The BaSIC Topology

B. Jayant Baliga • Ajit Kanale

The BaSIC Topology

A Revolutionary Power Device Control Strategy

Springer

B. Jayant Baliga
Electrical and Computer Engineering Department
North Carolina State University
Raleigh, NC, USA

Ajit Kanale
Electrical and Computer Engineering Department
North Carolina State University
Raleigh, NC, USA

ISBN 978-3-031-86629-6 ISBN 978-3-031-86630-2 (eBook)
https://doi.org/10.1007/978-3-031-86630-2

This Springer imprint is published by the registered company Springer Nature Switzerland AG
The registered company address is: Gewerbestrasse 11, 6330 Cham, Switzerland

The authors would like to dedicate this book to the international wide-bandgap semiconductor power device community.

Preface

The adverse environmental impact resulting from carbon emissions into the atmosphere due to human activity is widely accepted worldwide. Carbon emissions have been shown to produce global warming that leads to worsening of forest fires and intensification of hurricanes. The rise in sea levels due to melting of ice at the polar caps is threatening small island nations and coastal communities around the globe.

We have become increasingly dependent on the use of electrical power in our daily lives. It powers the appliances in our homes and the machinery in our factories. Nearly 70% of the electricity is generated using fossil fuels, primarily coal and natural gas. The burning of fossil fuels is responsible for a large amount of carbon dioxide emissions around the world in industrialized nations.

Mitigation of global warming can be accomplished by replacement of fossil fuels for the generation of electricity with renewable energy sources such as solar and wind power. These installations require power electronics to convert the generated power to a well-regulated 60 Hz AC power that can be distributed to consumers and the industry. Due to the relatively high power levels involved, the power semiconductor devices used in these applications must have high voltage and current-handling capability.

We have also become highly dependent upon the use of gasoline-powered cars and trucks for satisfying our transportation needs. More than 600 billion gallons of gasoline are consumed worldwide annually with 19.4 pounds of carbon dioxide emitted per gallon into the atmosphere. This is a major cause for global warming.

Carbon dioxide emissions from automobiles can be eliminated by deployment of electric vehicles. Power semiconductor devices are a key component in the adjustable speed motor drives required to control the electric motors.

The silicon-based insulated gate bipolar transistor (IGBT) was created in the early 1980s for applications in all sectors of the economy [1]. The device structure, its physics of operation, and extensive applications in all sectors of the economy have been documented [2]. It is the dominant power device technology used for all renewable energy generation as well as in hybrid-electric and all-electric cars and trucks.

In 1979, it was proposed that power device performance could be significantly enhanced by replacing silicon with wide-bandgap semiconductors [3]. This theory was published in 1982 leading to Baliga's figure of merit (BFOM) for power devices [4]. The BFOM predicted 1000-fold reduction in the specific on-state resistance of the drift region when replacing silicon with silicon carbide (SiC) and gallium nitride (GaN). This theoretical prediction was validated in the 1990s by the fabrication of high-performance SiC Schottky diodes and SiC power MOSFETs at the Power Semiconductor Research Center, North Carolina State University (NCSU). This achievement encouraged worldwide support for the preparation of low-defect-density SiC wafers and the creation of reliable SiC power device structures. SiC JBS rectifiers were commercialized in 2005 followed by SiC power MOSFETs in 2010. The physics and design of these devices have been described in a recent book [5].

The successful growth of gallium nitride layers on silicon substrates enabled creation of high-performance lateral GaN high-electron-mobility transistors (HEMTs). They were commercialized in 2015 providing another option for power electronics engineers. The physics and design of these devices have been described in a recent book [6].

The commercially available SiC power MOSFETs and GaN HEMT devices have excellent electrical characteristics for the replacement of silicon IGBT in electric vehicle and renewable energy applications. Their low on-state voltage drop and fast switching speed allow achieving superior efficiency when compared with silicon IGBTs. The power electronics can be operated at higher frequencies by using these devices. This shrinks the size of passive elements such as inductors and filters making the power electronics not only more compact but also less expensive.

Silicon IGBTs have been engineered by manufacturers to have short-circuit withstand time of at least 10 μs for its application in adjustable speed motor drives. This provides enough time to detect the onset of the short-circuit event and safely turn off the gate drive signal to the IGBT. A major roadblock for the application of the SiC- and GaN-based power switches has been found to be their poor short-circuit withstand time. Commercially available SiC power MOSFETs have short-circuit withstand times of only 3–4 μs, while the commercially available GaN HEMT devices have short-circuit withstand times of less than 0.3 μs. Reengineering the design of these devices leads to a substantial increase in the on-state resistance and switching power losses.

The **Ba**liga **S**hort-circuit **I**mprovement **C**oncept (BaSIC) was proposed [7] to increase the short-circuit withstand time for SiC and GaN transistors beyond 10 μs. This approach was thoroughly investigated during a 4-year period to complete a doctoral thesis at NCSU. It was compared with previously proposed alternatives for increasing the short-circuit withstand time. Previous reports in the literature discussed only the trade-off between improvements of short-circuit withstand time and increase in on-state resistance. In our work at NCSU, measurements of the switching losses were also included to provide a more comprehensive comparison of various approaches.

The monograph documents the benefits of using the BaSIC topology to enhance the short-circuit withstand capability of SiC and GaN power transistors. After an introductory chapter, the reader is familiarized with the short-circuit conditions for power transistors in the second chapter. The third chapter reviews various power transistor structures commercially fabricated from silicon, SiC, and GaN by the industry.

The **Ba**liga **S**hort-circuit **I**mprovement **C**oncept (BaSIC) is described in the fourth chapter. Its operating principles are explained for a generic case of a high-voltage power transistor whose short-circuit withstand capability is enhanced by using a nonlinear element in series. This combination is named the composite transistor. A key feature of the BaSIC topology is that it is tolerant to variations in manufacturing tolerances for the power transistor and the nonlinear element. A unique feature of the BaSIC topology is the ability to monitor the on-state current in the power transistor and onset of short-circuit events using a sense terminal. This is a valuable feature for exploitation by power electronics engineers for optimization of their chosen circuit configurations.

Chapters 5–7 discuss the utilization of a gate-source-shorted (GSS) depletion-mode MOSFET (DMM) as the nonlinear element for the BaSIC topology applied to SiC power MOSFETs. The use of a Si DMM is compared with the gate bias reduction approach in Chap. 6 and with the ballast resistor approach in Chap. 7. The BaSIC(DMM) topology is shown to create a much superior trade-off between increasing the short-circuit withstand time and the on-state resistance as well as the switching losses.

Chapter 8 demonstrates that the BaSIC topology can be scaled to achieve higher power-handling capability for SiC power MOSFETs. Excellent linear scaling of the power-handling capability is demonstrated by the paralleling of composite transistors. The sense terminal in the BaSIC topology is used to confirm very good current sharing between the composite transistors in the on-state and during the short-circuit event.

The experimental demonstration of the BaSIC topology with Si depletion-mode MOSFETs (DMMs) as the nonlinear element was constrained by the limited availability of commercial products. Chapter 9 describes a procedure for optimization of Si planar-gate (DMOS) and trench-gate (UMOS) devices to create a better trade-off curve between saturation current and on-state resistance. It is shown that a much superior trade-off curve becomes feasible for the BaSIC(DMM) topology by using the optimized Si MOSFET. The short-circuit withstand capability of the SiC power MOSFET can be increased beyond 10 μs using the optimized Si DMM device with less than 5% impact on the on-state resistance. These results should encourage the industry to manufacture the optimized Si depletion-mode MOSFETs (DMMs) for a high-volume application.

The BaSIC topology can also be implemented using widely available Si enhancement-mode (EMM) power MOSFETs as the nonlinear element. The DC gate bias applied to the Si EMM device can be programmed by the power electronics engineers to achieve a particular enhancement in the short-circuit capability within the least impact on increase in on-state resistance and switching losses. Experimental demonstration of this approach is described in Chap. 10. The BaSIC(EMM) topology is then compared with the gate bias reduction approach in Chap. 11 and with the ballast resistor approach in Chap. 12. The BaSIC(EMM) topology is shown to create a much superior trade-off between increasing the short-circuit withstand time and the on-state resistance as well as the switching losses.

A large variety of Si enhancement-mode (EMM) power MOSFETs are manufactured by the industry for a variety of applications. A selection methodology is provided in Chap. 13 for identification of the best Si EMM device for the BaSIC(EMM) topology. This selection methodology is applied to find a Si EMM device that extends the short-circuit withstand capability of a SiC power MOSFET beyond 10 μs with less than 5% increase in its on-state resistance.

The BaSIC(DMM) and BaSIC(EMM) topologies are compared in Chap. 14. It is shown that they are complementary with each other and operate in different portions of the same trade-off curve between short-circuit withstand capability and on-state resistance.

Power transistors are also required to tolerate unclamped inductive load switching conditions in applications. It is shown in Chap. 15 that the BaSIC topology does not compromise this capability for the SiC power MOSFETs.

In motor drive applications, the power transistor may be subjected to multiple short-circuit events during its lifetime. It has been shown that SiC power MOSFETs may survive single short-circuit events if the short-circuit time is below its short-circuit failure time. However, these events melt the chip aluminum top metallization. It can then penetrate through the inter-metal dielectric leading to a short circuit between the gate and source electrodes after multiple short-circuit pulses. It is shown in Chap. 16 that this failure mode can be eliminated using the BaSIC topology. It is the only method shown to provide this important benefit.

The DESAT protection method for short-circuit protection was created for Si IGBTs and is widely commercialized. It is based on sensing the collector voltage of the IGBT, which increases during the short-circuit event because of the large increase in current. The voltage at the collector terminal of the IGBT regularly switches between the low on-state value and the high DC bus supply value. The DESAT circuit has to be isolated from the high voltages at the IGBT collector terminal by using a high-voltage fast-recovery diode. The DESAT method for IGBTs requires a blanking time of 3 μs to avoid false triggering. The short-circuit response time is typically 5 μs, which dictates the need for short-circuit withstand times of over 10 μs for commercially available IGBTs optimized for motor drive applications.

It is demonstrated in Chap. 17 that the commercially available DESAT gate drivers for Si IGBTs can be utilized for the BaSIC topology with the sense node voltage. The sense node always remains below 20 V because it operates in the gate drive loop of the power transistor. This eliminates the need for a high-voltage isolation diode and shortens the blanking interval to only 0.5 μs. It is shown in this chapter that a SiC power MOSFET can be protected during the short-circuit event within 1 μs.

Although the BaSIC topology was conceived for improvement of short-circuit capability of SiC power MOSFETs, it can be used to improve the performance of Si IGBTs as well. The industry manufactures type 1 Si IGBTs with short-circuit capability of over 10 μs for motor drive applications. The short-circuit capability is achieved by trading off a larger on-state voltage drop for these devices. The industry also manufactures type 2 IGBT, which has poor short-circuit capability to reduce the on-state voltage drop in applications such as induction cooktops. It is shown in Chaps. 18 and 19 that the BaSIC(DMM) and BaSIC(EMM) topologies can achieve the desired 10 μs short-circuit withstand time using type 2 IGBTs. The composite transistor created using this approach is shown to offer superior performance to the Si type 1 IGBTs.

Gallium nitride high-electron-mobility transistors (HEMTs) have become commercially available for motor drive applications. These devices have very poor short-circuit withstand times of less than 0.3 μs. It is demonstrated in Chap. 20 that the short-circuit withstand time for a GaN HEMT device can be increased to 4 μs, an improvement by a factor of 15 times, by using the BaSIC(DMM) topology.

The ability to integrate the nonlinear element into the SiC power MOSFET or Si IGBT structure was proposed in the original BaSIC patent application [7]. The integration of a lateral SiC JFET structure into the SiC power MOSFET structure is described in Chap. 21. The characteristics of the lateral JFET are analyzed based on the doping concentration, thickness, and length of the drift region. Excellent short-circuit capability enhancement for the SiC power MOSFET is predicted by using the lateral integrated SiC JFET with only small impact on the on-state resistance.

The final Chap. 22 provides a brief summary of the benefits of the BaSIC topology.

The authors are hopeful that this monograph will encourage the industry to adopt the BaSIC topology when commercializing wide-bandgap semiconductor-based power devices. It is a particularly attractive approach for companies that manufacture both high-voltage SiC power MOSFETs and low-voltage Si MOSFETs. These devices can be co-packaged to provide an elegant product with excellent short-circuit capability and device current-sensing capability for their power electronics customers.

Raleigh, NC, USA B. Jayant Baliga
Raleigh, NC, USA Ajit Kanale
December 2024

References

1. B.J. Baliga, The creation of the IGBT, in *75th Anniversary of the Transistor* (IEEE Press Book, New York, 2023)
2. B.J. Baliga, *The IGBT Device*, 2nd edn. (Elsevier, Amsterdam, 2023)
3. B.J. Baliga, Optimum semiconductors for high voltage vertical channel field effect transistors. GE Class 2 Technical Report No. 79CRD186, Nov 1979
4. B.J. Baliga, Semiconductors for high voltage vertical channel field effect transistors, J. Appl. Phys. **53**, 1759–1764 (1982)
5. B.J. Baliga, *Modern Silicon Carbide Power Devices* (World Scientific Press, Singapore, 2024)
6. B.J. Baliga, *Gallium Nitride and Silicon Carbide Power Devices* (World Scientific Press, Singapore, 2017)
7. B.J. Baliga, Power devices having tunable saturation current clamps therein that support improved short-circuit capability. U.S. Patent # 12,094,932, Issued 17 Sept 2024

Competing Interests The authors have no competing interests to declare that are relevant to the content of this manuscript.

Contents

About the Authors

B. Jayant Baliga received his M.S. and Ph.D. in electrical engineering from Rensselaer Polytechnic Institute, Troy, NY, in 1971 and 1974, respectively. He spent 15 years at the General Electric R&D Center, Schenectady, NY, where he led their power device effort and was bestowed the highest rank of Coolidge Fellow for inventing, developing, and commercializing the IGBT. It is now utilized in all sectors of the economy, including electric vehicles and renewable energy generation. He derived the Baliga's-Figure-of-Merit (BFOM) in 1979, which is widely used for wide bandgap semiconductor power devices. He joined North Carolina State University (NCSU) in 1988 as a Full Professor and was promoted to Distinguished University Professor in 1997 and subsequently to Progress Energy Distinguished University Professor. He has founded four start-up companies with successful products while at NCSU. Dr. Baliga has authored 28 books and over 750 publications in international journals and conference digests. He is the author of the textbook *Fundamentals of Power Semiconductor Devices* (Springer, 2018), now in its second edition, as well as three Springer monographs: *Advanced High Voltage Power Device Concepts* (2012), *Advanced Power MOSFET Concepts* (2010), and *Advanced Power Rectifier Concepts* (2009). In 2011, President Obama honored him with the National Medal of Technology and Innovation at the White House and he is also the recipient of the highest IEEE recognition, the Medal of Honor in 2014, the Global Energy Prize in 2015, and the Millennium Technology Prize in 2024. He holds

124 U.S. patents and was inducted into the National Inventors Hall of Fame as the sole inventor of the insulated-gate bipolar transistor (IGBT) in 2016.

Ajit Kanale is a Power Applications Engineer in the Advanced Power Applications Group at Wolfspeed Inc. He graduated with a Bachelor's degree in Avionics from the Indian Institute of Space Science and Technology, Thiruvananthapuram, India. In 2012, he was appointed as a Satellite Engineer at the Indian Space Research Organization's satellite mission control facility in Bangalore, where he managed remote sensing and scientific satellite missions. In 2016, he moved to Raleigh, NC, to pursue graduate studies at North Carolina State University. He obtained his Master's degree in Electrical Engineering in 2019, with a thesis on short-circuit characterization of the 1.2 kV Silicon Carbide JBSFET device under the joint guidance of Prof. B. Jayant Baliga and Prof. Subhashish Bhattacharya. From 2018 to 2022, under the guidance of the same advisors, he pursued doctoral research on BaSIC topology to improve short-circuit ruggedness of power devices. In addition, he also assisted in characterizing several innovative 650 V and 1.2 kV SiC power MOSFET and JBSFET device designs, including the split gate, buffered gate, hexagonal-cell MOSFETs, and the bidirectional FET (BiDFET).

Chapter 1
Introduction

Power devices need to serve applications that work over a wide range of power levels. The large variation in operating voltages and currents is illustrated in Fig. 1.1 [1]. Automotive applications in conventional gasoline powered vehicles operate from the 14 V DC battery for running the air-conditioning, power windows, etc. They require power devices with blocking voltages below 30 V.

Power supplies for microprocessors and memory chips in laptops and personal computers operate from a 19 V DC bus. The power devices for this application are designed with blocking voltages <30 V.

Telecommunication applications and display drives need devices with higher blocking voltages with small current ratings. The large number of devices needed for these applications can be integrated on a single chip.

Lamp ballasts for compact fluorescent lamps (CFLs) require power devices with blocking voltages up to 900 V. A similar voltage rating is needed for robotics and numerical controls used in factory automation.

Two-thirds of the worldwide electrical power is consumed for driving motors for multifold applications. Industrial applications include pumping of fluids and gases, and conveyor belts for transporting materials. Household applications include the heat pumps for air-conditioning of buildings and refrigeration of food to delay spoilage.

The increasing deployment of electric vehicles requires adjustable speed motor drives. These devices can range from electric buses and trams for urban transit to high-speed electric trains to travel between cities and countries. The power handling capability of power devices for this application can extend to multimegawatt.

In recent years, the transition from gasoline power cars and trucks to electric or hybrid electric vehicles has been accelerating. This application requires power devices with blocking voltages up to 2000 V and current handling capability of several hundred amperes.

The transmission of power from fossil-fuel based generation facilities to cities and industries is done with high voltage DC (HVDC) lines. These lines must handle

B. J. Baliga, A. Kanale, *The BaSIC Topology*,
https://doi.org/10.1007/978-3-031-86630-2_1

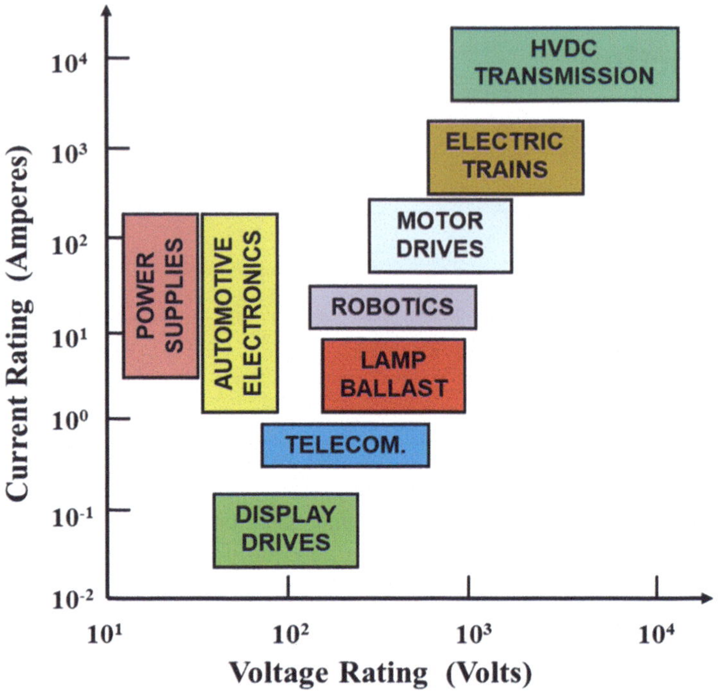

Fig. 1.1 Power device applications extending over a broad spectrum of voltage and current levels

gigawatts of power. This holds true for power generation from renewable energy sources such as large solar and wind farms.

1.1 Motor Drive Application

As mentioned above, a large number of industrial, commercial, and transportation applications require delivery of power to the windings of motors. Electric trains require operation of motors over a wide range of frequencies in order to alter the speed of the locomotive. The most commonly used topology converts the constant frequency input AC power to a DC bus voltage and then uses an inverter stage to produce the variable frequency output power [2] that controls the speed of the motor.

The circuit diagram for commonly used H-bridge topology is shown in Fig. 1.2. Four IGBTs are used with fly-back rectifiers in the inverter stage to deliver the variable frequency power to the motor winding. A pulse-width modulation (PWM) scheme is used to generate the variable frequency AC voltage waveform that is fed to the motor windings [3]. When IGBT-1 and IGBT-4 are simultaneously turned-on by application of the on-state gate bias, current flowing through the motor winding from left to right increases gradually during the on-time for these devices.

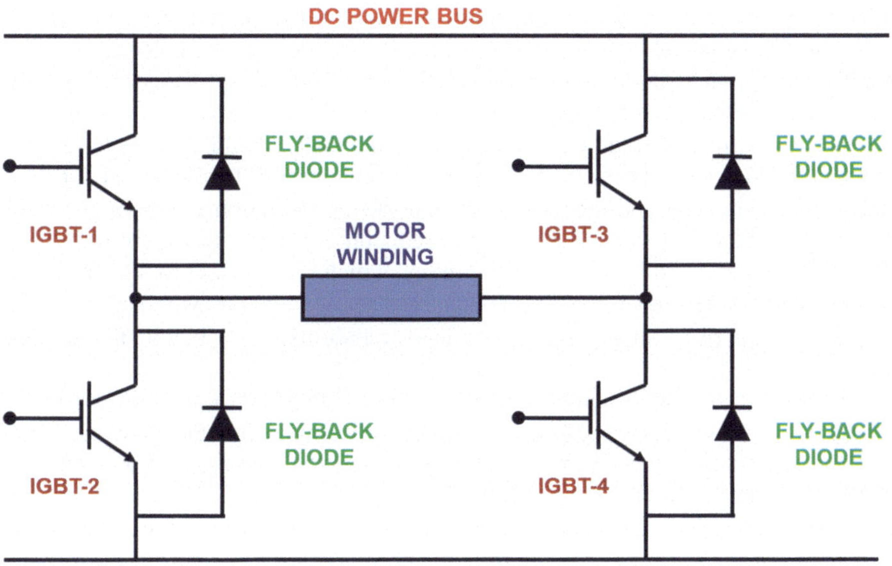

Fig. 1.2 H-bridge topology for driving motor winding

Subsequently, when IGBT-2 and IGBT-3 are simultaneously turned-on by application of the on-state gate bias, current flowing through the motor winding from left to right gradually decreases during the on-time for these devices. This phenomenon can be used to generate a sinusoidal waveform of variable frequency.

1.2 Short-Circuit Withstand Capability

Motor windings are susceptible to failure as short circuits due to rupture of the insulation. This reduces the motor winding impedance to zero. The DC bus voltage is then imposed across those IGBTs that are operating with the on-state gate bias applied to them. This produces a rapid increase in the current until it is limited by the saturation current of the device at the on-state gate bias. This current is usually more than 10 times the on-state current level resulting in high power dissipation because the device is supporting the DC bus voltage. The heat generated within the chip cannot be removed by the package due to the short time interval. The adiabatic heating raises the junction temperature until the device fails.

Silicon IGBTs are the most widely used power device for motor drive applications. This is partly due to engineering their design so that they have a short-circuit withstand time of more than 10 μs. This allows detection of the short circuit followed by turning-off the gate drive voltage in <10 μs. However, the 10 μs short-circuit withstand capability is achieved by trading off the on-state and switching power losses. Consequently, manufacturers have commercialized two types of

IGBTs. The first Type 1 devices are tailored for motor drive applications with 10 μs short-circuit withstand capability. The second Type 2 devices are tailored for other applications, such as induction cooktops, where no short-circuit withstand capability is required. The Type 2 devices have lower power losses than the Type 1 devices.

Recently, silicon carbide power MOSFETs and GaN HEMT devices have been commercialized with the goal of replacing the silicon IGBTs. They can substantially reduce on-state and switching losses allowing operation of the motor drive circuits at higher frequencies. This allows reducing the size and weight of passive elements (inductors and capacitors) and filters, which is important for electric vehicle applications. However, these wide band gap semiconductor switches have been found to have poor short-circuit withstand capability, well below the desired 10 μs value.

The short-circuit withstand time for SiC power MOSFETs can be increased by modifications of the device cell structure or by reducing the on-state gate bias. Both of these approaches have been found to significantly increase the on-state and switching power losses as discussed in this book.

Special gate drive circuits have been proposed and demonstrated to allow detection of the short circuit in <1 μs so that SiC power MOSFETs can survive a short-circuit event. However, the peak short-circuit current occurs during this time. It has been found that SiC power MOSFETs are prone to failure after repetitive short-circuit events of even a short duration due to melting of the aluminum source contact metallization as discussed in this book.

1.3 The BaSIC Topology

The **Ba**liga **S**hort-circuit **I**mprovement **C**oncept (BaSIC) was proposed [4] to address the poor short-circuit withstand time of commercially available SiC power MOSFETs. It was later applied to achieve good short-circuit withstand capability for the Type 2 IGBTs which reduces losses in motor drive applications. In addition, it was demonstrated to significantly extend the short-circuit time for GaN HEMT devices. Furthermore, it was shown to solve the problem of repetitive short-circuit failure in SiC power MOSFETs.

The purpose of this book is to introduce power electronics engineers to the BaSIC topology. They can take advantage of this idea to design motor drives with SiC power MOSFETs that can tolerate short-circuit events.

The BaSIC topology has an inherent current sensing capability. This feature can be used to monitor the current flow through each of the power switches within the power electronic circuit. Sensing the current through each of the transistors allows designer to adjust the control algorithm to maximize the output power and avoid reliability problems due to excessive currents in one device. The discrete current sensors used in power modules can be eliminated to reduce cost and size of the circuits.

1.4 Summary

A revolutionary concept called the BaSIC topology is discussed in detail in this book. Its application to SiC power MOSFETs is demonstrated to allow increasing the short-circuit withstand time to over 10 μs with <5% increase in the on-state resistance. The BaSIC topology is shown to prevent repetitive short-circuit failure of SiC power MOSFETs. It provides a current sensing capability that can be taken advantage by power circuit designers for circuit optimization.

The BaSIC topology is shown in this book to significantly enhance the short-circuit capability of GaN HEMT devices. It has also been applied to silicon IGBTs to allow using type 2 devices, which are not designed with 10 μs short-circuit withstand capability, in motor drive circuits.

References

1. B.J. Baliga, *Fundamentals of Power Semiconductor Devices*, 2nd edn. (Springer Science, New York, 2019)
2. J.D. van Wyk, Power electronic converters for drives, in *Power Electronics and Variable Frequency Drives*, ed. by B.K. Bose, (IEEE Press, 1997), pp. 81–137
3. J. Holtz, Pulse width modulation for electronic power conversion, in *Power Electronics and Variable Frequency Drives*, ed. by B.K. Bose, (IEEE Press, 1997), pp. 138–208
4. B.J. Baliga, *Power Devices Having Tunable Saturation Current Clamps Therein that Support Improved Short-Circuit Capability*, U. S. Patent 12,094,932 (17 Sept 2024)

Chapter 2
Short-Circuit Conditions

Inadequate short-circuit withstand capability is a major issue that could impede the commercialization of SiC power MOSFETs and GaN HEMT devices for automotive and industrial motor drive applications. As explained in Chap. 1, this problem occurs when the motor winding undergoes a short circuit due to failure of its insulation while the power transistor is in its on-state with full gate bias applied to deliver current and power to the motor winding from a DC link power supply. The transistor must support the DC supply voltage while operating in its current saturation mode at a current level that is typically more than 10 times the on-state current level.

Silicon IGBTs have been commercialized with designs that can tolerate the short-circuit conditions for at least 10 μs. Commercial IGBT gate control circuits based on the DESAT principle are available that can detect the on-set of the short-circuit condition and turn-off the gate drive voltage in <10 μs to prevent destructive failure of these devices.

This chapter describes the basic short-circuit operating conditions. Analytical solutions for the short-circuit failure time are derived based on adiabatic heating conditions for the semiconductor chips [1]. The analysis allows relating the short-circuit withstand time to the basic transistor device properties.

2.1 Short-Circuit Analysis

Under short-circuit conditions due to failure of a motor winding in the voltage source inverter topology, the power transistor is connected directly across the DC bus while the on-state gate bias is being applied. The circuit diagram that represents the short-circuit condition is shown in Fig. 2.1. The short circuit across the load is represented by the switch. Closing the switch connects the DC power supply (V_{DC}) directly to the drain of the transistor while its on-state gate bias (V_{GS}) is still being applied.

B. J. Baliga, A. Kanale, *The BaSIC Topology*,
https://doi.org/10.1007/978-3-031-86630-2_2

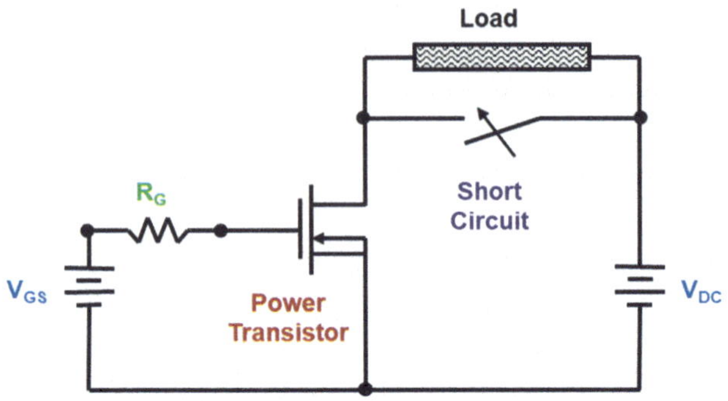

Fig. 2.1 Circuit representing short-circuit conditions for a power transistor

It would be desirable to detect the short-circuit condition and turn-off the transistor using feedback to the gate drive control circuit before the device undergoes destructive failure. The typical DESAT gate control circuit developed for commercial silicon IGBTs can require close to 10 μs to detect and respond to the occurrence of the short-circuit conditions. For this reason, commercially available silicon IGBTs for motor drive applications are designed to withstand short-circuit conditions for more than 10 μs. This criterion has now become the expected performance for all power transistors used in motor drive applications.

The waveforms for the gate voltage and drain current and drain voltage for the power transistor are illustrated in Fig. 2.2 during the short-circuit event. The drain voltage (v_D) becomes equal to the DC supply voltage while the gate voltage (v_G) remains at the on-state value (V_{GS}). The current flowing through the transistor (i_D) must be limited by its drain saturation current due to zero load impedance during the short-circuit event. The drain current is shown to have a constant value during the short-circuit event in the figure. In practice, the short-circuit current increases to a peak value within a few μs and then decreases with time due to the increase in junction temperature.

The power being dissipated by the transistor per unit area (power density) during the short-circuit event is therefore given by:

$$p_D = J_{D,SAT} V_{DC} \tag{2.1}$$

where V_{DC} is the DC supply voltage and $J_{D,SAT}$ is the drain saturation current density determined by the gate bias voltage (V_G).

The duration of the short-circuit event before device failure is too short for removal of heat via the package and heat sink. Consequently, the maximum junction temperature (T_M) increases under adiabatic heating conditions. The rate of rise of the temperature is determined by the semiconductor volume within which the power dissipation is occurring, which is given by the product of the device active area and the drift layer thickness (t_D). The rise in temperature is given by:

$$\Delta T(t) = \frac{p_D t}{t_D C_V} = \frac{J_{D,SAT} V_{DC} t}{t_D C_V} \tag{2.2}$$

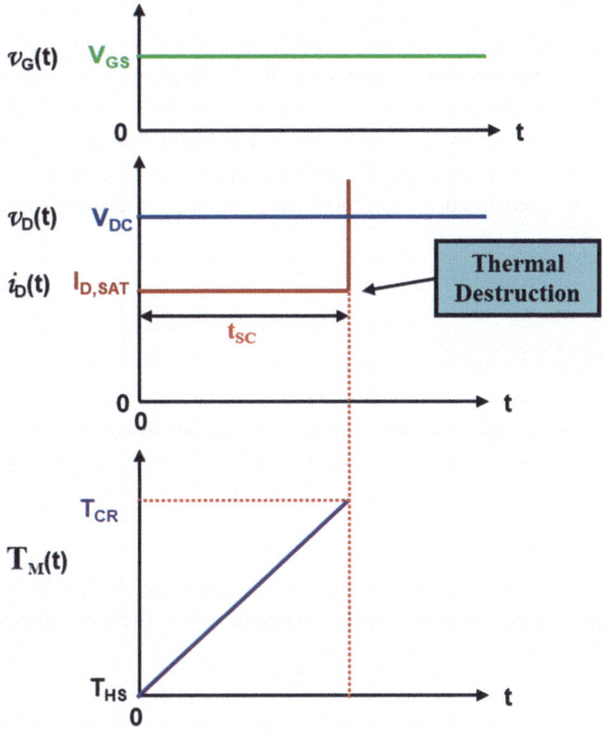

Fig. 2.2 Transistor waveforms and temperature rise during a short-circuit event

where C_V is the volumetric specific heat of the semiconductor material.

Under the assumptions used for the analytical model, the temperature increases linearly as a function of time, as shown in Fig. 2.2 by the purple line. In practice, the saturated current does not remain constant during the short-circuit event because the temperature rise produces a decrease in the saturation current due to the reduction of channel mobility as the temperature rises.

The destruction of the transistor occurs when the temperature reaches a critical value (T_{CR}). The mechanism postulated for the destructive failure of the SiC power MOSFETs is turn-on of the parasitic NPN transistor [2]. A second short-circuit failure mechanism that has been postulated is penetration of the source metallization through the junctions at the critical temperature [3]. The latch-up of the parasitic thyristor in silicon IGBTs initiates their failure.

From Eq. (2.2), the short-circuit withstand time for the transistor can be obtained using:

$$t_{SC} = \frac{(T_{CR} - T_{HS})t_D C_V}{J_{D,SAT} V_{DC}} \tag{2.3}$$

where T_{HS} is the initial (heat sink) temperature. This expression shows that the short-circuit withstand time is inversely proportional to the saturation current level. It is therefore clear that methods for reducing the saturation current level are required to achieve the goal of extending the short-circuit time to more than 10 μs.

From this expression, it can also be concluded that the short-circuit withstand time will decrease with increasing DC power supply voltage. The DC supply voltage (V_{DC}) is usually two-thirds of the blocking voltage rating for transistors in motor drive applications.

2.2 Summary

The short-circuit withstand capability of transistors has become a critical parameter for many applications, especially motor drives for electric vehicles. Commercially available silicon Type 1 IGBTs designed for this application have a short-circuit withstand capability of over 10 μs. This allows the modern DESAT gate control circuits to detect the short-circuit event and turn-off the gate drive voltage in time to prevent destructive failure of the devices. This attribute of the Si IGBT must be replicated for any transistors that replace them in motor drive applications.

References

1. B.J. Baliga, *Fundamentals of Power Semiconductor Devices*, 2nd edn. (Springer-Science, 2019)
2. G. Romano et al., A comprehensive study of short-circuit ruggedness of silicon carbide power MOSFETs. IEEE J. Emerg. Sel. Top. Power Electron. **4**, 978–987 (2016)
3. K. Han, A. Kanale, B.J. Baliga, B. Ballard, A. Morgan, D.C. Hopkins, New short circuit failure mechanism for 1.2kV 4H-SiC MOSFETs and JBSFETs, in *IEEE 6th Workshop on Wide Bandgap Power Devices and Applications (WiPDA)* (2018), pp. 108–113

Chapter 3
Power Devices

As mentioned in Chap. 1, this book is focused on power devices used for applications with blocking voltage ratings ranging from 600 to 1700 V. These applications have been served by silicon IGBTs during the last 40 years. During the past decade, silicon carbide power MOSFETs and gallium nitride HEMT devices have been commercialized to supplement the silicon IGBTs. The basic structure of these devices is described in this chapter for the convenience of readers. A detailed description of the physics of operation of the silicon IGBT is available in a textbook [1]. Its multifarious applications are thoroughly explored in another reference [2]. A detailed description and analysis of the silicon carbide MOSFETs and gallium nitride HEMT devices are also available in another textbook [3].

3.1 Silicon IGBT

The Insulated Gate Bipolar Transistor (IGBT), developed in the 1980s by combining the physics of operation of bipolar transistors and MOSFETs [4], has become the most widely used silicon device for power switching applications with high voltages (>300 V). The structure of the IGBT is deceptively similar to that for the power MOSFET as shown in Fig. 3.1 if viewed as a mere replacement of the N^+ substrate with a P^+ substrate. However, this substitution creates a four-layer parasitic thyristor which can latch up resulting in destructive failure due to loss of gate control. Fortunately, the parasitic thyristor can be defeated by the addition of the P^+ region within the cell [1]. The benefit of the P^+ substrate is the injection of minority carriers into the N-drift region resulting in greatly reducing its resistance. This has enabled the development of high voltage IGBT products with high current carrying capability. The main disadvantage of the Si IGBT is a slow switching speed that increases the turn-off power losses. This problem was solved by using electron irradiation to reduce the lifetime in the drift region [2]. Extensive application of the

B. J. Baliga, A. Kanale, *The BaSIC Topology*,
https://doi.org/10.1007/978-3-031-86630-2_3

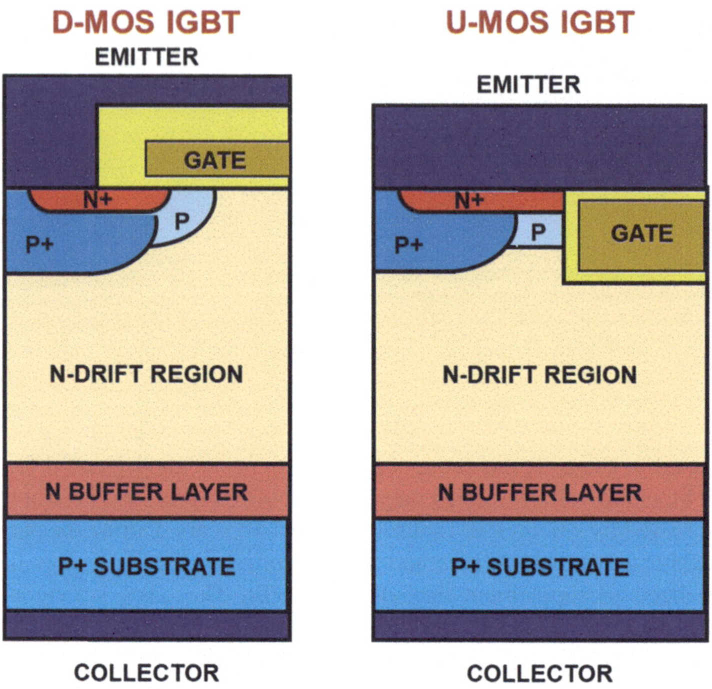

Fig. 3.1 Silicon IGBT device structures

silicon IGBT has occurred in all segments of the economy in the last 40 years with enormous benefits to society [2].

3.1.1 On-State Voltage Drop

The on-state voltage drop for the silicon IGBT can be obtained by using the P-i-N rectifier/MOSFET model [1]. In this model, the on-state voltage drop for the IGBT structure is obtained by adding the voltage drops across the P-i-N and MOSFET portions of the device:

$$V_{\mathrm{F,IGBT}} = \frac{2kT}{q}\ln\left[\frac{J_{\mathrm{C}}W_{\mathrm{N}}}{4qD_{\mathrm{a}}n_{\mathrm{i}}F\left(W_{\mathrm{N}}/2L_{\mathrm{a}}\right)}\right] + \frac{W_{\mathrm{Cell}}L_{\mathrm{CH}}J_{\mathrm{CH}}}{2\mu_{\mathrm{ni}}C_{\mathrm{OX}}\left(V_{\mathrm{G}}-V_{\mathrm{TH}}\right)} \tag{3.1}$$

In this equation, k is Boltzmann's constant, T is the absolute temperature, q is the electron charge, J_{C} is the on-state collector current density, W_{N} is the width of the N-drift layer plus the N buffer layer, D_{a} is the ambipolar diffusion constant, n_{i} is the intrinsic concentration, L_{a} is the ambipolar diffusion length, W_{cell} is the cell width with two channels, L_{CH} is the channel length, μ_{ni} is the inversion channel mobility,

C_{OX} is the specific gate oxide capacitance, V_G is the gate drive voltage, and V_{TH} is the threshold voltage. The specific gate oxide capacitance C_{OX} can be calculated by dividing the oxide dielectric constant with the gate oxide thickness. The function $F(W_N/2L_a)$ can be computed using formulae provided in the textbook [1]. From this equation, it can be surmised that a smaller on-state voltage drop is achieved by reducing the channel length and increasing the gate drive voltage.

3.1.2 Collector Saturation Current Density

The short-circuit withstand capability is determined by the device saturation current level under short-circuit conditions as discussed in Chap. 2. A smaller saturation current produces an increase in the short-circuit withstand time. In the case of the silicon IGBT, the collector saturation current density is related to the device cell parameters by

$$J_{C,SAT} = \frac{\mu_{ni} C_{OX} W_{cell}}{L_{CH} \left(1 - \alpha_{PNP}\right)} \left(V_G - V_{TH}\right)^2 \tag{3.2}$$

where α_{PNP} is the current gain of the IGBT internal NPN transistor. From this equation, it can be surmised that the saturation current can be reduced by increasing the channel length and reducing the gate drive voltage.

3.1.3 Trade-Off Curve Using Channel Length

It is obvious based on fundamental operating physics of the silicon IGBT that there is a trade-off between increasing the short-circuit withstand time and maintaining a low on-state voltage drop. This trade-off is inherent in the design of the Type-2 IGBTs that are commercially available with short-circuit withstand times of more than 10 μs.

The case of a 1200 V asymmetric blocking silicon IGBT is considered here as an example. This device was analytically modelled as described in the textbook with the above equations. The device structure has an N-drift region thickness of 100 μm with doping concentration of 5×10^{13} cm^{-3}. The N buffer layer has a width of 10 μm with doping concentration of 5×10^{16} cm^{-3}. The MOSFET was modelled with an inversion layer mobility of 250 cm^2/V-s and gate oxide thickness of 500 Å. The high-level lifetime in the drift region was chosen as 2 μs. A gate drive voltage of 15 V was assumed with a threshold voltage of 5 V. The on-state voltage drop was computed at an on-state current density of 100 A/cm^2. The trade-off curve between reducing the collector saturation current and the increase in on-state voltage drop obtained by varying the channel length is shown in Fig. 3.2. The collector current

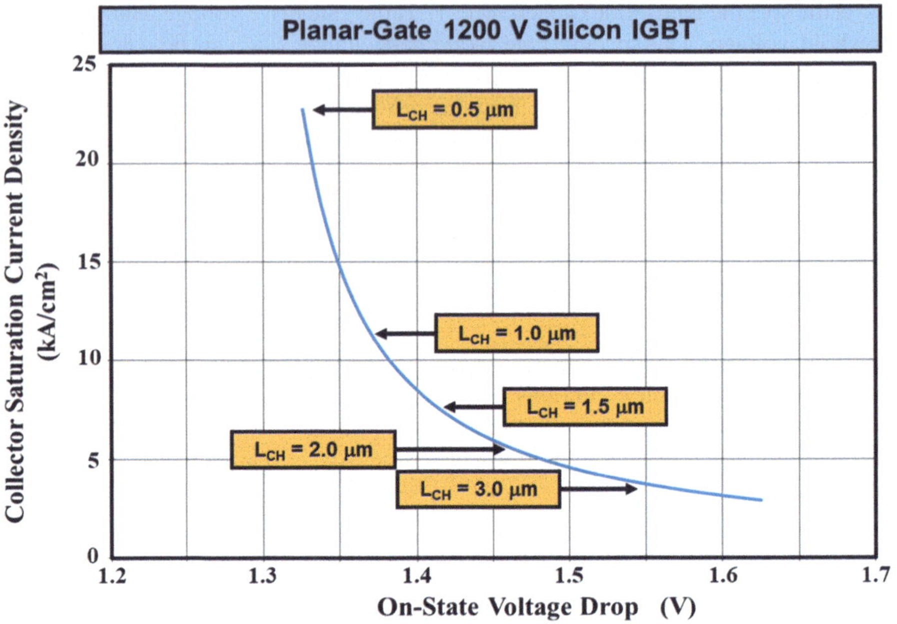

Fig. 3.2 Trade-off curve for 1200 V asymmetric silicon IGBT device obtained by changing the channel length

density can be reduced from over 20,000 A/cm^2 to <5000 A/cm^2 by increasing the channel length. This produces an increase in on-state voltage drop from 1.3 to 1.6 V.

3.1.4 Trade-Off Curve Using Gate Drive Voltage

The trade-off curve can also be generated by varying the gate drive voltage. This can be illustrated by using the same asymmetric IGBT device structure described above with a fixed channel length of 1 µm. The on-state voltage drop increases when the gate drive voltage is reduced in accordance with Eq. (3.1) due to an increase in the channel resistance for the internal MOSFET of the IGBT. The collector saturation current decreases when the gate drive voltage is reduced in accordance with Eq. (3.2). The resulting trade-off curve is shown in Fig. 3.3 with some particular gate drive voltages marked on the graph. The collector saturation current can be reduced below 4000 A/cm^2 by decreasing the gate drive voltage to <11 V with an increase in on-state voltage drop from 1.37 to 1.44 V.

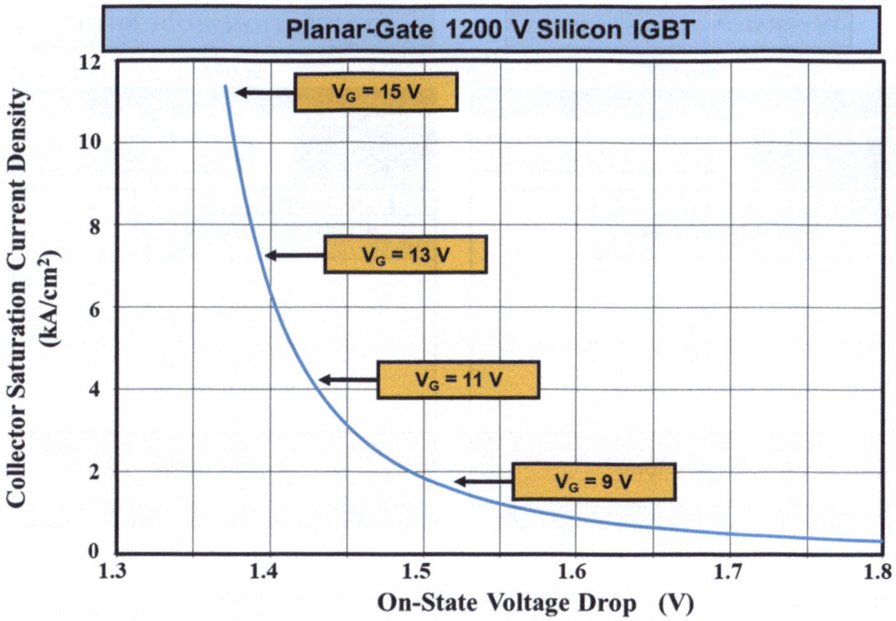

Fig. 3.3 Trade-off curve for 1200 V asymmetric silicon IGBT device obtained by changing the gate drive voltage

3.2 Silicon Carbide Power MOSFET

A detailed analysis of the silicon carbide power MOSFET structures is provided in the textbook [1]. It is shown that the conventional silicon power MOSFET structure could not be directly adopted for the silicon carbide devices due to various reasons. Novel silicon carbide power MOSFET structures [5] that prevent P-base reach-through breakdown and high electric field in the gate oxide were required. These devices, called shielded channel silicon carbide power MOSFETs are illustrated in Fig. 3.4.

The inversion-mode structure has a P-base region and a P^+ shielding region under the N^+ source and P-base regions. The accumulation-mode structure has an N-base region with the P^+ shielding region under the N^+ source and the N-base regions. The P-base and N-base regions can be formed by adjusting the energy and dose of the ion implants during device fabrication. The drain voltage is supported across the junction between the P^+ shielding region and the N-drift region in the voltage blocking mode.

The P^+ shielding region creates a JFET region where a potential barrier is created in the blocking mode at location A. The gap between the P^+ shielding regions must be optimized during the design of the MOSFET cell structure to simultaneously obtain a low specific on-resistance and low electric field in the gate oxide [6].

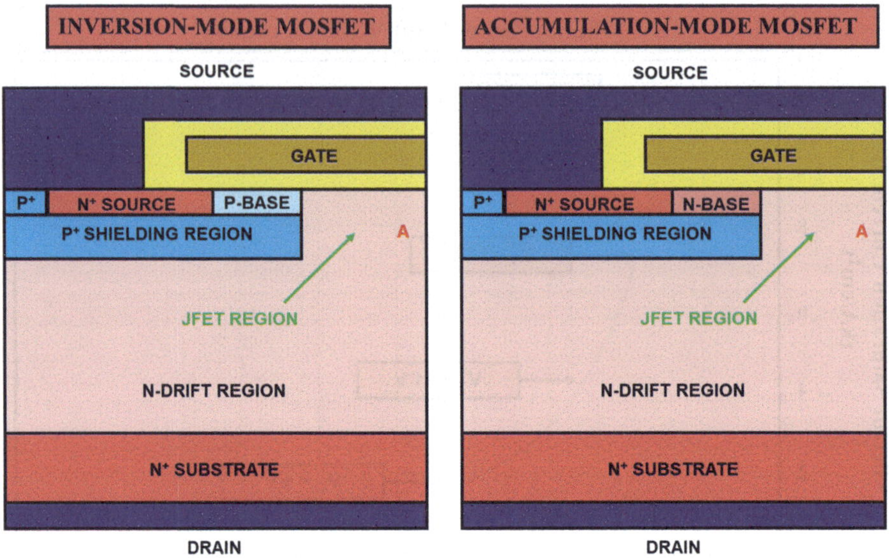

Fig. 3.4 Silicon carbide shielded channel planar power MOSFET structures

The minimum P-base thickness for the shielded inversion-mode 4H-SiC power MOSFET is no longer constrained by the reach-through limitation due to the shielding provided by the P^+ shielding region. This allows reducing the channel length to below 1 μm. In addition, the P-base doping concentration can be reduced to achieve threshold voltages below 3 V. The channel resistance contribution is reduced using these attributes of the shielded channel structure.

The built-in potential of the junction between the sub-surface P^+ shielding region and the N-base region will completely deplete the N-base region if its doping concentration and thickness are appropriately chosen in the case of the accumulation-mode shielded channel MOSFET structure. A potential barrier is created for electrons in the channel that enables normally-off operation for the accumulation-mode shielded channel MOSFET structure at zero gate bias. The mobility of electrons in the accumulation layer has been found to larger than the inversion channel mobility. This allows obtaining a lower specific on-resistance for the accumulation-mode devices because the channel resistance is dominant for devices with ratings of 600–1700 V [6].

3.2.1 On-State Resistance

The planar SiC power MOSFET structure contains many resistance components in series during current conduction after application of the gate drive voltage. The channel resistance is a dominant component for devices with ratings of 600–1700 V. The specific channel resistance is given by:

$$R_{\mathrm{CH.sp}} = \frac{L_{\mathrm{CH}}\, p}{\mu_{\mathrm{CH}} C_{\mathrm{OX}} \left(V_{\mathrm{G}} - V_{\mathrm{TH}}\right)} \tag{3.3}$$

where p is the cell pitch ($p = W_{\mathrm{Cell}}/2$). Based on the above equation, from the cell design viewpoint, the channel resistance can be reduced by decreasing the channel length and cell pitch and by reducing the gate oxide thickness to make the gate oxide capacitance larger. When a larger gate drive voltage is applied, the channel resistance decreases, leading to a smaller total specific on-resistance.

3.2.2 Drain Saturation Current Density

As previously pointed out, the short-circuit withstand capability is determined by the device saturation current level under short-circuit conditions. A smaller saturation current produces an increase in the short-circuit withstand time. The drain saturation current is determined by the basic power MOSFET cell design parameters:

$$J_{\mathrm{D,SAT}} = \frac{\mu_{\mathrm{CH}} C_{\mathrm{OX}}}{L_{\mathrm{CH}} W_{\mathrm{Cell}}} \left(V_{\mathrm{G}} - V_{\mathrm{TH}}\right)^2 \tag{3.4}$$

where μ_{CH} is the channel mobility, C_{OX} is the specific gate oxide capacitance, L_{CH} is the channel length, W_{Cell} is the MOSFET cell width, and V_{TH} is the threshold voltage.

Based on Eq. (3.4), the drain saturation current can be reduced by increasing the channel length and cell pitch and by increasing the gate oxide thickness to reduce the gate oxide capacitance. It can be concluded that a trade-off is required between decreasing the drain saturation current to improve the short-circuit withstand time and an increase in specific on-resistance for the SiC power MOSFET.

3.2.3 Trade-Off Curve Using Channel Length

Based on the fundamental operating physics of the silicon carbide power MOSFET, a trade-off is necessary between increasing the short-circuit withstand time and achieving a low specific on-resistance. One approach to perform this trade-off is by variation of the channel length. It is worth pointing out that the cell width also increases when increasing the channel length.

The trade-off curve obtained by varying the channel length is shown in Fig. 3.5 for the case of a 1.2 kV inversion channel 4H-SiC shielded power MOSFET. A channel mobility of 15 cm^2/V-s was used in the analytical model with gate oxide thickness of 500 Å. The saturated current density is large (25,000 A/cm^2) for a typical channel length of 0.5 μm, leading to a measured short-circuit withstand time of 4.8 μs [7]. Based on applying Eq. (2.3), the short-circuit withstand time can be made larger than 10 μs if the drain saturation current density is reduced to 12,000 A/cm^2. It can be concluded that a channel length of 1.0 μm will achieve this goal. This will produce a 40% increase in the specific on-resistance from 2.5 to 3.5 mΩ-cm^2.

3.2.4 Trade-Off Curve Using Gate Drive Voltage

The drain saturation current density of the shielded SiC MOSFET structure can also be reduced by decreasing the gate bias voltage per Eq. (3.4). However, the reduction of the gate bias voltage increases the channel resistance per Eq. (3.3), which increases the total specific on-resistance of the MOSFET. A trade-off between the reduction of drain saturation current density and increase in on-resistance is

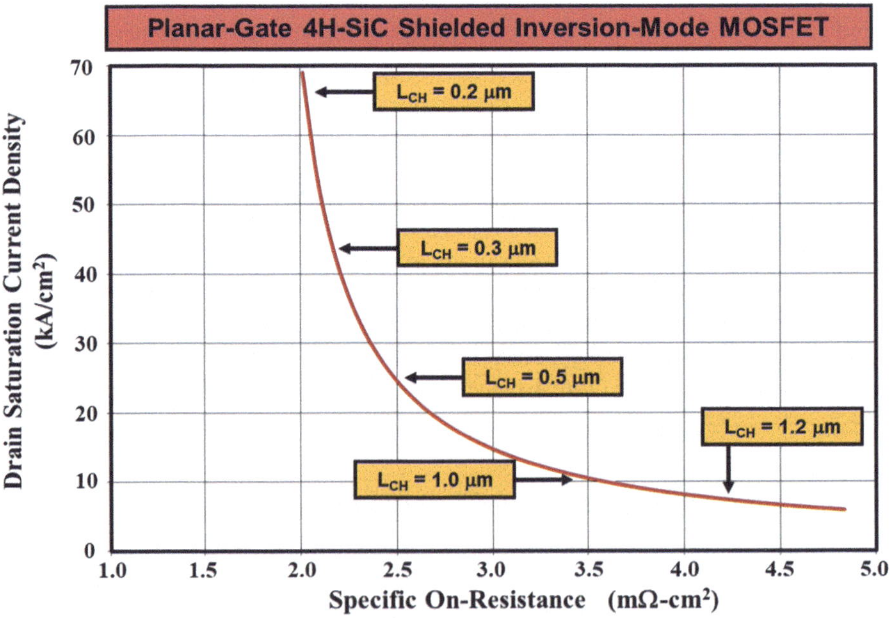

Fig. 3.5 Trade-off curve between drain saturation current and specific on-resistance for shielded channel SiC planar-gate power MOSFET: Channel length as parameter

therefore necessary. Reduction of the gate bias has also been found to increase the switching losses.

The trade-off curve obtained for the 1.2 kV SiC shielded power MOSFET structure described in the previous section was obtained using the analytical models by varying the gate bias. The results are displayed in Fig. 3.6. The drain saturation current could be reduced from 24,000 A/cm^2 to <5000 A/cm^2 by reducing the gate bias from 20 to 10 V. This results in an increase in specific on-resistance by 43% from 2.50 to 3.58 mΩ-cm^2.

The trade-off curve between raising the short-circuit withstand time and the increase in the on-resistance has been experimentally evaluated by reducing the gate drive voltage. The measurements were performed on a typical 1.2 kV SiC planar-gate power MOSFET product (C2M0280120D [8]). A summary of the measured data is given in Table 3.1 [9]. The peak short-circuit current, which is determined by the drain saturation current, was found to reduce by 60% when the gate bias was reduced from 20 to 15 V. This produced an increase in the short-circuit withstand time by 87.5% from 4.8 to 9.0 μs. The on-resistance was found to increase by 66.2% when the gate bias is reduced from 20 to 15 V. In addition, the switching losses increased by about 10% when the gate bias was reduced from 20 to 15 V. Although this approach may be a convenient method to achieve a longer short-circuit time, it has a significant detrimental impact on the power losses and efficiency of power circuits.

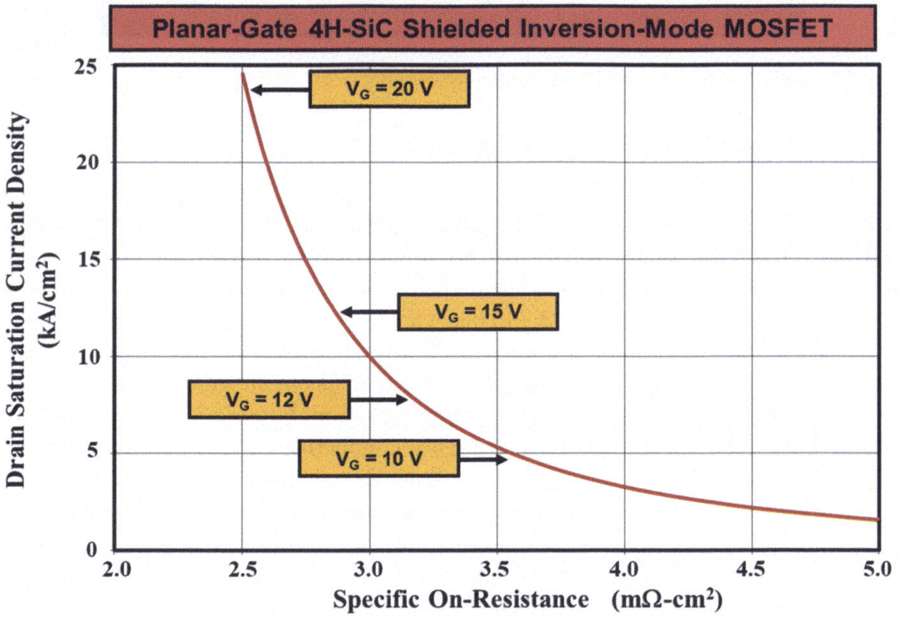

Fig. 3.6 Trade-off curve between drain saturation current and specific on-resistance for shielded channel SiC planar-gate power MOSFET: Gate bias as parameter

Table 3.1 Measured improvement in short-circuit withstand time for a typical commercial SiC power MOSFET

Gate bias (V)	On-resistance (mΩ) (% Increase)	Peak short-circuit current (A)	Short-circuit withstand time (μs) (% Increase)
20	346	52	4.8
18	423 (22.3%)	40 (77%)	6.8 (41.7%)
15	575 (66.2%)	31 (60%)	9.0 (87.5%)

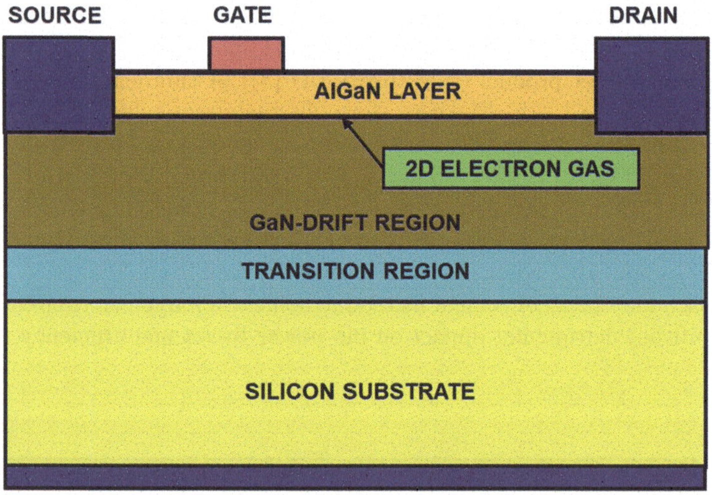

Fig. 3.7 Gallium nitride lateral HEMT structure

3.3 Gallium Nitride HEMT Device

The commercialization of gallium nitride high electron mobility transistors (HEMTs) has been greatly accelerated by the successful growth of high quality gallium nitride layers on silicon substrates. This technology required optimization of a transition layer as shown in Fig. 3.7 to take care of the lattice mismatch between the materials. A thin aluminum-gallium-nitride layer is grown on top of the gallium nitride layer to create the two-dimensional electron gas in the gallium nitride layer. The electrons in the 2D electron gas exhibit a high mobility (~2000 $cm^2/V\text{-}s$) and charge (~10^{13} cm^{-2}). This produces a highly conductive current path through the GaN drift region with low specific on-resistance despite the lateral device structure.

Field plates are applied to the gate and drain electrodes (not shown in the figure) to suppress the high local electric fields at their edges. This allows the GaN HEMT structure to support a large voltage between the drain and source with an average electric field of about 1 MV/cm. A 600 V blocking voltage can therefore be achieved with a gate-drain spacing of 6 μm. A highly interdigitated structure with source, drain, and gate fingers must be designed and fabricated to achieve the desired low on-resistance for a power device.

The GaN HEMT power devices have been found to have extremely poor short-circuit withstand times of under 1 μs as discussed later in the book. Segmentation of the channel has been used to increase the short-circuit withstand time from 0.3 to 3 μs with 38% increase in the on-state resistance [10].

3.4 Summary

The basic structures for power devices have been described in this chapter. The silicon IGBT has become the workhorse of the power electronics community for the past 40 years. The trade-off curve between its on-state voltage drop and its short-circuit withstand capability has been analytically modelled by varying the channel length and gate drive voltage.

Wideband gap semiconductor-based power switches have been commercialized during the past decade. The shielded channel silicon carbide power MOSFET structure is described in this chapter. The trade-off curves between on-resistance and short-circuit withstand capability for this device have been provided by varying the channel length and gate drive voltage.

The gallium nitride HEMT structure is also described in this chapter. It has been found to exhibit very poor short-circuit withstand capability. Segmentation of its channel has been used to extend the short-circuit withstand time while trading-off an increase in its on-resistance.

References

1. B.J. Baliga, *Fundamentals of Power Semiconductor Devices*, 2nd edn. (Springer Science, New York, 2019)
2. B.J. Baliga, *The IGBT Device: Physics, Design, and Applications of the Insulated Gate Bipolar Transistor*, 2nd edn. (Elsevier Press, 2023)
3. B.J. Baliga, *Gallium Nitride and Silicon Carbide Power Devices* (World Scientific Press, Singapore, 2017)
4. B.J. Baliga, Evolution of MOS-bipolar power semiconductor technology. Proc. IEEE **76**, 409–418 (1988)
5. B.J. Baliga, *Silicon Carbide Semiconductor Devices having Buried Silicon Carbide Conduction Barrier Layers Therein*, U. S. Patent 5,543,637 (6 Aug 1996)
6. B.J. Baliga, *Modern Silicon Carbide Power Devices* (World Scientific Press, Singapore, 2024)
7. A. Kanale, B.J. Baliga, Enhancing short circuit capability of 1.2 kV SiC power MOSFETs using a gate-source shorted Si depletion-mode MOSFET in series with the source, in *IEEE Int. Conf. Power Electronics and Drive Systems*, Paper 827 (2019), pp. 1–6
8. Wolfspeed product page, https://www.wolfspeed.com/c2m0280120d
9. A. Kanale, B.J. Baliga, Comparison of current suppression methods to enhance short circuit capability of 1.2 kV SiC power MOSFETs: a new approach using a series-connected,

gate-source-shorted Si depletion-mode MOSFET vs reduced gate bias operation, in *IEEE Industrial Electronics Conf.* (2019), pp. 5114–5119
10. D. Bisi, D. Cruse, P. Zuk, P. Parikh, U. Mishra, Short-circuit capability with GaN HEMTs: invited, in *2022 IEEE International Reliability Physics Symposium (IRPS)* (2022), pp. 1–7

Chapter 4
The BaSIC Topology

The BaSIC topology is a new approach for controlling the switching of power transistors. It was initially proposed to improve the short-circuit withstand time for silicon carbide power MOSFETs [1]. The BaSIC topology is based on using a non-linear resistance in series with the source of the silicon carbide MOSFETs. Commercially available depletion mode and enhancement mode silicon MOSFETs are the most convenient devices to form the non-linear element.

In general, the BaSIC topology can be used to increase the short-circuit withstand time for any power device. Experimental demonstrations of the BaSIC topology have been successfully performed with silicon IGBTs and gallium nitride HEMT devices, as well as silicon carbide power MOSFETs, as discussed in various chapters in this book.

The BaSIC topology has an inherent current sensing capability. This feature can be utilized to monitor the current flowing through various devices in a power electronic converter. This feature is useful for examination of current distribution among paralleled devices when increasing the power handling capability. An important benefit of utilizing the BaSIC topology is preventing failure of devices under repetitive short-circuit events. It is the only approach that has been shown to achieve this outcome.

The BaSIC topology allows detection of the short-circuit event by monitoring the voltage across the non-linear element. This voltage can be fed to the commercially available DESAT protection circuits for silicon IGBTs without the need for a high voltage isolation diode. A shorter response time can be achieved with this approach leading to less stress on the power devices during the short-circuit event.

This chapter reviews the operating principles of the BaSIC topology. This lays the groundwork for its practical implementation as described in later chapters. The impact of its implementation on the on-state and switching performance of the power switch are discussed to allow minimization of any adverse influence on power circuit operation.

B. J. Baliga, A. Kanale, *The BaSIC Topology*,
https://doi.org/10.1007/978-3-031-86630-2_4

4.1 The Baliga Short-circuit Improvement Concept

The **Bal**iga **S**hort-circuit **I**mprovement **C**oncept (BaSIC) is based on utilizing a non-linear resistance in series with the source electrode of the silicon carbide power MOSFET. In order to generalize the discussion, the power device is referred to as a high voltage transistor labeled Q_{HV} and the non-linear resistance is referred to as a low voltage transistor labeled Q_{LV} in this chapter.

A schematic diagram of the BaSIC topology is shown in Fig. 4.1. The high voltage transistor Q_{HV} has terminals labeled D for drain, S for source, and G for gate analogous to the silicon carbide power MOSFET. The non-linear resistance Q_{LV} has terminals labeled T_1 and T_2. One of its terminals T_2 is connected to the source terminal S of the high voltage transistor Q_{HV}. The gate drive voltage is supplied to the gate G of the high voltage transistor Q_{HV} in the same manner as done for silicon carbide power MOSFETs. The gate supply voltage is grounded as usual and thus connected to terminal T_1 of the non-linear resistance Q_{LV}. Consequently, the non-linear resistance is located in the gate drive loop and experiences only low voltages (below 30 V) during power circuit operation.

The on-state current in the high voltage transistor Q_{HV} flows directly through the non-linear resistance Q_{LV}. As explained in the next section, the non-linear resistance is designed to operate in its linear regime at the on-state current rating of the high voltage transistor. The voltage developed across the Q_{LV} is consequently proportional to the current in the high voltage transistor Q_{HV}. The voltage at terminal T_2 of Q_{LV}, labeled V_{SENSE} in Fig. 4.1, can therefore be utilized to monitor the current flowing through the high voltage transistor Q_{HV}. This is an important benefit of using the BaSIC topology because it allows power circuit designers to avoid excessive stress

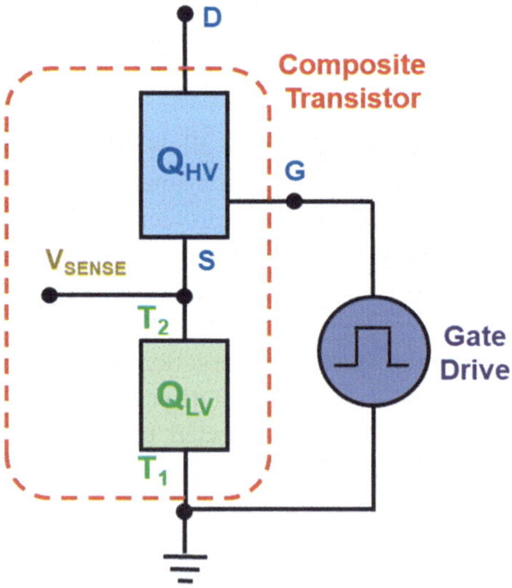

Fig. 4.1 The Baliga Short-circuit Improvement Concept (BaSIC)

on some transistors within the power circuit. It avoids the need for adding additional current sensors, such as Rogowski coils.

4.1.1 Non-Linear Resistance

A key property of the non-linear resistance Q_{LV} is that the resistance must increase with increasing current flowing through it. The non-linear resistance Q_{LV} should preferably limit the current when the voltage across it becomes larger than 5–10 V. A typical characteristic for the non-linear resistance Q_{LV} is illustrated in Fig. 4.2. It must be designed to operate in its linear regime at the on-state current rating of the high voltage power transistor Q_{HV}. This operating point is labeled A in Fig. 4.2. The voltage across the non-linear resistance $V_{ON}(Q_{LV})$ is proportional to the current flowing through it $I_{ON}(Q_{LV})$. The voltage $V_{ON}(Q_{LV})$ can therefore be used to sense the current flowing through Q_{LV} which is equal to the current flowing through the power transistor Q_{HV}.

The current through the power transistor tries to increase to its current saturation value at the gate drive voltage during the short-circuit event. The current saturation value for silicon carbide power transistors is typically 10 times the on-state value. The current saturation value $I_{SAT}(Q_{LV})$ for the non-linear resistance is designed to be about 3 times larger than the on-state current level and consequently 3.5 times smaller than the drain saturation current of the high voltage power transistor. Consequently, the non-linear resistance Q_{LV} is pushed to operation in the current saturation regime as indicated by point C in Fig. 4.2. The voltage across the non-linear resistance $V_{SC}(Q_{LV})$ at point C is well below the gate drive voltage $V_{GS,ON}$ and is in the range of 5–8 V. This behavior is discussed in a subsequent section.

The current saturation value for the non-linear element is chosen to be about 3 times the on-state current rating of the power device Q_{HV}. This avoids disruption of normal power circuit operation where the transistor current can spike to twice the on-state level due to reverse recovery of antiparallel diodes. The current limiting

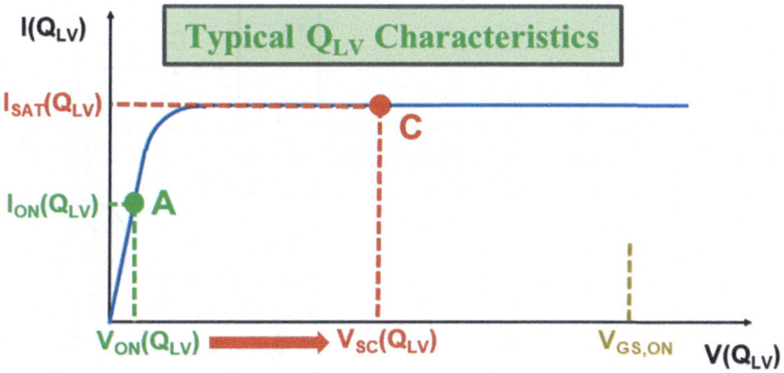

Fig. 4.2 Typical characteristics of the non-linear resistance Q_{LV}

value for the non-linear resistance is typically 3.5 times smaller than the saturation current level of the power transistor. This increases the short-circuit withstand time by a proportionate amount.

4.2 BaSIC Topology: Normal Operation

The BaSIC topology must be engineered to increase the short-circuit withstand time of power transistors without compromising their on-state and switching performance during normal operation. The characteristics of the non-linear resistance must be chosen using these considerations. The various states of operation for the BaSIC topology are discussed in this section.

4.2.1 On-State Operation

A schematic diagram of the BaSIC topology is shown in Fig. 4.3 pertinent to operation under normal power circuit conditions with the power transistor in its on-state. The gate drive voltage is switched to +20 V in the on-state for the power transistor Q_{HV}. In the on-state, the power transistor conducts current at its typical current rating. This current flows through the non-linear resistance Q_{LV}. The non-linear resistance Q_{LV} is designed to operate in its linear region at point A shown in Fig. 4.2 at the on-state rated current of the power transistor Q_{HV}.

The voltages developed across the non-linear resistance Q_{LV} and across the gate-source terminals of the power transistor Q_{HV} are indicated in Fig. 4.3. The

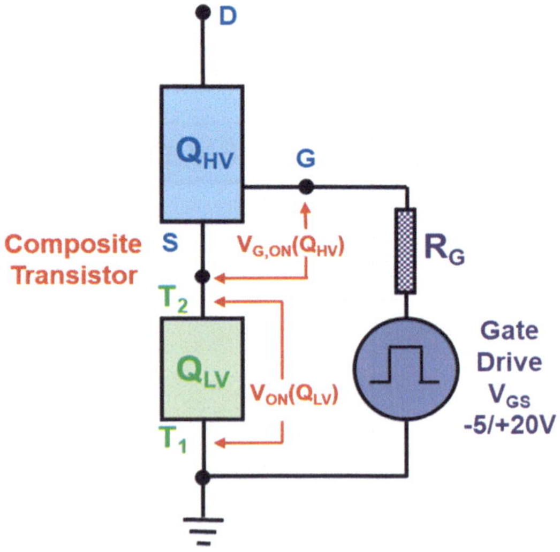

Fig. 4.3 The BaSIC topology under on-state operating conditions

gate-source voltage $V_{G,ON}(Q_{HV})$ is reduced from the gate drive voltage V_{GS} by the voltage $V_{ON}(Q_{LV})$ developed across the non-linear resistance. This will increase on-resistance for a power transistor Q_{HV} such as a silicon carbide power MOSFET. The voltage developed across the non-linear resistance $V_{ON}(Q_{LV})$ must be kept below 1 V in order to minimize the adverse impact of on-state power losses.

4.2.2 Blocking State Operation

The gate drive voltage is reduced to zero or most often to −5 V when controlling power transistors such as silicon carbide power MOSFETs. The power transistor Q_{HV} can then support a high voltage with a small leakage current. This leakage current flows through the non-linear resistance Q_{LV} in its linear regime with a very small voltage drop across it due to its small resistance. It does not experience any high voltages during the blocking mode of operation for the BaSIC topology.

4.2.3 Switching Operation

The turn-on and turn-off of the power transistor Q_{HV} are achieved by toggling the gate drive voltage between −5 V and +20 V. The switching transients for typical power MOSFETs are discussed in detail in the textbook [2]. The switching time intervals during which the currents and voltages are large produce the significant power losses. These voltage transition time intervals become larger when the gate drive voltage is reduced. In addition, the transconductance is reduced by the presence of a resistance in series with the source of the power transistor. A smaller tranconductance slows down the current transients. Both of these phenomena increase the switching power losses. The adverse impact on switching performance can be minimized by reducing the resistance of the non-linear resistance in its linear regime.

4.2.4 On-State Characteristics

The operation of the high voltage power transistor Q_{HV} and the non-linear resistance Q_{LV} under normal on-state conditions can be explained with the aid of Fig. 4.4. The output characteristics of the non-linear resistance Q_{LV} are shown in the left-hand side together with the output characteristics of the power transistor Q_{HV} in the right-hand side. The power transistor has much larger saturation current $I_{SAT}(Q_{HV})$ than the saturation current $I_{SAT}(Q_{LV})$ of the non-linear resistance. The total on-state power loss is the sum of the power dissipated in the power transistor Q_{HV} and the non-linear resistance Q_{LV}. This is close to the on-state power loss of the power transistor

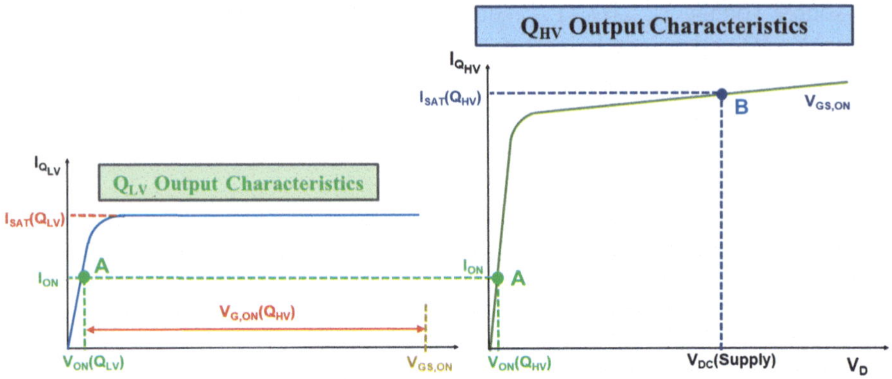

Fig. 4.4 Operating principle of the BaSIC topology: Normal operation

Q_{HV} if the non-linear resistance is designed with a resistance much smaller than that of the power transistor Q_{HV}.

The gate drive voltage for the power transistor under on-state operation is shown in Fig. 4.4 as $Q_{G,ON}(Q_{HV})$. It is close to the gate drive voltage $V_{GS,ON}$ if the non-linear resistance has a low resistance in the on-state.

4.3 BaSIC Topology: Short-Circuit Operation

The pertinent voltages are shown in Fig. 4.5 when the BaSIC topology is operating under short-circuits conditions. When a short circuit occurs across the load with on-state gate bias applied to the stand-alone power transistor Q_{HV}, its current increases to the saturation current $I_{SAT}(Q_{HV})$. It then operates at point B shown in Fig. 4.4 while supporting the DC supply voltage V_{DC}(Supply). The large power dissipation under these conditions leads to failure of the power transistor in a small time interval.

In contrast, when a short-circuit event occurs with the BaSIC topology, the current increases until it becomes limited by the saturation current of the non-linear resistance $I_{SAT}(Q_{LV})$. The voltage across the non-linear resistance $V_{SC}(Q_{LV})$, then increases until it reaches an operating point C as shown in Fig. 4.6. This reduces the gate drive voltage to $V_{G,SC}(Q_{HV})$ for the power transistor. Its output characteristic then shifts from the green line to the purple line as indicated by the red arrow with a reduced saturation current. An important feature of the BaSIC topology is that the value for $V_{SC}(Q_{LV})$ *automatically adjusts* until the new saturation current of the power transistor matches the saturation current of $I_{SAT}(Q_{LV})$ of Q_{LV}. No external influence is required in the BaSIC topology to achieve this outcome. The smaller value of the saturation current at operating point C increases the short-circuit withstand time by the ratio of $I_{SAT}(Q_{HV})$ and $I_{SAT}(Q_{LV})$.

Fig. 4.5 The BaSIC topology under short-circuit operating conditions

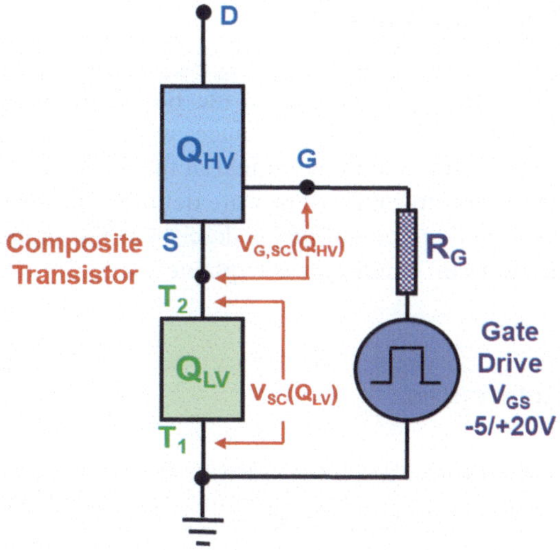

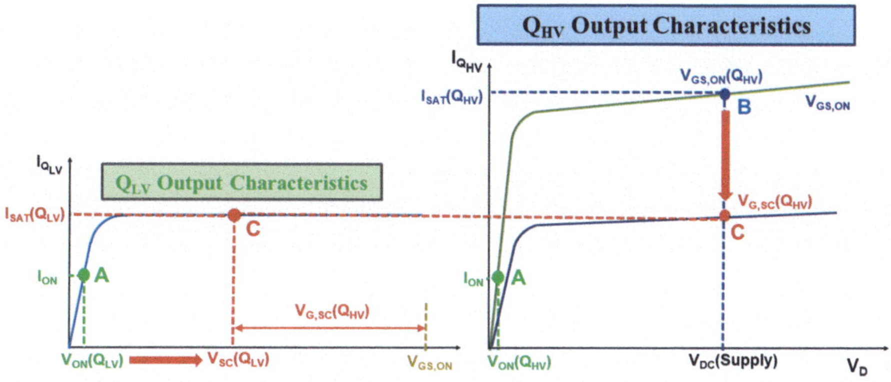

Fig. 4.6 Operating principle of the BaSIC topology: Short-circuit conditions

The power dissipation in the non-linear resistance Q_{LV} under short-circuit conditions is much smaller than that in the power transistor because the voltage drop across it $V_{SC}(Q_{LV})$ is in the range of 5–8 V. It does not undergo a large temperature rise in contrast to the high voltage transistor Q_{HV}. In contrast, the voltage supported by the power transistor Q_{HV} is typically 800 V for a 1200 V rated device. Its temperature does increase substantially during the short-circuit event. The rate of rise in the temperature is reduced by ratio of $I_{SAT}(Q_{HV})$ and $I_{SAT}(Q_{LV})$. This extends the short-circuit withstand time allowing the DESAT circuit to respond in time to safely turn-off the high voltage transistor.

4.4 Summary

The operating principles of the BaSIC topology are explained in this chapter. It employs a non-linear resistance, whose resistance increases with current level, in series with the source terminal of the power transistor. The optimum characteristics for the non-linear resistance are defined. The operation of the BaSIC topology under normal circuit conditions is described. The unique behavior of this topology under short-circuit conditions is explained.

References

1. B.J. Baliga, *Power Devices having Tunable Saturation Current Clamps Therein that Support Improved Short-Circuit Capability*, U. S. Patent 12,094,932 (17 Sept 2024)
2. B.J. Baliga, *Fundamentals of Power Semiconductor Devices*, 2nd edn. (Springer, 2019)

Chapter 5
BaSIC(DMM) for SiC Power MOSFETs

The efficacy of the BaSIC topology was first demonstrated for enhancing the short-circuit capability of silicon carbide power MOSFETs. These devices have been found to exhibit short-circuit withstand times of <5 μs [1, 2]. The short-circuit withstand time for the SiC power MOSFETs can be improved by using a gate-source-shorted (GSS) silicon depletion mode MOSFET (DMM) as the non-linear current limiting element [3]. A search for silicon depletion mode power MOSFETs produced only one device with the characteristics suitable for the BaSIC topology. It is an IXYS 100 V rated Si depletion-mode MOSFET with on-resistance of 64 mΩ at zero gate bias, i.e., with gate shorted to the source terminal. This device is referred to as the silicon DMM in this chapter, and the implemented composite transistor structure is referred to CM in the BaSIC(DMM) topology.

Trade-off curves between the short-circuit withstand time and the increase in on-resistance and switching energy loss achieved with the BaSIC(DMM) topology were obtained using a 1.2 kV SiC power MOSFETs with datasheet on-resistance of 280 mΩ. These results are compared in subsequent chapters with two other short-circuit ruggedness improvement techniques: (a) using reduced gate drive voltage and (b) using a series resistance.

5.1 Short-Circuit Configuration: Stand-Alone SiC Power MOSFET

The circuit configuration for performing the short-circuit test on the stand-alone SiC power MOSFET is shown in Fig. 5.1 with the device labeled M. It consists of a DC power supply with voltage V_{DC}, a capacitor C_{DC} across the DC power supply to deliver the short-circuit current into the SiC power MOSFET while maintaining the DC voltage, and the gate drive with voltage V_{GS} with series resistor $R_{G(ext)}$. The gate drive voltage for the SiC power MOSFET was set at 20 V in accordance with the

B. J. Baliga, A. Kanale, *The BaSIC Topology*,
https://doi.org/10.1007/978-3-031-86630-2_5

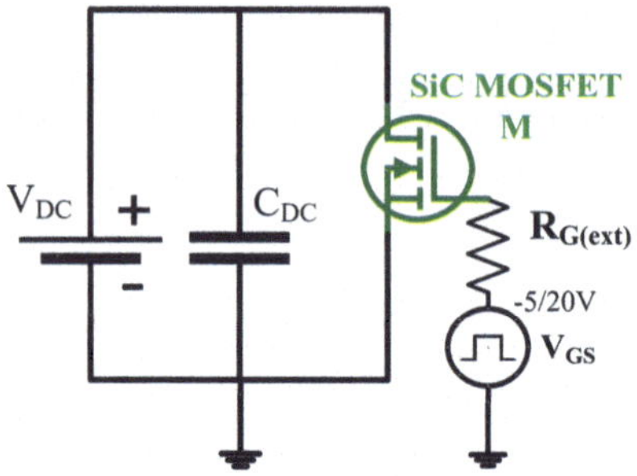

Fig. 5.1 Short-circuit circuit with stand-alone SiC power MOSFET M

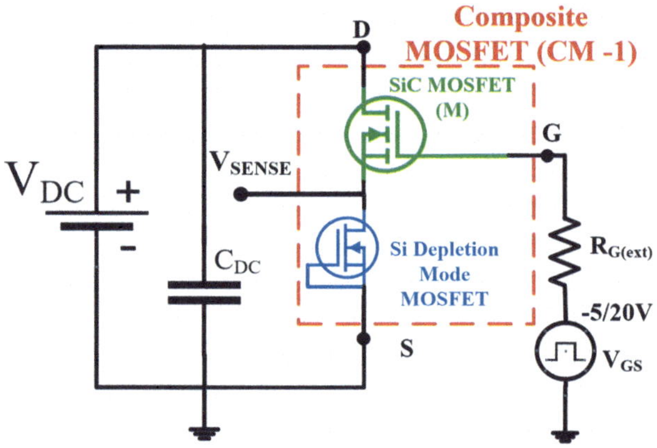

Fig. 5.2 Short-circuit circuit with BaSIC(DMM) topology: Case CM-1 with single DMM

datasheet [4]. The DC supply voltage was set at 800 V to match the power MOSFET switching conditions specified in the datasheet.

The short-circuit tests were performed by initially using a short duration gate pulse of 1 μs to avoid damaging the SiC power MOSFET. The width of the gate pulse was gradually increased until the device failed while capturing the drain current waveform. This allowed determination of the short-circuit withstand capability of the stand-alone SiC power MOSFET.

5.2 Short-Circuit Configuration: BaSIC(DMM) Topology

The short-circuit tests for the BaSIC(DMM) topology were performed for two cases. In the first case, a single silicon gate-source-shorted (GSS) DMM was connected in series with the source electrode of the SiC power MOSFET. This configuration is shown in Fig. 5.2 with the composite MOSFET labeled as CM-1. It is worth pointing out that the gate drive voltage is applied to the gate of the SiC power MOSFET as usually done in its power circuit applications. The silicon DMM device only serves as the non-linear resistance Q_{LV} with no gate bias applied to its gate terminal.

The output characteristics of the silicon DMM device at 25 °C in its linear regime of operation are provided in its datasheet [5]. They are reproduced here in Fig. 5.3 for discussion. It is appropriate to use the 25 °C data because the silicon DMM does not undergo switching losses under normal circuit operating conditions preventing a significant increase in its junction temperature. The operating point A for silicon DMM device at the SiC power MOSFET rated current of 6 A is marked with the green dot using the characteristic with zero gate bias. This point corresponds to the operating point A shown in Fig. 4.4 for the non-linear resistance Q_{LV}. The voltage drop across the silicon DMM device is <0.3 V. This minimizes the reduction in gate drive voltage for the SiC power MOSFET and makes the on-resistance of the composite MOSFET close to that of the stand-alone SiC power MOSFET.

The output characteristics of the silicon DMM device at 100 °C in its current saturation regime of operation are also provided in its datasheet [4]. They are reproduced here in Fig. 5.4 for discussion. It is appropriate to use the 100 °C data because the silicon DMM undergoes some adiabatic heating during the short-circuit event. Its voltage rises to about 8 V with a high short-circuit current. The resulting power dissipation produces an increase in its junction temperature that is estimated to reach 100 °C. As an example, the operating point C for this device is marked with

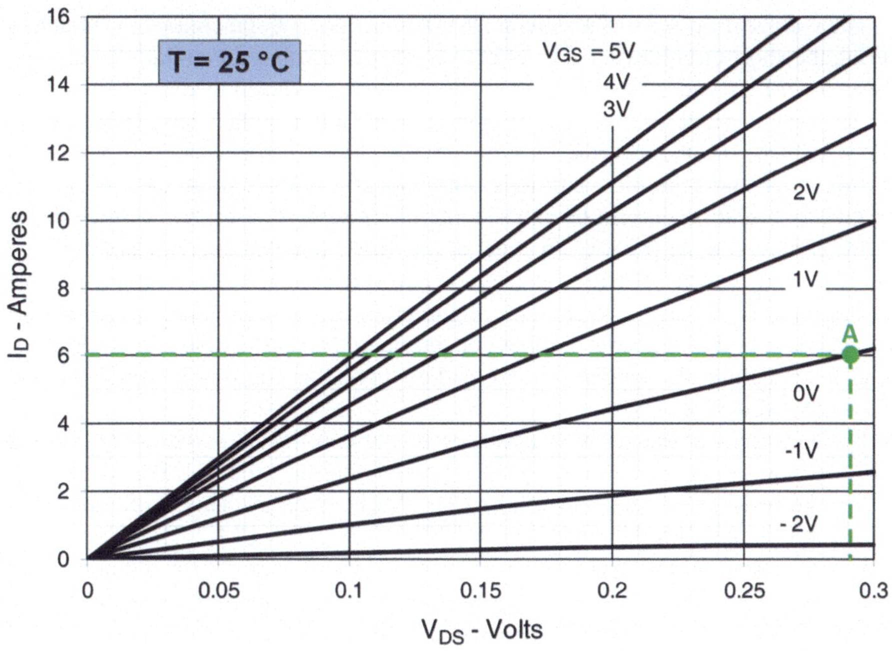

Fig. 5.3 Output characteristics of the silicon DMM device in the linear region

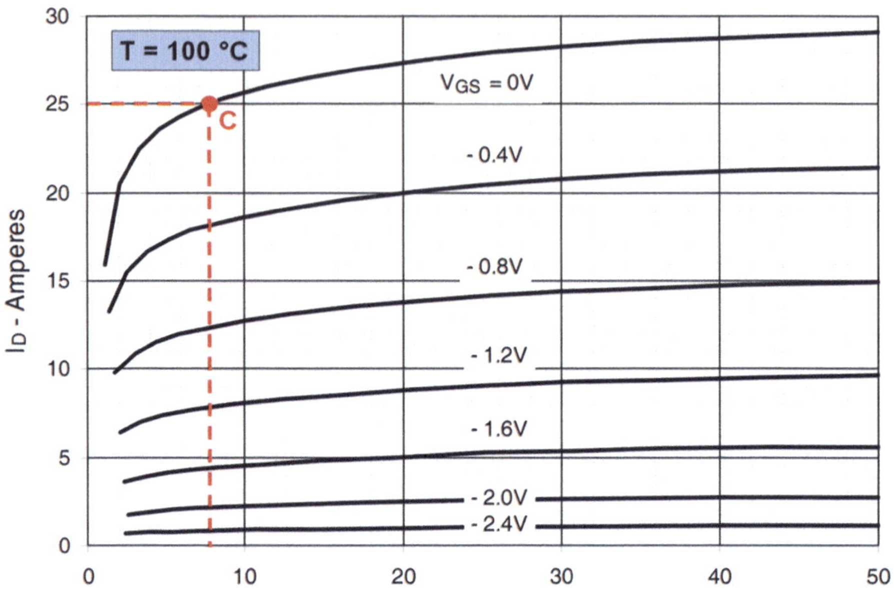

Fig. 5.4 Output characteristics of the silicon DMM device in the current saturation region

the red dot. This point corresponds to the operating point C shown in Fig. 4.4 for the non-linear resistance Q_{LV}.

The voltage drop across the silicon DMM device $V_{SC}(Q_{LV})$ increases to about 8 V during the short-circuit event. The effective gate drive voltage $V_{SC}(Q_{HV})$ for the SiC power MOSFET under short-circuit conditions is then reduced to about 12 V, which brings down its saturation current until it automatically matches that for the silicon DMM device. In other words, the voltage $V_{SC}(Q_{LV})$ developed across the silicon DMM under short-circuit conditions automatically adjusts until the SiC power MOSFET saturation current becomes equal to the saturation current of the silicon DMM at its operating point C. In general, the saturation characteristics of both the silicon carbide power MOSFET and silicon DMM device will vary from part to part for the same device due to manufacturing tolerances. The automatic adjustment of the operating point C ensures seamless operation of the BaSIC topology without the need for matching the characteristics of the silicon carbide power MOSFET and silicon DMM device.

It is beneficial to reduce the resistance of the non-linear resistance Q_{LV} in the BaSIC topology. This can be achieved by paralleling the silicon DMM devices to reduce the resistance in half for the BaSIC(DMM) topology. This configuration is shown in Fig. 5.5 with the composite MOSFET labeled as CM-2. However, paralleling the silicon DMM devices increases the saturation current for the non-linear resistance Q_{LV}, which shortens the short-circuit withstand time. The composite MOSFET CM-2 was therefore evaluated to create a trade-off curve for the BaSIC(DMM) topology.

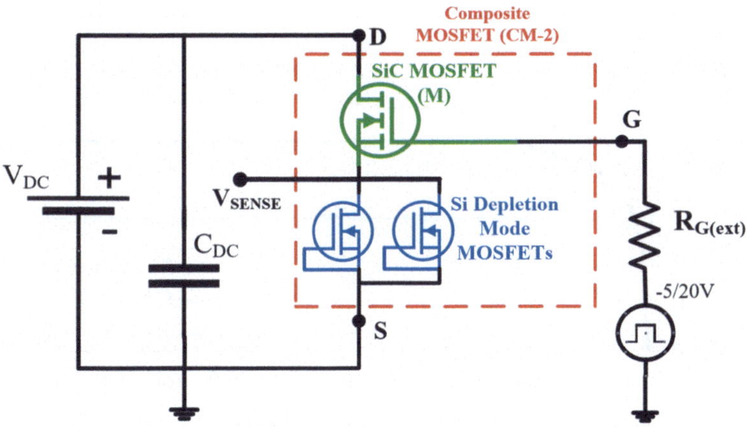

Fig. 5.5 Short-circuit circuit with BaSIC(DMM) topology: Case CM-2 with two DMM devices in parallel

5.3 High Voltage Output Characteristics

As explained in Chap. 2, the power transistor operates in its current saturation mode while supporting a high voltage equal to the DC supply voltage for the load during the short-circuit conditions. This operating point is part of the power transistor's output characteristics. Knowledge of the high-voltage, high-current output characteristics is therefore invaluable for understanding the short-circuit capability of power transistors. However, these characteristics are not provided by device manufacturers in the datasheets of high voltage power transistors.

A novel method for extracting the high-voltage, high-current output characteristics of power transistors was developed by using the short-circuit test configuration [6]. The short-circuit test configuration shown in Fig. 5.1 was used with a small pulse width of 1.5 μs. During this time, the current increases to a peak value that is representative of the saturation current at the DC supply voltage. The DC supply voltage can be varied in small steps to generate the high-voltage, high-current output characteristic for the chosen gate bias. The experimentally extracted high-voltage, high-current output characteristic at a gate bias of 20 V for the stand-alone SiC power MOSFET M is shown in Fig. 5.6 with the red line. The current increases gradually up to 400 V and then saturates a value of 52 A.

The same test methodology was used to extract the high-voltage, high-current output characteristics of the composite MOSFETs CM-1 and CM-2. These characteristics are shown in Fig. 5.6 by the green and blue lines. The saturation current for the composite MOSFET CM-1 with single silicon DMM device is 23 A. This corresponds to operating point C in Fig. 5.4. The saturation current for the composite MOSFET CM-2 with two silicon DMM devices in parallel is 45 A, which is twice that for the single silicon DMM case as expected.

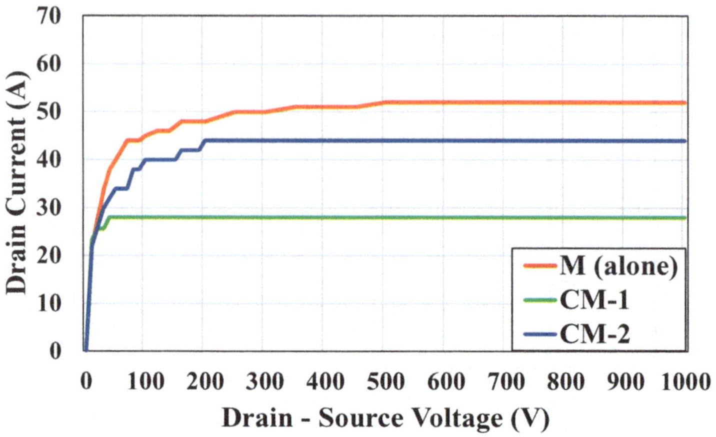

Fig. 5.6 High-voltage, high-current output characteristics: Stand-alone SiC power MOSFET M, composite MOSFET CM-1 and composite MOSFET CM-2

The test methodology for acquiring the high-voltage, high-current output characteristics for power transistors described here can be applied to any power devices, such as silicon IGBTs. The important criterion for using this approach is that the power transistor should be able to survive the short-circuit conditions until its current reaches the peak value in a small time interval, such as 1.5 μs used for the above devices.

5.4 Short-Circuit Test Results

The short-circuit withstand time for the power transistor can be obtained by using the test configuration shown in Fig. 5.1. The short-circuit pulse width is gradually increased while capturing the current waveform until the device fails. The DC supply voltage is monitored as shown by the black line in Fig. 5.7. It is important to make sure that the capacitor C_{DC} is adequate to maintain the voltage at the desired value, 800 V in this case.

The red lines in Fig. 5.7 are obtained for the stand-alone SiC power MOSFET M. It can be seen that the device survived pulses of 3.0 and 4.0 μs in width but failed when the pulse width was increased to 4.8 μs. The green lines were obtained for the composite MOSFET CM-1. It survived a pulse width of 6 and 7 μs but failed when the pulse width was increased to 7.9 μs. The blue lines were obtained for the composite MOSFET CM-2. It survived a pulse width of 5 μs but failed when the pulse width was increased to 5.3 μs. The short-circuit withstand time increases in inverse proportion to the short-circuit current level as expected from the theory provided in Chap. 2.

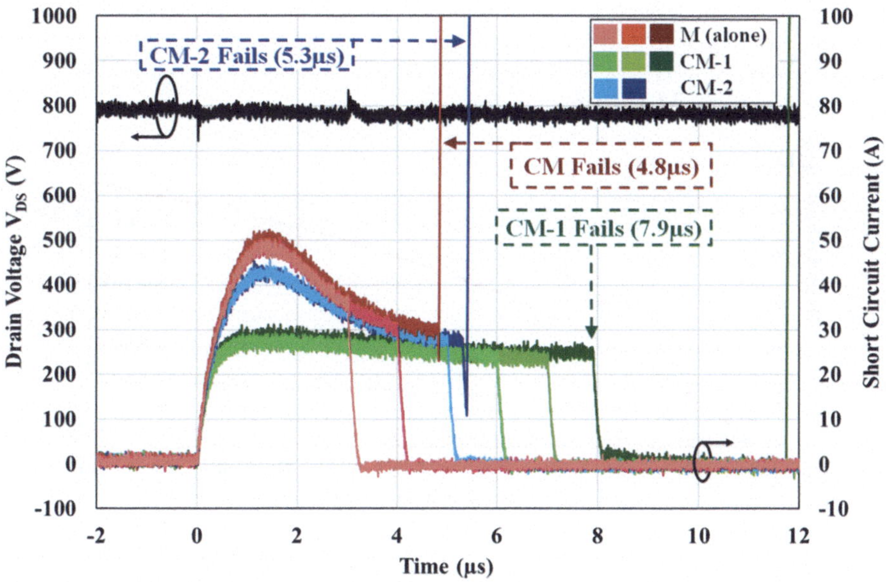

Fig. 5.7 Measured short-circuit waveforms: Stand-alone SiC power MOSFET M, composite MOSFET CM-1 and composite MOSFET CM-2

These short-circuit test results provide validation of the BaSIC topology idea. This topology was able to reduce the short-circuit current level and significantly increase the short-circuit withstand time for the SiC power MOSFET.

5.5 Impact on Normal Circuit Operation

As discussed in Chap. 4, the BaSIC topology is based on the addition of a non-linear resistance in series with the source electrode of the power transistor. The impact of the resistance of this element on the on-state resistance and switching behavior of the SiC power MOSFET must be quantified. Trade-off curves between enhancing the short-circuit time and an increase on-resistance and switching losses must be developed to fully comprehend the value of the BaSIC topology.

5.5.1 Impact on On-Resistance

The output characteristics of the three configurations discussed above were obtained at lower drain voltages by using a curve tracer. The results for stand-alone SiC power MOSFET M are shown in Fig. 5.8 for several gate bias values. The on-state operating point for this device at its rated current of 6 A is highlighted with the red

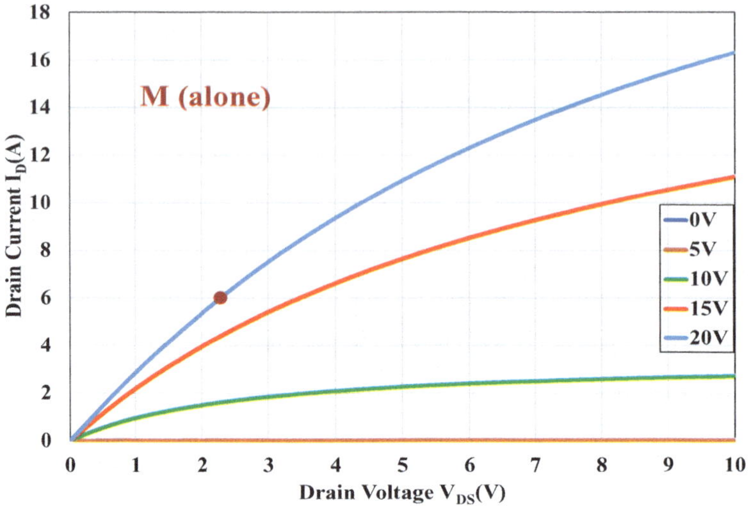

Fig. 5.8 Measured output characteristics of the stand-alone SiC power MOSFET M

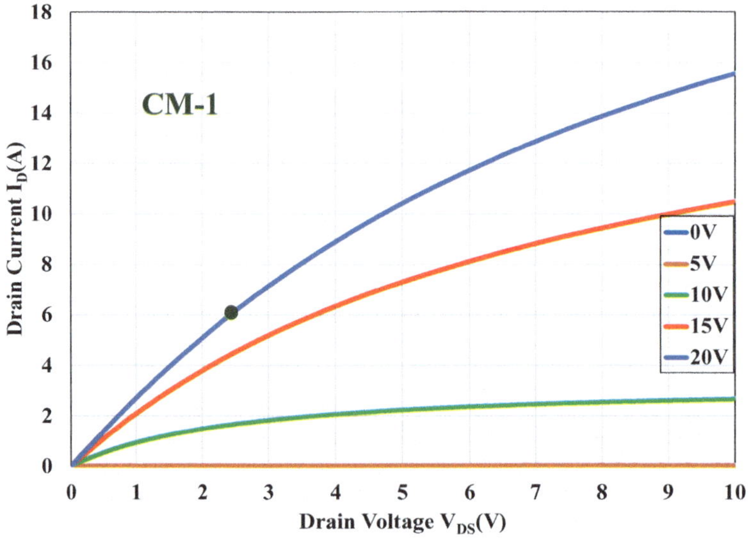

Fig. 5.9 Measured output characteristics of the composite MOSFET CM-1

dot. The on-resistance at this operating point is 346 mΩ. The typical value cited in the datasheet for this transistor is 280 mΩ with a maximum value of 370 mΩ. The measured values fall within the datasheet parametric variation of 32% above the typical value.

The output characteristics obtained for the composite MOSFETs CM-1 and CM-2 are shown in Figs. 5.9 and 5.10 for several gate bias values. The on-state

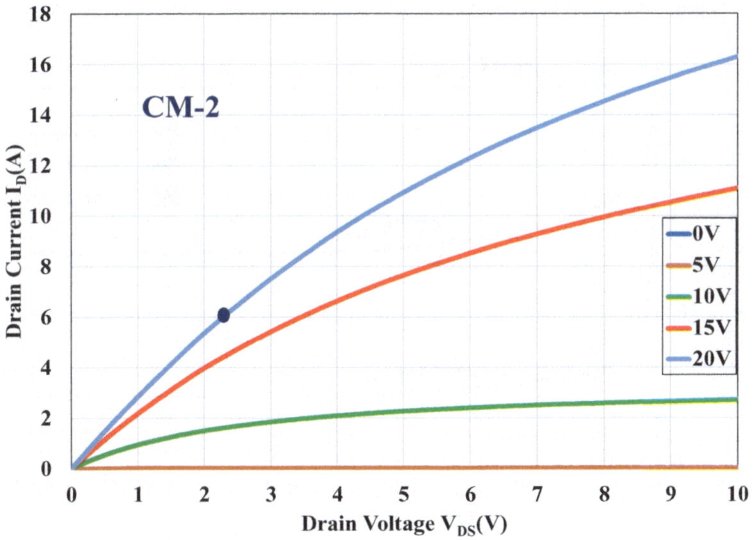

Fig. 5.10 Measured output characteristics of the composite MOSFET CM-2

operating point for each device at its rated current of 6 A is highlighted with the green and blue dot. The on-resistance at this operating point is 405 mΩ for CM-1. This is consistent with the additional silicon DMM resistance of about 60 mΩ. The on-resistance at this operating point is 381 mΩ for CM-2. This is consistent with the additional silicon DMM resistance of about 30 mΩ when connected in parallel. The on-resistance is increased by 17% for the CM-1 case and by 10% for the CM-2 case. It is worth pointing out that these changes are much smaller the 32% device-to-device variation when supplied by the manufacturer.

5.5.2 Impact on Capacitances

The capacitances of a power MOSFET determine its switching performance [7]. The input, output, and reverse transfer capacitances of the three configurations were therefore obtained. These measured characteristics are displayed in Fig. 5.11. The red lines apply for the stand-alone SiC power MOSFET. The measured values for the C_{ISS}, C_{OSS}, and C_{RSS} are very close to values provided in the datasheet for the device.

The green and blue lines in Fig. 5.11 apply to the composite MOSFET CM-1 and CM-2. They overlap the lines for the stand-alone SiC power MOSFET. This demonstrates that the BaSIC topology does not alter the capacitances.

The reason for this can be explained by using the equivalent circuit for the BaSIC topology shown in Fig. 5.12. The SiC power MOSFET has its usual gate-source $C_{GS,SiC}$, drain-source $C_{DS,SiC}$, and gate-drain $C_{GD,SiC}$ capacitances. In contrast, the

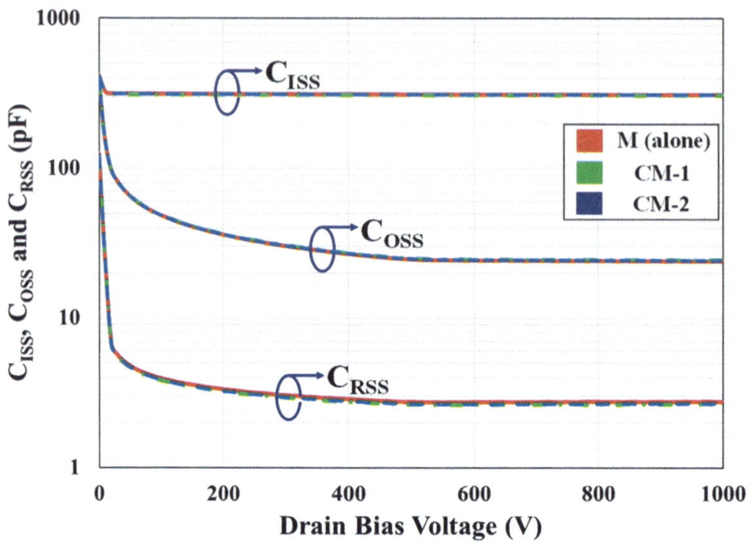

Fig. 5.11 Measured capacitances of the stand-alone SiC power MOSFET M, composite MOSFET CM-1, and composite MOSFET CM-2

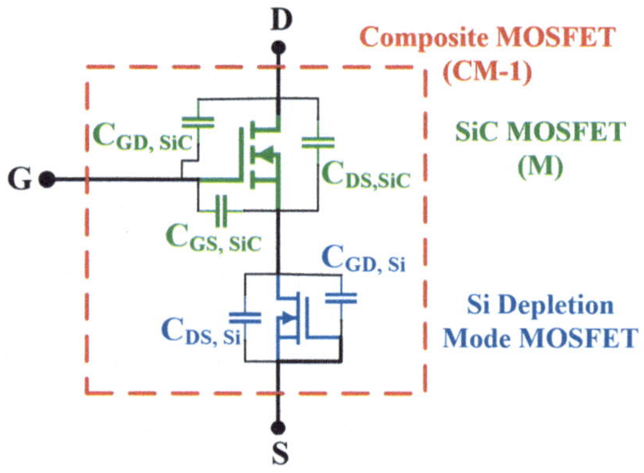

Fig. 5.12 Equivalent circuit model for the capacitances of the composite MOSFET CM-1

silicon DMM device has only its usual drain-source $C_{\mathrm{DS,SiC}}$, and gate-drain $C_{\mathrm{GD,SiC}}$ capacitances. The gate-source $C_{\mathrm{GS,SiC}}$ capacitance is eliminated by the shorting of its gate and source.

The input capacitance of the composite MOSFET obtained by using the above equivalent circuit is as follows:

$$C_{\mathrm{ISS}}(\text{Composite MOSFET}) = \frac{C_{\mathrm{ISS}}(\mathrm{SiC})C_{\mathrm{OSS}}(\mathrm{Si})}{C_{\mathrm{ISS}}(\mathrm{SiC}) + C_{\mathrm{OSS}}(\mathrm{Si})} \approx C_{\mathrm{ISS}}(\mathrm{SiC}) \quad (5.1)$$

The value of C_{OSS}(Si) for the DMM device has a very large value of 10,000 pF at small drain voltages as shown in Fig. 5.13. This value is much bigger than the C_{ISS}(SiC) of 260 pF for the SiC power MOSFET. Consequently, the C_{ISS} of the composite MOSFET becomes essentially equal to that of the stand-alone SiC power MOSFET as indicated in the equation.

The output capacitance of the composite MOSFET obtained by using the above equivalent circuit is as follows:

$$C_{OSS}\left(\text{Composite MOSFET}\right) = \frac{C_{OSS}\left(\text{SiC}\right)C_{OSS}\left(\text{Si}\right)}{C_{OSS}\left(\text{SiC}\right)+C_{OSS}\left(\text{Si}\right)} \approx C_{OSS}\left(\text{SiC}\right) \quad (5.2)$$

The value of C_{OSS}(Si) has a very large value of 10,000 pF at small drain voltages as shown in Fig. 5.13. This value is much bigger than the C_{OSS}(SiC) of 23 pF for the SiC power MOSFET at large drain bias values as given in its datasheet. Consequently, the C_{OSS} of the composite MOSFET becomes essentially equal to that of the stand-alone SiC power MOSFET as indicated in the equation.

The reverse transfer capacitance, C_{RSS}, for the composite MOSFET is measured with gate of the SiC MOSFET shorted to the source of the Si DMM device per standard practice in the industry. The shorting of the gate and source terminals of the composite MOSFET eliminates any influence of the Si DMM on the capacitance. Consequently, the C_{RSS} for the composite device is equal to that of the SiC MOSFET.

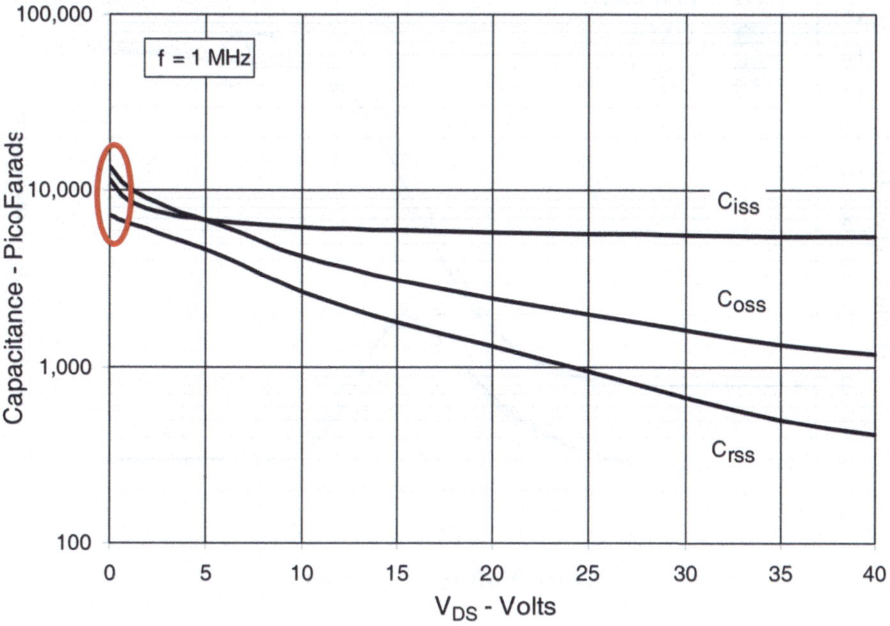

Fig. 5.13 Capacitances of the silicon DMM device

From the above discussion, it is apparent that the BaSIC topology does not alter the capacitances of the SiC power MOSFET. This is a favorable outcome from the point of view of the switching behavior.

5.5.3 Impact on Transconductance

The transconductance g_M of the power transistor plays a role in the switching transient for the drain voltage and current during operation with an inductive load [6]. Its value can be ascertained from the transfer characteristics of the device. The measured transfer characteristics for the three configurations at a drain bias of 20 V are shown in Fig. 5.14. The transconductance, extracted by taking a derivative of the drain current with respect to the gate voltage, is also shown in the figure. The value of the transconductance can be compared for the three cases at the same drain current level of 6 A.

The stand-alone SiC power MOSFET has a transconductance of 3.35 S, which is close to the value in its datasheet. The transconductances of the composite MOSFET CM-1 and CM-2 are 2.86 and 3.02 S. The reduction in the transconductance is associated with the resistance of the non-linear resistance Q_{LV}:

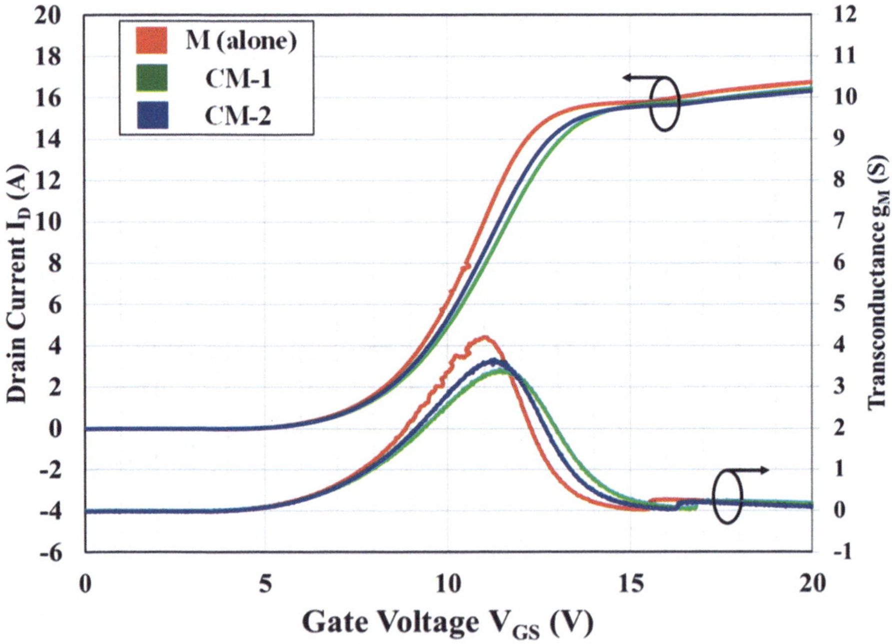

Fig. 5.14 Transconductances for the stand-alone SiC power MOSFET M, the composite MOSFET CM-1, and the composite MOSFET CM-2

$$g_{\mathrm{M}}\left(\text{with series}\, R\right) = \frac{g_{\mathrm{M}}\left(\mathrm{SiC}\right)}{1 + R_{\mathrm{S}} g_{\mathrm{M}}\left(SiC\right)} \tag{5.3}$$

The transconductance value predicted by this equation is 2.9 S for the case of a series resistance R_S of 60 mΩ. This matches the measured value for the composite MOSFET CM-1 where the silicon DMM has a resistance of 60 mΩ. The transconductance value predicted by this equation is 3.05 S for the case of a series resistance R_S of 30 mΩ. This matches the measured value for the composite MOSFET CM-2 where the two paralleled silicon DMMs have a resistance of 30 mΩ. The impact of the reduction of the transconductance on the switching losses is discussed in the next section.

5.5.4 *Impact on Switching Losses*

The double-pulse test is commonly used to evaluate the switching power losses of power transistors. Its basic circuit diagram is shown in Fig. 5.15 where the power MOSFET is controlling current flow from the DC supply to the load inductance. When the power MOSFET is initially turned-on by closing switch S_1, current I_L flows from the DC power supply through the load inductance. This current increases linearly with time during the first pulse duration until it reaches the rated current for the power transistor. When the gate voltage is turned-off by opening switch S_1 and closing switch S_2, the drain voltage of the power MOSFET first increases until it becomes one diode voltage drop larger than the DC supply voltage. This forward

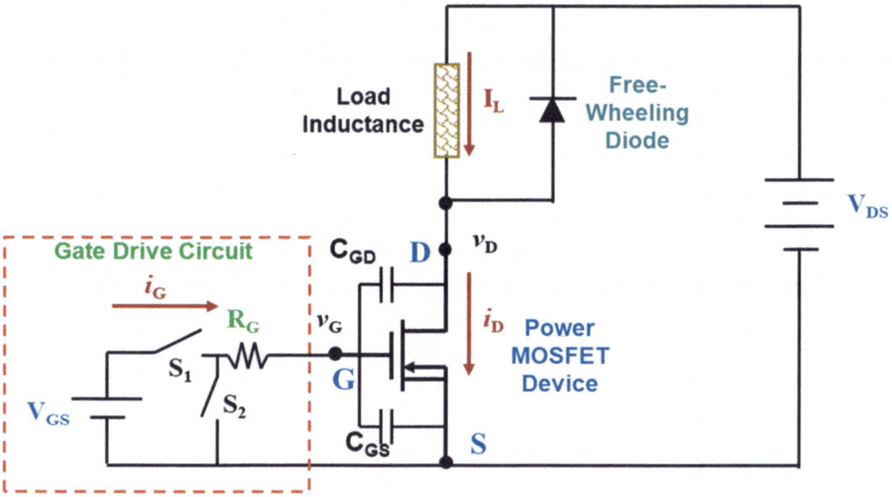

Fig. 5.15 Double-pulse test configuration for evaluation of switching losses for power transistors

biases the freewheeling diode. The inductor current then transfers from the power MOSFET to the freewheeling diode, allowing its current i_D to reduce to zero. The waveforms for this circuit are analyzed in the textbook with analytical expressions derived for the various portion of the voltage and current transients [6].

A second gate pulse is applied after a short time interval to turn-on the power MOSFET again. This allows transfer of the inductor current from the freewheeling diode to the power MOSFET. The diode undergoes its reverse recovery process at this switching event. A large reverse recovery current is observed for silicon Pi-N rectifiers [6]. For this reason, the double-pulse testing of SiC power MOSFETs was performed using SiC JBS rectifiers as freewheeling diodes. The turn-on current was observed to have an overshoot during turn-on even in this case due to the high dV_D/dt at the drain terminal of the power MOSFET and the capacitance of the SiC JBS diode.

The rate of change of voltage for a power MOSFET during its turn-on is given by [6]:

$$\frac{dV_D}{dt} = \frac{V_{GS} - V_{TH}}{R_G C_{GD}(V_D)} - \frac{I_D}{R_G C_{GD}(V_D) g_M} \tag{5.4}$$

Based on this expression, the dV_D/dt decreases when the transconductance g_M becomes smaller. A smaller dV_D/dt produces greater switching losses. It can therefore be expected that the switching losses for the composite MOSFET CM-1 and CM-2 will be larger than for the stand-alone SiC power MOSFET because they have smaller transconductance values as discussed in Sect. 5.5.3.

The measured turn-on transients for the three transistor configurations are shown in Fig. 5.16. It can be observed that the dV_D/dt is decreasing from the stand-alone SiC power MOSFET to the composite MOSFET CM-2 to the composite MOSFET CM-1. Although there is considerable oscillation in the current waveform due to stray inductances in the test circuit, the energy loss per cycle can be extracted by integration of the instantaneous drain current and voltage. The turn-on loss for the stand-alone SiC power MOSFET, the composite MOSFET CM-2, and the composite MOSFET CM-1 were found to be 157, 181, and 196 μJ. The composite MOSFETs exhibit larger turn-on losses due to their smaller transconductances as concluded from Eq. (5.4).

The measured turn-off transients for the three transistor configurations are shown in Fig. 5.17. Very little difference between the three cases is observed. The turn-off loss for the stand-alone SiC power MOSFET, the composite MOSFET CM-2, and the composite MOSFET CM-1 were found to be 46, 44, and 47 μJ. The total energy loss per cycle for the stand-alone SiC power MOSFET, the composite MOSFET CM-2, and the composite MOSFET CM-1 were 203, 225, and 240 μJ. The energy loss per cycle for the composite MOSFETs CM-2 and CM-1 is larger than that for the stand-alone SiC power MOSFET by 11 and 18%.

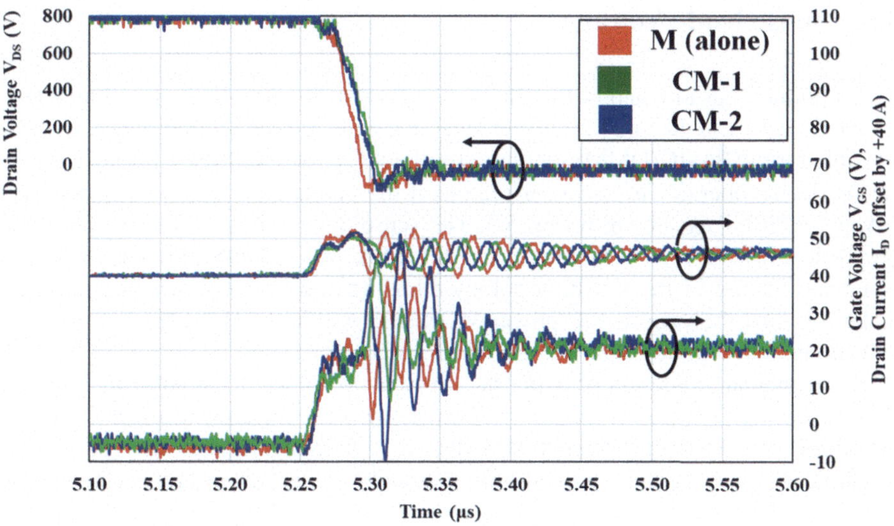

Fig. 5.16 Turn-on transients for the stand-alone SiC power MOSFET M, the composite MOSFET CM-1, and the composite MOSFET CM-2

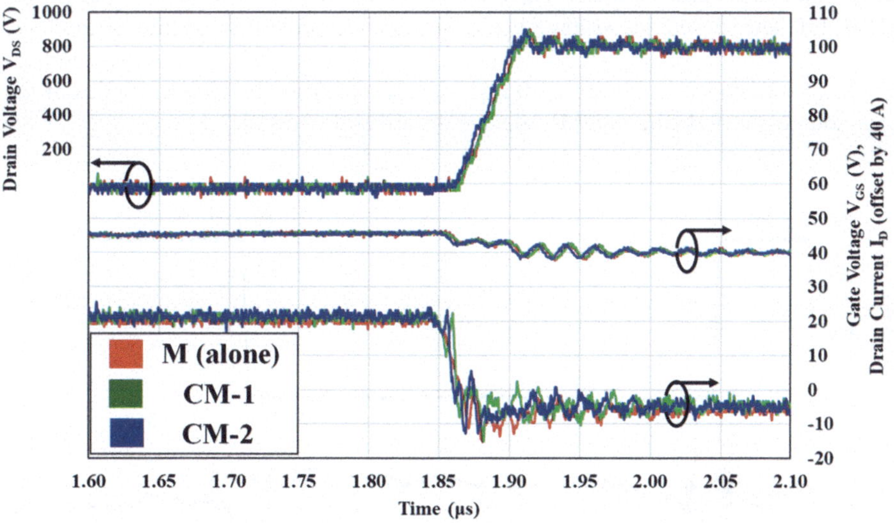

Fig. 5.17 Turn-off transients for the stand-alone SiC power MOSFET M, the composite MOSFET CM-1, and the composite MOSFET CM-2

In conclusion, the double-pulse tests demonstrate that the composite MOSFETs have larger switching losses than the stand-alone SiC power MOSFET. However, the increase in the switching losses is modest.

5.6 Sense Node Voltage

It was pointed out in Chap. 4 that the voltage across the non-linear resistance Q_{LV} can be used for monitoring the current level in the SiC power MOSFET Q_{HV}. This is possible for the BaSIC(DMM) topology for the SiC power MOSFET because the Si DMM device operates in its linear regime at the on-state current level of the SiC power MOSFET as shown in Fig. 5.3. The sense node voltage V_{SENSE} is therefore proportional to the current flowing through the SiC power MOSFET. This is a very beneficial feature of the BaSIC topology allowing power electronics engineers to make sure that the various SiC power MOSFETs within the H-bridge or other circuits do not exceed their reliable operating limits.

Experimental validation of using the sense node voltage to monitor the current flowing through the power transistor was obtained for the case of the SiC power MOSFET with a single silicon DMM device connected in series with its source terminal. The waveforms obtained with the double-pulse test circuit are shown in Fig. 5.18. The current starts ramping up after application of the first gate drive pulse as shown by the red line. It reaches a value of 6 A at 2.5 μs. The sense node voltage V_{SENSE} is shown by the blue line. There is considerable noise in the circuit implemented using discrete devices connected together with wires. This introduces some stray inductance that produces oscillations at the turn-on time of the SiC power MOSFET. The sense node voltage V_{SENSE} increases in proportion to the current level and reaches a value of about 0.36 V when the current has a value of 6 A. The equivalent sense resistance is therefore 60 mΩ as expected from the characteristics of the gate-source-shorted silicon DMM device described in Sect. 5.5.1.

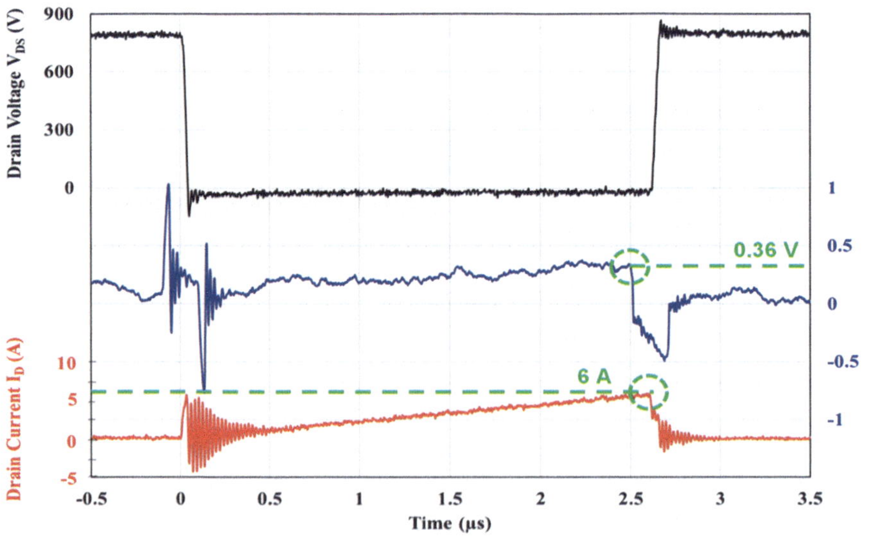

Fig. 5.18 Sense node voltage for the composite MOSFET CM-1 during normal operation

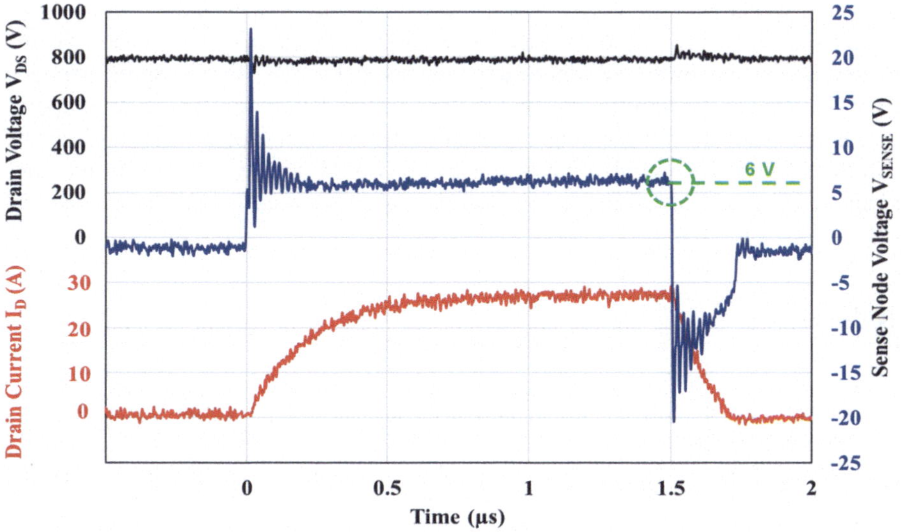

Fig. 5.19 Sense node voltage for the composite MOSFET CM-1 during short-circuit operation

The sense node voltage V_{SENSE} can be monitored to detect the short-circuit event as well. This was demonstrated for the case of the SiC power MOSFET with a single silicon DMM device connected in series with its source terminal. The resulting waveforms are shown in Fig. 5.19. The sense node voltage has a value of 6 V at the end of the short-circuit pulse when the current reaches a value of 28 A. The sense node voltage therefore changes from 0.36 V in the normal on-state operating mode to 6 V under short-circuit conditions. This large change can be easily detected by the control circuits to identify the occurrence of the short circuit and then trigger turning-off the gate drive voltage. This is discussed in a later chapter.

A significant amount of oscillation in the sense node voltage V_{SENSE} can be noted at the beginning of the short-circuit event. This noise settles down in about 0.25 μs. Blanking of the control circuit for this duration will be required when using the BaSIC topology. This time interval is much smaller than the 3 μs blanking time needed for the conventional IGBT DESAT gate control circuits. This difference is due to the sense node voltage V_{SENSE} operating at low voltages compared with the DESAT circuit where the drain terminal voltage of the power transistor is being monitored. The oscillations in the sense node voltage V_{SENSE} can be reduced if the power transistor and silicon DMM are co-packaged in a 4-terminal module to minimize the stray inductance.

5.7 Summary

The benefits of using the BaSIC topology is demonstrated in this chapter for the case of a silicon carbide power MOSFET by using a gate-source-shorted silicon depletion mode MOSFET as the non-linear resistance. An increase in the short-circuit withstand time by 65% was observed. The BaSIC topology increased the on-state resistance by 17% and the switching losses by 18%. This trade-off between improving the short-circuit withstand time and degradation of normal circuit operation is much superior to previous reported methods. Even superior trade-off results can be achieved by optimization of the silicon DMM device as shown in the later chapter.

References

1. A. Kadavelugu, E. Aeloiza, C. Belcastro, Short-circuit performance of multi-chip SiC MOSFET modules, in *IEEE Workshop on Wide Bandgap Power Devices and Applications* (2017), pp. 285–290
2. X. Zhang, G. Sheh, L. Gant, S. Banerjee, In-depth study of short-circuit robustness and protection of 1200V SiC MOSFETs, in *Power Electronics and Intelligent Motion Conference* (2018), pp. 866–872
3. A. Kanale, B.J. Baliga, Enhancing short-circuit capability of 1.2 kV 4H-SiC power MOSFETs using a gate-source shorted Si depletion-mode MOSFET in series with the source, in *2019 IEEE 13th International Conference on Power Electronics and Drive Systems (PEDS)* (2019), pp. 1–6
4. 1.2 kV SiC power MOSFET datasheet: Wolfspeed product page, https://www.wolfspeed.com/c2m0280120d
5. IXYS Power Semiconductor product page, https://www.littelfuse.com/products/power-semiconductors/ixys-power-semiconductors-and-ics.aspx
6. A. Kanale, K. Han, B.J. Baliga, S. Bhattacharya, Superior short circuit performance of 1.2kV SiC JBSFETs compared to 1.2kV SiC MOSFETs, in *12th European Conference on Silicon Carbide and Related Materials*, Paper WE.P.MO9 (2018)
7. B.J. Baliga, *Fundamentals of Power Semiconductor Devices*, 2nd edn. (Springer Science, New York, 2019)

Chapter 6
BaSIC(DMM) Topology Versus Gate Bias Reduction

The efficacy of the BaSIC topology for improving the short-circuit withstand time for SiC power MOSFETs was demonstrated in the previous chapter by using a gate-source-shorted silicon depletion mode MOSFET (DMM) in series with the source electrode. It is important to compare the BaSIC(DMM) topology with the alternative approach of reducing the gate bias. Reducing the gate bias for a power MOSFET brings down its saturation current level. This will increase the short-circuit withstand time. However, this approach also produces an increase in the on-resistance and switching losses. A comparison of the BaSIC(DMM) topology with gate bias reduction is performed in this chapter by using trade-off curves for the two approaches.

6.1 Gate Bias Reduction Physics

The on-state resistance of SiC power MOSFETs is dependent on the applied gate voltage as shown in their datasheets by vendors. The SiC power MOSFET structure has many series resistance components during current conduction [1]. Among these components, the channel resistance is dominant component for devices with blocking voltage ratings of 600–1700 V.

The specific channel resistance is given by [1]:

$$R_{\text{CH,sp}} = \frac{L_{\text{CH}}\, p}{\mu_{\text{CH}} C_{\text{OX}} \left(V_{\text{G}} - V_{\text{TH}}\right)} \tag{6.1}$$

where L_{CH} is the channel length, p is the cell pitch ($p = W_{Cell}/2$), μ_{CH} is the channel mobility, C_{OX} is the specific gate oxide capacitance, V_G is the gate bias, and V_{TH} is the threshold voltage. Based on the above equation, the channel resistance becomes

B. J. Baliga, A. Kanale, *The BaSIC Topology*,
https://doi.org/10.1007/978-3-031-86630-2_6

larger when the gate drive voltage is reduced leading to an increase in the total specific on-resistance.

6.2 Drain Saturation Current Density

The short-circuit withstand capability is determined by the device saturation current level under short-circuit conditions as explained in Chap. 2. A reduction of the saturation current produces an increase in the short-circuit withstand time.

The drain saturation current for a power MOSFET is given by [1]:

$$J_{\mathrm{D,SAT}} = \frac{\mu_{\mathrm{CH}} C_{\mathrm{OX}}}{L_{\mathrm{CH}} W_{\mathrm{Cell}}} \left(V_{\mathrm{G}} - V_{\mathrm{TH}}\right)^2 \tag{6.2}$$

Based on this relationship, the drain saturation current can be reduced by decreasing the gate drive voltage used in the on-state. Consequently, the short-circuit withstand time can be increased by reducing the gate drive voltage.

6.3 Switching Losses

The rate of change of drain voltage during the turn-on of a power MOSFET is given by [1]:

$$\frac{\mathrm{d}V_{\mathrm{D}}}{\mathrm{d}t} = \frac{V_{\mathrm{GS}} - V_{\mathrm{TH}}}{R_{\mathrm{G}} C_{\mathrm{GD}}\left(V_{\mathrm{D}}\right)} - \frac{I_{\mathrm{D}}}{R_{\mathrm{G}} C_{\mathrm{GD}}\left(V_{\mathrm{D}}\right) g_{\mathrm{M}}} \tag{6.3}$$

Based on this expression, the $\mathrm{d}V_{\mathrm{D}}/\mathrm{d}t$ will decrease when the gate bias is reduced, which will produce greater switching losses.

6.4 Theoretical Trade-Off

The discussion in the previous section shows that a trade-off between decreasing the drain saturation current and an increase in the on-state resistance must be performed when the gate bias is reduced.

The trade-off curve obtained for the 1.2 kV SiC shielded power MOSFET structure described in Chap. 3 can be obtained using the above analytical models while varying the gate bias. The results are shown in Fig. 6.1. The drain saturation current is reduced from 24,000 A/cm^2 to <5000 A/cm^2 by decreasing the gate bias from 20

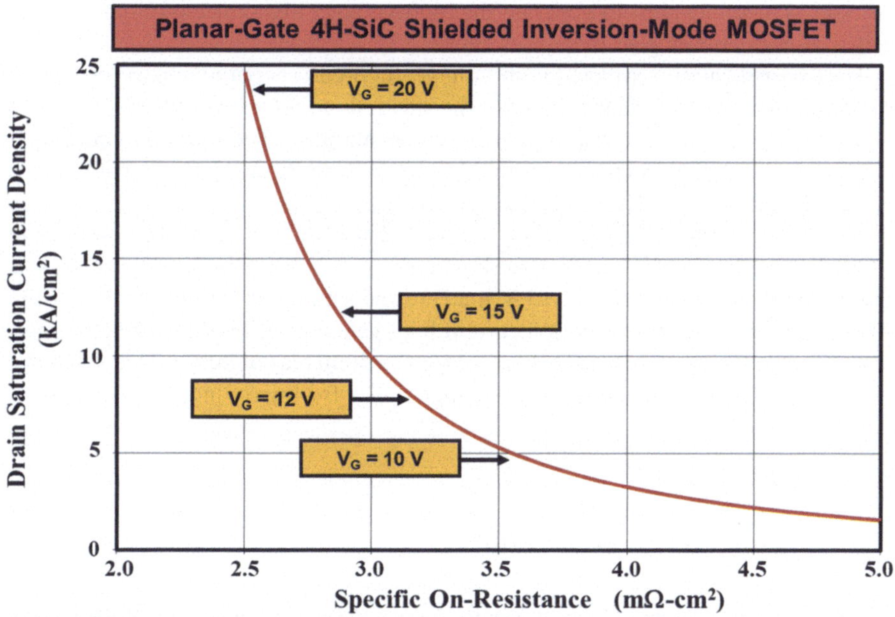

Fig. 6.1 Trade-off curve between drain saturation current and specific on-resistance for shielded channel SiC planar-gate power MOSFET: Gate bias as parameter

to 10 V. This produces an increase in specific on-resistance from 2.50 to 3.58 mΩ-cm^2 and an increase of 43% in the on-resistance.

6.5 Experimental Short-Circuit Test Results

The measurements were performed on a typical 1.2 kV SiC planar-gate power MOSFET product (C2M0280120D [2]), which is the same device used for experimental validation of the BaSIC(DMM) topology in the previous chapter [3]. The gate bias values were chosen so that the short-circuit conditions for this approach were similar to the two DMM device cases CM-1 and CM-2.

6.5.1 High-Voltage, High-Current Output Characteristics

The high-voltage, high-current output characteristics of the SiC power MOSFET were extracted by using the short-circuit test configuration [4] with a small pulse width of 1.5 μs. During this time, the current increases to a peak value that is representative of the saturation current at the DC supply voltage. The DC supply voltage was varied in small steps to generate the high-voltage, high-current output characteristic at each chosen gate bias.

The experimentally extracted high-voltage, high-current output characteristics at gate bias values of 15, 18, and 20 V for the stand-alone SiC power MOSFET M are shown in Fig. 6.2 with the red, brown, and magenta lines. The saturation current at a drain bias of 800 V decreases from 52 A to 34 A to 28 A when the gate bias is reduced from 20 V to 18 V to 15 V, respectively.

The high-voltage, high-current output characteristics for the composite MOSFETs CM-1 and CM-2 are also shown in Fig. 6.2. It can be seen that the saturation current for a gate bias of 15 V matches the saturation current for composite MOSFET CM-1 and the saturation current for a gate bias of 18 V matches the saturation current for composite MOSFET CM-2 at a drain bias of 800 V. This is helpful when making a comparison of the two approaches for improving the short-circuit withstand time.

6.5.2 *Short-Circuit Test Results*

The short-circuit withstand times for the stand-alone SiC power MOSFET were obtained by using the test configuration shown in Fig. 5.1. The short-circuit pulse width is gradually increased while capturing the current waveform until the device fails. The DC supply voltage was monitored as shown by the black line in Fig. 6.3. It remains constant at the desired value of 800 V.

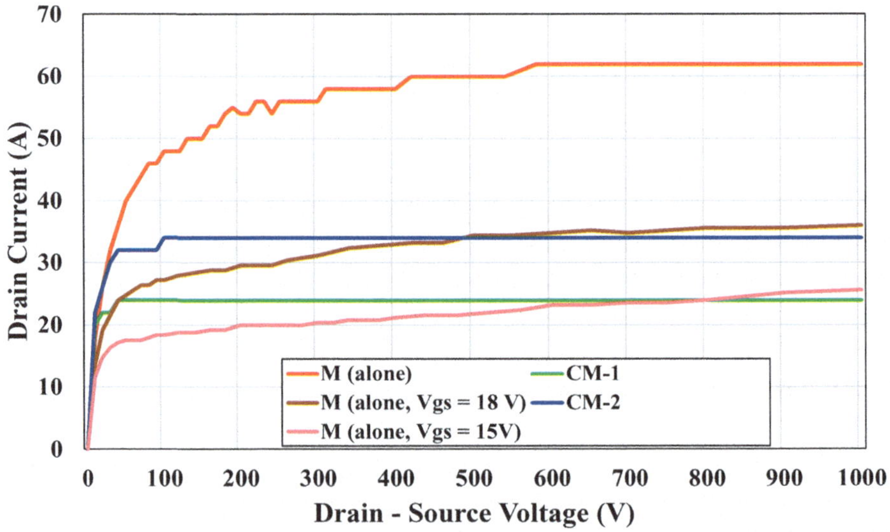

Fig. 6.2 High-voltage, high-current output characteristics: Stand-alone SiC power MOSFET M, composite MOSFET CM-1, and composite MOSFET CM-2

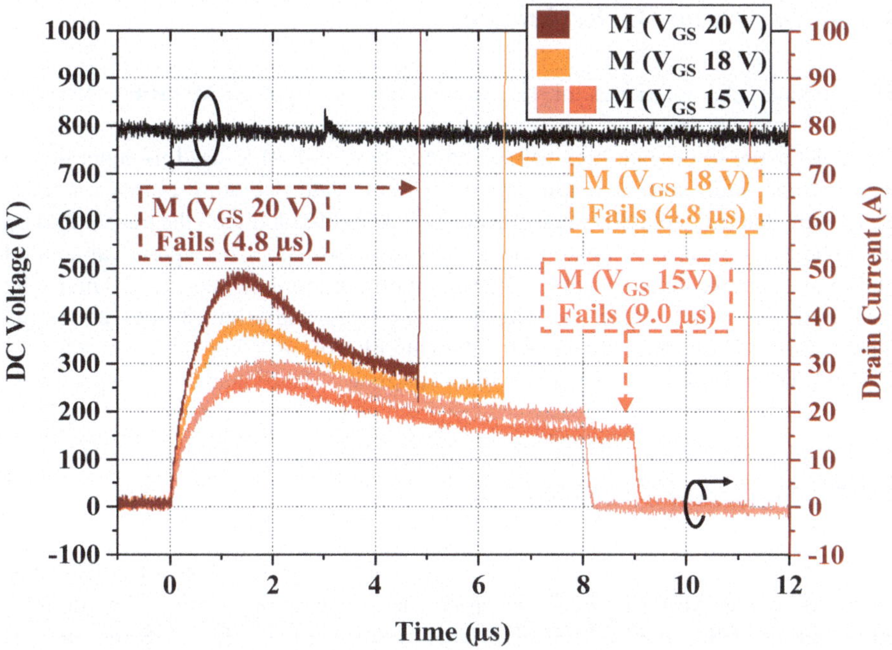

Fig. 6.3 Measured short-circuit waveforms for stand-alone SiC power MOSFET M, with various gate bias values

The stand-alone SiC power MOSFET M failed when the pulse width was increased to 4.8 µs when a gate bias of 20 V was applied. The short-circuit current was reduced as expected with decrease of the gate bias. The device failed at a pulse width of 6.3 µs when the gate bias was reduced to 18 V, and at a pulse width of 9.0 µs for a gate bias of 15 V.

These short-circuit test results demonstrate that reducing the gate bias brings down the short-circuit current level. It is possible to significantly increase the short-circuit withstand time for the SiC power MOSFET when the gate bias is reduced to 15 V.

6.6 Normal Circuit Operation

The impact of reducing the gate bias on the on-state resistance and switching behavior of the SiC power MOSFET were quantified. Trade-off curves between enhancing the short-circuit time and an increase on-resistance and switching losses were then developed for comparison with the BaSIC(DMM) topology.

6.6.1 Impact on On-Resistance

The on-state characteristics of the stand-alone SiC power MOSFET M were obtained at lower drain voltages by using a curve tracer. The results are shown in Fig. 6.4 for the three gate bias values: 15 V, 18 V, and 20 V used to generate the high-voltage, high-current output curves. The on-state operating points for these cases were determined at the rated current of 6 A highlighted with the purple dashed line. The on-resistance at a gate bias of 20 V was 346 mΩ. The typical value cited in the datasheet for this transistor is 280 mΩ with a maximum value of 370 mΩ. The measured values fall within the datasheet parametric variation of 32% above the typical value. The values obtained for the gate bias of 18 V and 15 V are 423 and 575 mΩ. These values are an increase in on-resistance of 22% and 66%, respectively.

6.6.2 Impact on Switching Losses

The double-pulse test was used to evaluate the switching power losses for the stand-alone SiC power MOSFET with the three cases of gate bias values. The double-pulse testing of SiC power MOSFETs was performed using SiC JBS rectifiers as freewheeling diodes. The turn-on current is observed to have an overshoot during turn-on even in this case due to the high dV_D/dt at the drain terminal of the power MOSFET and the capacitance of the SiC JBS diode.

The measured turn-on transients for the stand-alone SiC power MOSFET are shown in Fig. 6.5 for the three gate bias values. It can be observed that the dV_D/dt

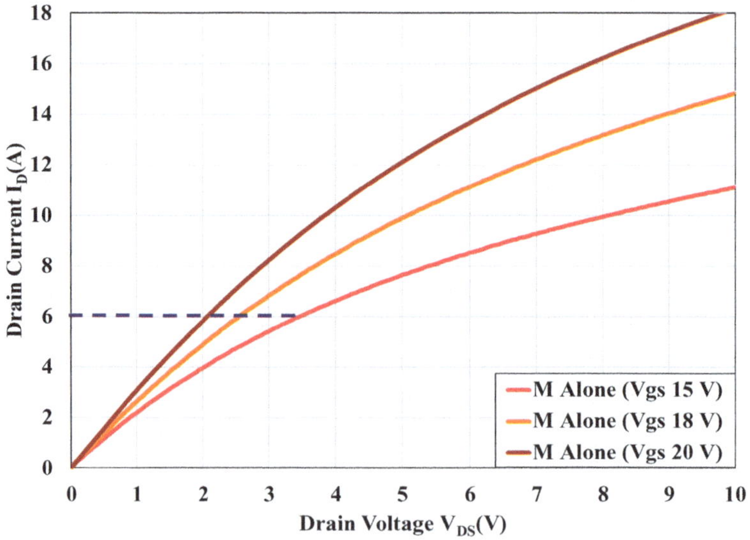

Fig. 6.4 Measured on-state characteristics of the stand-alone SiC power MOSFET M with various gate bias values

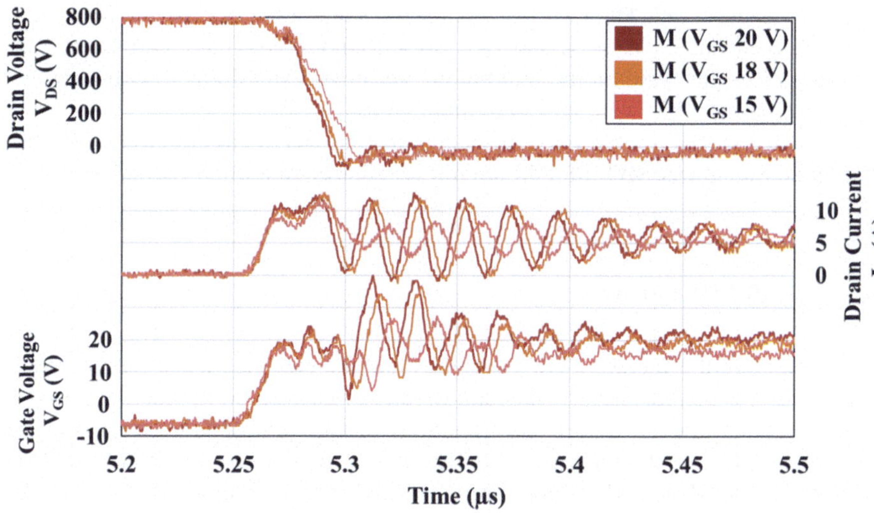

Fig. 6.5 Turn-on transients for the stand-alone SiC power MOSFET M with various gate bias values

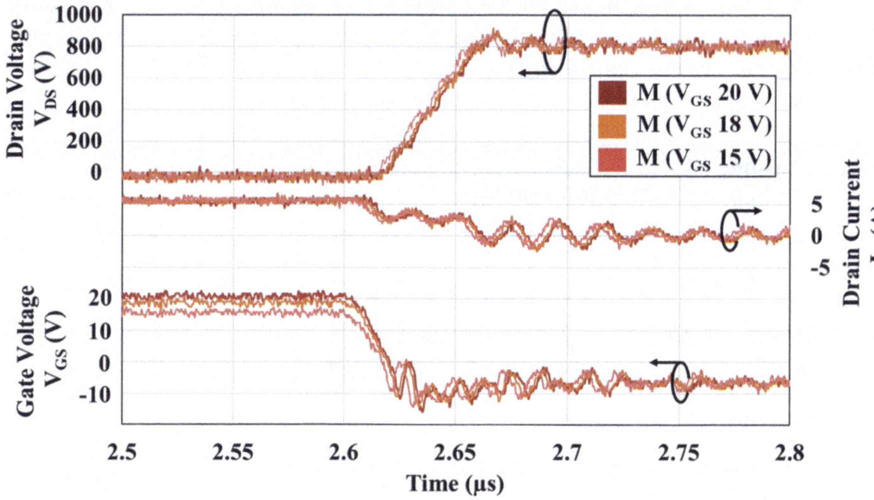

Fig. 6.6 Turn-off transients for the stand-alone SiC power MOSFET M with various gate bias values

decreases when the gate bias value is reduced. This is consistent with Eq. (6.3). A reduction of the dV_D/dt increases the turn-off time duration, which results in larger energy loss per cycle. The turn-on loss for the stand-alone SiC power MOSFET with gate bias values of 20 V, 18 V, and 15 V were found to be 157, 161, and 177 μJ, respectively.

The measured turn-off transients for the stand-alone SiC power MOSFET are shown in Fig. 6.6 for the three gate bias values. Very little difference between the three cases is observed. The turn-off loss for the stand-alone SiC power MOSFET M with gate bias of 20 V, 18 V, and 15 V were found to be 43, 42, and 43 μJ. The total energy loss per cycle for the stand-alone SiC power MOSFET M with gate bias of 20 V, 18 V, and 15 V were 203, 204, and 223 μJ.

In conclusion, the double-pulse tests demonstrate that switching losses for the stand-alone SiC power MOSFET M are not very sensitive to the gate bias value.

6.7 Trade-Off Curves

A comparison of the BaSIC(DMM) topology and the gate bias reduction approach for increasing the short-circuit withstand time can be achieved by making a trade-off curve between the increase in short-circuit withstand time and the increase in the on-resistance. This is shown in Fig. 6.7. The trade-off curve for the BaSIC(DMM) case lies to the left-hand side of that for the reducing gate bias approach demonstrating that it is a superior approach. A quantitative difference between the cases can be derived by choosing an increase in short-circuit withstand time of 65%. The increase in on-resistance for the BaSIC(DMM) topology is 17% compared with 44% when reducing the gate bias as shown by the purple arrows in the figure. This is a factor 2.6 times better performance with the BaSIC(DMM) topology. A much bigger difference is achievable by optimizing the silicon gate-source-shorted DMM device as discussed in the later chapter.

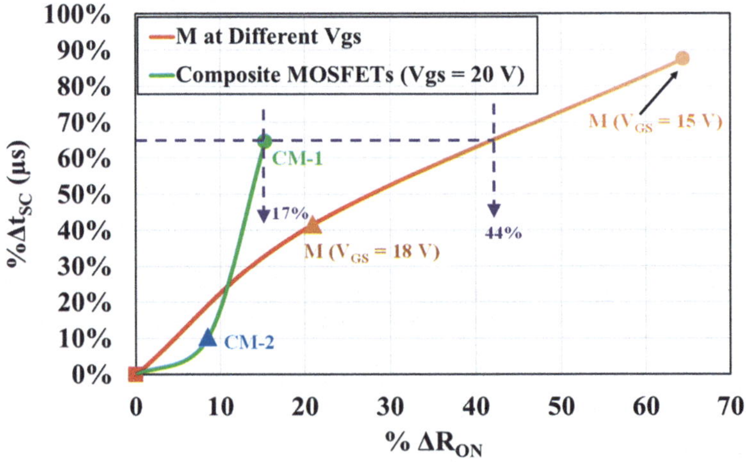

Fig. 6.7 Trade-off curves for the stand-alone SiC power MOSFET M with various gate bias values versus the BaSIC(DMM) topology

6.8 Summary

The performance of the BaSIC(DMM) topology for a silicon carbide power MOSFET has been compared with the gate bias reduction method in this chapter. The results obtained with the BaSIC(DMM) topology are shown to be far superior to those achievable by gate bias reduction.

References

1. B.J. Baliga, *Fundamentals of Power Semiconductor Devices*, 2nd edn. (Springer Science, New York, 2019)
2. Wolfspeed product page, https://www.wolfspeed.com/c2m0280120d
3. A. Kanale, B.J. Baliga, Comparison of current suppression methods to enhance short circuit capability of 1.2 kV SiC power MOSFETs: a new approach using a series-connected, gate-source-shorted Si depletion-mode MOSFET vs reduced gate bias operation, in *IEEE Industrial Electronics Conf.* (2019), pp. 5114–5119
4. A. Kanale, K. Han, B.J. Baliga, S. Bhattacharya, Superior short circuit performance of 1.2kV SiC JBSFETs compared to 1.2kV SiC MOSFETs, in *12th European Conference on Silicon Carbide and Related Materials*, Paper WE.P.MO9 (2018)

Chapter 7
BaSIC(DMM) Topology Versus Ballast Resistance

The BaSIC topology with a gate-source-shorted silicon depletion mode MOSFET (DMM) in series with the source electrode was shown to be effective for improving the short-circuit withstand time of a SiC power MOSFET in Chap. 5. An alternative approach to improving the short-circuit withstand time for the SiC power MOSFET is by connecting a ballast resistance with the source electrode in series. The voltage drop across the ballast resistance during the short-circuit event will reduce the gate drive voltage for the SiC power MOSFET and bring down its saturation current level. This will increase the short-circuit withstand time. However, this approach will obviously increase the on-resistance and also the switching losses. A comparison of the BaSIC(DMM) topology with ballast resistance approach is performed in this chapter by using trade-off curves for the two approaches [1].

7.1 Ballast Resistance Approach

A SiC power MOSFET with the ballast resistance approach is illustrated in Fig. 7.1. The ballast resistance R_S is connected to the source electrode of the SiC power MOSFET. The current flowing through the SiC MOSFET also flows through the ballast resistance due to their series connection. Its resistance will therefore add to the on-resistance of the SiC power MOSFET. The composite device for the SiC MOSFET with a ballast resistance will be referred to as a *MOSFET with Series Resistance (MSR)* in this chapter. During a short-circuit event, the voltage drop across the ballast resistance subtracts from the gate drive voltage V_{GS} because the ballast resistor is in the gate loop. This reduces the effective gate drive voltage for the SiC power MOSFET, which brings down its saturation current level. A longer short-circuit withstand time is achieved by reduction of the drain saturation current level.

B. J. Baliga, A. Kanale, *The BaSIC Topology*,
https://doi.org/10.1007/978-3-031-86630-2_7

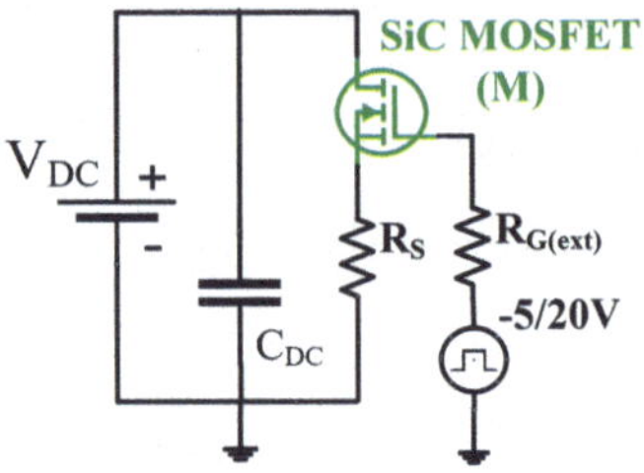

Fig. 7.1 The ballast resistance approach

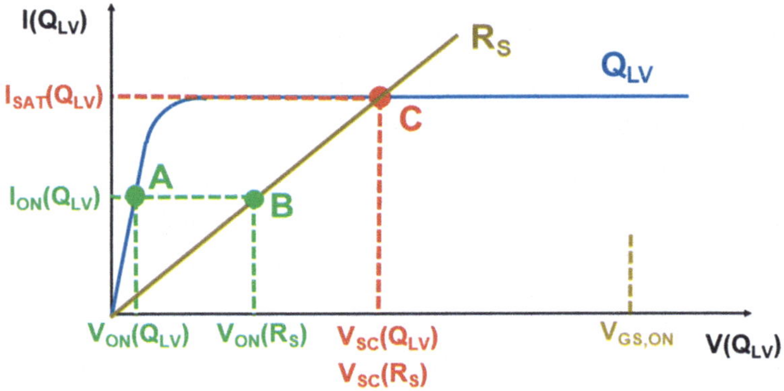

Fig. 7.2 Comparison of the characteristics of the non-linear resistance with the ballast resistance

7.2 Comparison of Characteristics

A perspective of the relative performance of the BaSIC topology with ballast resistance approach can be obtained by examination of the characteristics of the non-linear resistance Q_{LV} and the series ballast resistance R_S for achieving the same level of the drain saturation current for the SiC power MOSFET during the short-circuit event.

The *i-v* characteristics of the non-linear resistance Q_{LV} and the series ballast resistance R_S are superimposed in Fig. 7.2 to allow comparison of these approaches. The same voltage drop $V_{SC}(Q_{LV})$ and $V_{SC}(R_S)$ occurs across both cases under short-circuit conditions at point C. The drain saturation current level for the SiC power MOSFET is then decided by the difference between the gate drive voltage V_{GS} and this voltage. Consequently, both approaches will produce the same short-circuit withstand time.

The on-state operating point for the two approaches has the same on-state current level $I_{ON}(Q_{LV})$. The non-linear resistance operates at point A with a voltage drop $V_{ON}(Q_{LV})$ while the series ballast resistance operates at point B with a voltage drop $V_{ON}(R_S)$. From this illustration, it becomes obvious that the series ballast resistance value R_S must be much larger than the resistance of the non-linear resistance Q_{LV} in the on-state mode to achieve the same short-circuit withstand time for the SiC power MOSFET.

7.3 Experimental Short-Circuit Results

Comparison of the BaSIC(DMM) topology with the series ballast resistance was performed using the same 1.2 kV SiC planar-gate power MOSFET product (C2M0280120D [2]) used for experimental validation of the BaSIC(DMM) topology in the previous chapter. The series ballast resistance values were chosen so that the short-circuit conditions for this approach were similar to the two DMM device cases CM-1 and CM-2.

7.3.1 High Voltage High Current Output Characteristics

The series ballast resistance R_S values were chosen such that the drain saturation current level matches that observed for the BaSIC(DMM) topology composite MOSFET CM-1 and CM-2. This required extraction of the high voltage high current output characteristics of the SiC power MOSFET with series resistance (MRS) using the short-circuit test configuration [3] with a small pulse width of 1.5 μs. During this time, the current increases to a peak value that is representative of the saturation current at the DC supply voltage. The DC supply voltage was varied in small steps to generate the high voltage high current output characteristic at each chosen gate bias.

The values for the series ballast resistance were varied until the drain saturation current level at a drain bias of 800 V matched that previously observed for the BaSIC(DMM) topology composite MOSFETs CM-1 and CM-2. It was found that a series ballast resistance value of 200 mΩ produced a high voltage high current output characteristic shown by the green dashed line in Fig. 7.3. The drain saturation current at a drain bias of 800 V then matches that observed for composite MOSFET CM-1 as indicated by the purple circle. Similarly, it was found that a series ballast resistance value of 50 mΩ produced a high voltage high current output characteristic shown by the blue dashed line in Fig. 7.3. The drain saturation current at a drain bias of 800 V then matches that observed for composite MOSFET CM-2 as indicated by the purple circle. This is helpful when making a comparison of the two approaches for improving the short-circuit withstand time.

7.3.2 Short-Circuit Test Results

The short-circuit withstand time for the SiC power MOSFET with the series ballast resistance was obtained by using the test configuration shown in Fig. 7.1. The short-circuit pulse width is gradually increased while capturing the current waveform until the device fails. The DC supply voltage was monitored as shown by the black line in Fig. 7.4. It remains constant at the desired value of 800 V.

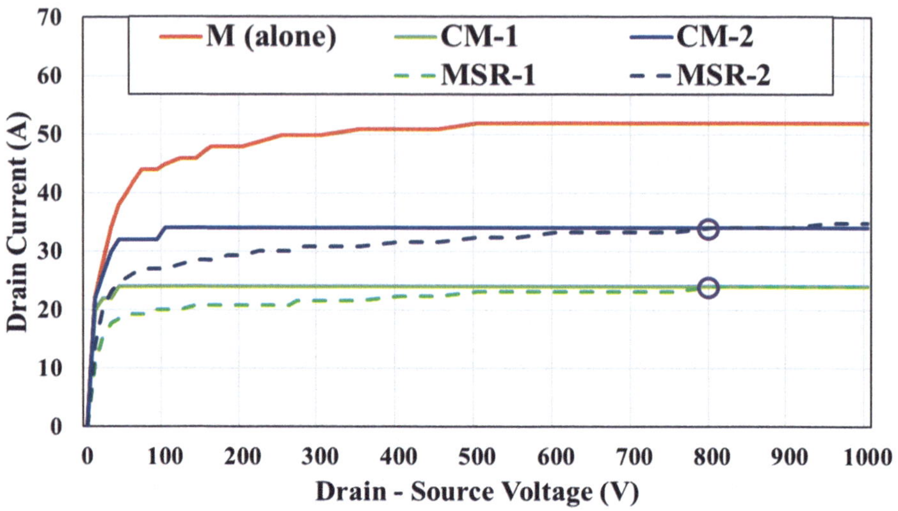

Fig. 7.3 High voltage high current output characteristics: Stand-alone SiC power MOSFET M, composite MOSFETs CM-1 and CM-2c and the MOSFET with series resistance MSR-1 and MSR-2

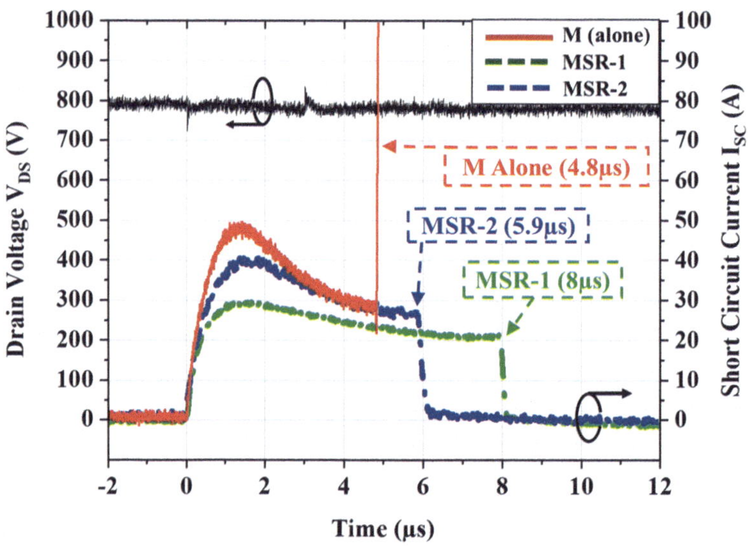

Fig. 7.4 Measured short-circuit waveforms for stand-alone SiC power MOSFET M, and MOSFET with series resistance MSR-1 and MSR-2

The stand-alone SiC power MOSFET M failed when the pulse width was increased to 4.8 μs when a gate bias of 20 V was applied. The short-circuit current was reduced as expected when the series ballast resistance was added. The device failed at a pulse width of 8.0 μs for the MSR-1 case with a series resistance of

200 mΩ. The MOSFET failed at a pulse width of 5.9 μs for the MSR-2 case with a series resistance of 50 mΩ.

These short-circuit test results demonstrate that the short-circuit withstand time for the SiC power MOSFET can be significantly increased by putting a series ballast resistance at its source electrode.

7.4 Normal Circuit Operation

The impact of series ballast resistance on the on-state resistance and switching behavior of the SiC power MOSFET was quantified. Trade-off curves between enhancing the short-circuit time and an increase on-resistance and switching losses were then developed for comparison with the BaSIC(DMM) topology.

7.4.1 Impact on On-Resistance

The on-state characteristics of the SiC power MOSFETs with series ballast resistance MSR-1 and MSR-2 were obtained at lower drain voltages by using a curve tracer. The results are shown in Fig. 7.5 for a gate bias of 20 V. The on-state operating point for these cases was determined at the rated current of 6 A highlighted with the purple dashed line and the dots. The on-resistance at a gate bias of 20 V was 346 mΩ for the stand-alone SiC power MOSFET. The typical value cited in the datasheet for this transistor is 280 mΩ with a maximum value of 370 mΩ. The measured values falls within the datasheet parametric variation of 32% above the typical

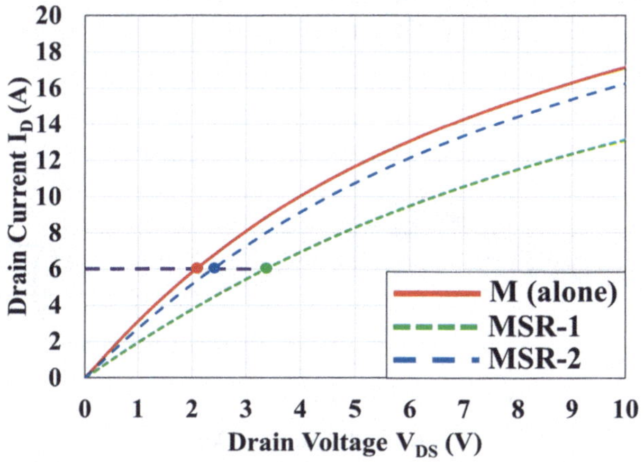

Fig. 7.5 Measured on-state characteristics of the stand-alone SiC power MOSFET M, and MOSFET with series resistance MSR-1 and MSR-2

value. The values obtained for the MSR-1 and MSR-2 cases are 550 and 397 mΩ as expected from adding the series ballast resistance values to that of the SiC power MOSFET. The on-resistances for MSR-1 and MSR-2 are of 59% and 15% larger than for the stand-alone SiC power MOSFET.

7.4.2 Impact on Transconductance

The transconductance g_M of the power MOSFET plays a role in the switching transient for the drain voltage during operation with an inductive load. Its value can be extracted from the transfer characteristics of the device. The transfer characteristics for the MSR-1 and MSR-2 configurations were measured at a drain bias of 20 V and are compared with the stand-alone SiC power MOSFET in Fig. 7.6. The transconductance, extracted by taking a derivative of the drain current with respect to the gate voltage, is also shown in the figure. The value of the transconductance can be compared for the three cases at the same drain current level of 6 A indicated by the purple dashed line. It is given by the values at the three points indicated on the transconductance curves.

The stand-alone SiC power MOSFET has a transconductance of 3.35 S, which is close to the value in its datasheet. The transconductances of the MOSFETs with

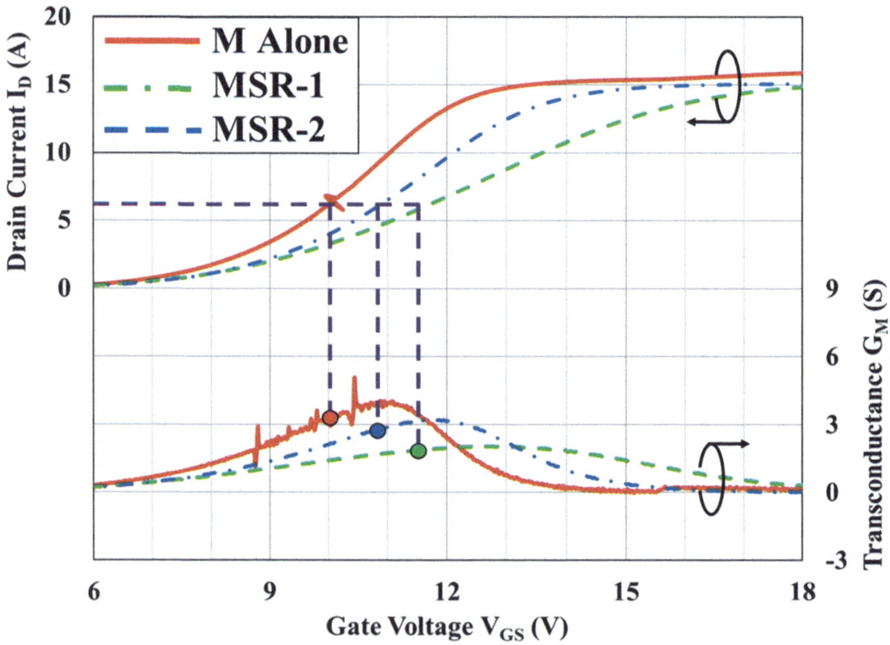

Fig. 7.6 Transconductances for the stand-alone SiC power MOSFET M, and with the series ballast resistance MSR-1 and MSR-2

series resistance MSR-1 and MSR-2 are 2.2 and 2.5 S. The reduction in the transconductance is associated with the resistance of the ballast resistance as given in Eq. (5.3). The transconductance value predicted by this equation is 2.0 S for the case of a series resistance R_S of 200 mΩ which matches the measured value for the MSR-1. The transconductance value predicted by this equation is 2.8 S for the case of a series resistance R_S of 50 mΩ. This matches the measured value for the MSR-2. The impact of the reduction of the transconductance on the switching losses is discussed in the next section.

7.4.3 Impact on Switching Losses

Double-pulse test was used to evaluate the switching power losses for the SiC power MOSFET with series ballast resistances. The double-pulse testing of SiC power MOSFETs was performed using SiC JBS rectifiers as freewheeling diodes. The turn-on current is observed to have an overshoot during turn-on even in this case due to the high dV_D/dt at the drain terminal of the power MOSFET and the capacitance of the SiC JBS diode.

The measured turn-on transients for the stand-alone SiC power MOSFET and the MOSFETs with the series resistances MSR-1 and MSR-2 are shown in Fig. 7.7. It can be observed that the dV_D/dt decreases when the series resistance is added. This

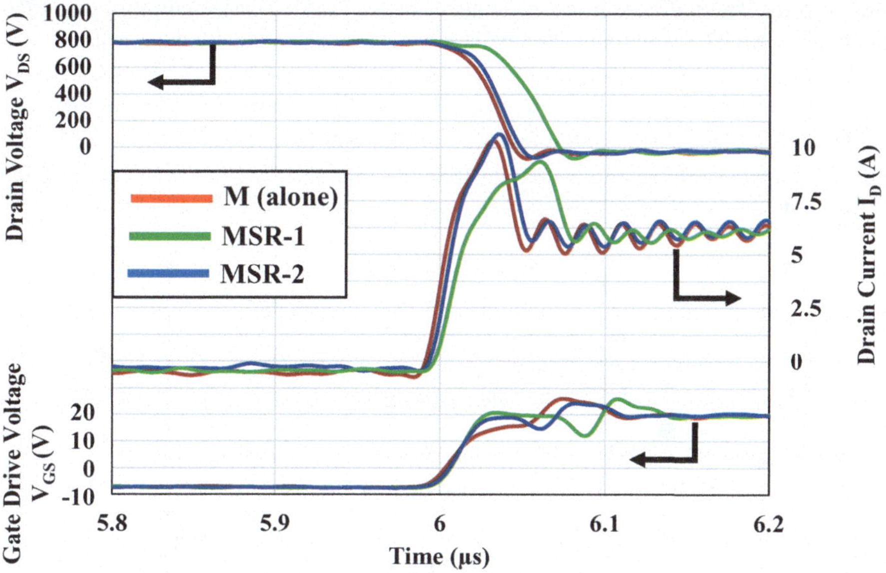

Fig. 7.7 Turn-on transients for the stand-alone SiC power MOSFET M and MOSFET with series resistances MSR-1 and MSR-2

is consistent with Eq. (6.3) due to the reduction of the transconductance. A reduction of the dV_D/dt increases the turn-on time duration, which results in larger energy loss per cycle. The turn-on loss for the stand-alone SiC power MOSFET with gate bias values of 20 V was found to be 157 μJ. The turn-on losses for the MOSFETs with series resistances MSR-1 and MSR-2 were 246 and 196 μJ, respectively.

Very little difference between the measured turn-off transients for the stand-alone SiC power MOSFET and the MOSFETs with series resistances MSR-1 and MSR-2 was observed. The turn-off losses for the devices were 46, 42, and 46 μJ. The total energy loss per cycle for the stand-alone SiC power MOSFET M and the MOSFETs with series resistances MSR-1 and MSR-2 were 203, 288, and 248 μJ. In conclusion, the double-pulse tests demonstrate that switching losses get worse in the presence of the series ballast resistance.

7.5 Trade-Off Curves

A comparison of the BaSIC(DMM) topology and the MOSFET with series ballast resistance (MSR) approach can be performed by making a trade-off curve between the increase in short-circuit withstand time and the increase in the on-resistance. The graphs for the two methods are shown in Fig. 7.8. The trade-off curve for the BaSIC(DMM) case lies to the left-hand side of that for the series ballast resistance approach for the higher short-circuit withstand time range. A quantitative difference between the two approaches can be derived by choosing a short-circuit withstand time of 8 μs. The increase in on-resistance for the BaSIC(DMM) topology is 17% compared with 62% with the series ballast resistance as shown by the purple arrows in the figure. This is a factor 3.6 times better performance with the BaSIC(DMM) topology. A much bigger difference is achievable by optimizing the silicon gate-source-shorted DMM device as discussed in the later chapter.

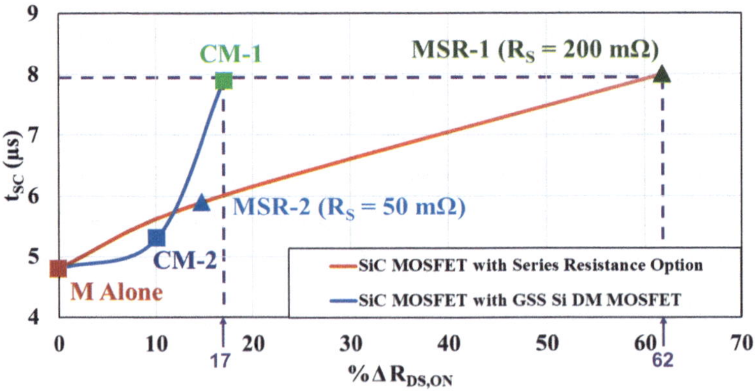

Fig. 7.8 Trade-off curves for the stand-alone SiC power MOSFET M with series ballast resistances versus the BaSIC(DMM) topology: On-resistance comparison

Similarly, the trade-off curve between short-circuit withstand time and switching energy loss can be used to compare the two approaches. These graphs are shown in Fig. 7.9 for the BaSIC(DMM) topology and the MOSFET with series resistance (MSR) approach. A quantitative difference between the two approaches can be derived by choosing a short-circuit withstand time of 8 μs. The increase in total switching energy loss for the BaSIC(DMM) topology is 22% compared with 42% with the series ballast resistance as shown by the purple arrows in the figure. This is a factor 1.9 times better performance with the BaSIC(DMM) topology. A much bigger difference is achievable by optimizing the silicon gate-source-shorted DMM device as discussed in the later chapter.

7.6 Summary

The performance of the BaSIC(DMM) topology for a silicon carbide power MOSFET has been compared with the option of putting a ballast resistance in series with the source electrode of the MOSFET in this chapter. The results obtained with the BaSIC(DMM) topology are shown to far superior to those achievable by ballast resistance approach.

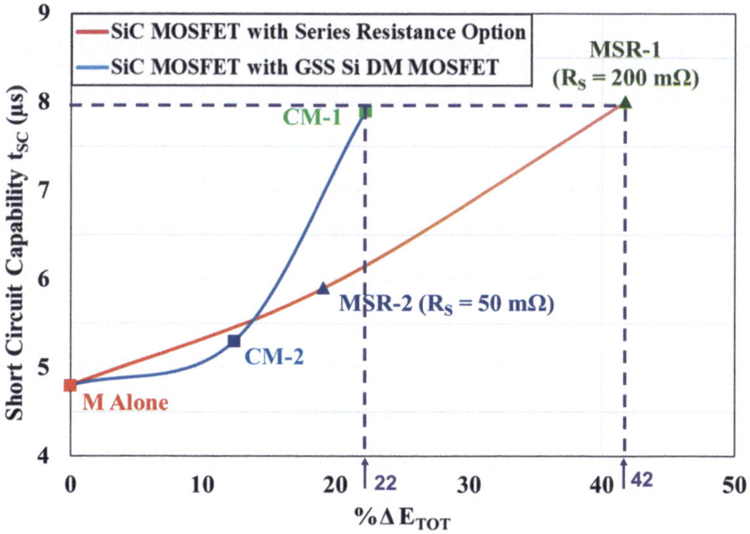

Fig. 7.9 Trade-off curves for the stand-alone SiC power MOSFET M with series ballast resistances versus the BaSIC(DMM) topology: switching loss comparison

References

1. A. Kanale, B.J. Baliga, Comparison of current suppression methods to enhance short circuit capability of 1.2 kV SiC power MOSFETs: a new approach using a series-connected, gate-source-shorted Si depletion-mode MOSFET vs use of a series resistance, in *IEEE Workshop on Wide Bandgap Power Devices and Applications* (2019), pp. 53–58
2. Wolfspeed product page, https://www.wolfspeed.com/c2m0280120d
3. A. Kanale, K. Han, B. J. Baliga, S. Bhattacharya, Superior short circuit performance of 1.2kV SiC JBSFETs compared to 1.2kV SiC MOSFETs, in *12th European Conference on Silicon Carbide and Related Materials*, Paper WE.P.MO9 (2018)

Chapter 8
Scaling BaSIC(DMM) Topology

The BaSIC topology with a gate-source-shorted silicon depletion mode MOSFET (DMM) in series with the source electrode was shown to be effective for improving the short-circuit withstand time of a SiC power MOSFET in Chap. 5. These results were gathered by using a single SiC power MOSFET with current rating of 6 A. Many applications require greater levels of current. This is commonly achieved by either paralleling of discrete devices or by the fabrication of multi-chip modules.

The on-resistance of the SiC power MOSFET has a positive temperature coefficient. This is favorable for the paralleling of devices. When the current through one of the devices increases, the larger power dissipation produces a rise of its on-resistance. This diverts the current flow to other devices that are in parallel. However, even with this feature, the paralleling of SiC power MOSFETs for increasing the current and power handling capability has been reported to be problematic.

A matching of the individually packaged 1.2 kV SiC power MOSFET devices has been found to be necessary [1]. It was found that a mismatch in the threshold voltage produces non-uniform current sharing during switching. The influence of device and circuit mismatch has also been examined [2]. It was found that mismatch in on-resistance influences on-state current distribution, while mismatch in threshold voltage influences transient behavior. Circuit loop inductance can also play a role. The addition of an inductor in series with the source electrode of paralleled SiC power MOSFETs has been explored to improve current sharing during switching [2]. The screening of SiC power MOSFETs to improve paralleled operation has also been undertaken [3, 4]. The gate-drain transfer characteristics of the SiC power MOSFETs were matched to improve the transient current sharing. A comprehensive review of the issues and proposed solutions for the current imbalance during the on-state and switching transients has been published [5]. These studies demonstrate that special effort is required to achieve good scaling of static and dynamic characteristics for SiC power MOSFETs.

Based on the previous work on paralleling SiC power MOSFETs, it is important to determine the behavior of the BaSIC topology when the composite transistors are

B. J. Baliga, A. Kanale, *The BaSIC Topology*,
https://doi.org/10.1007/978-3-031-86630-2_8

paralleled. This chapter describes the results of a study conducted for the BaSIC topology using a typical SiC power MOSFET product with the gate-source-shorted silicon depletion mode MOSFET in series [6].

8.1 BaSIC(DMM) On-Resistance Temperature Dependence

As pointed out earlier, it is important that the power transistor displays a positive temperature coefficient for the on-resistance to ensure good current sharing between paralleled devices. In the case of the BaSIC(DMM) topology, the on-resistance of the composite MOSFET is the sum of the resistance of the SiC power MOSFET and the silicon DMM device. These devices exhibit different temperature dependence behavior.

The temperature dependence of the on-resistance for the typical 1.2 kV SiC power MOSFET product C2M0280120D [7] is available in its datasheet and shown in Fig. 8.1 by the blue line. The on-resistance has been normalized to its room temperature (25 °C) value. The on-resistance increases by a factor of 1.85× from 25 to 150 °C.

The temperature dependence of the on-resistance for the typical silicon DMM device IXTH16N10D2 [8] is available in its datasheet and shown in Fig. 8.1 by the green line. Unlike the SiC power MOSFET, its on-resistance does not increase monotonically with temperature. A slight reduction of on-resistance is observed up to 85 °C followed by an increase to its 25 °C value when the temperature reaches 150 °C.

The temperature dependence of the on-resistance for the composite MOSFET in the BaSIC(DMM) topology can be ascertained by the addition of the on-resistance

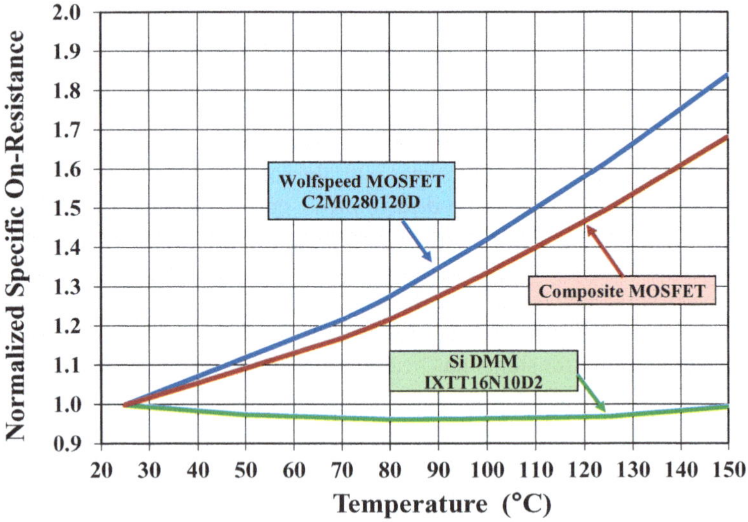

Fig. 8.1 Temperature dependence of the on-resistance of the BaSIC topology

of the two internal devices. The variation of the on-resistance for the composite MOSFET is shown by the red line in Fig. 8.1. It can be seen that the composite MOSFET exhibits a monotonic rise with temperature by a factor of 1.68× from 25 to 150 °C. This indicates that the current should share well between paralleled composite MOSFETs in the BaSIC(DMM) topology.

8.2 Testing Configuration

A compact testing configuration was designed for investigation of the paralleling of the composite MOSFETs in the BaSIC(DMM) topology. Commercially available devices in discrete TO-220 packages were used. The gate and source of the silicon DMM device were soldered together. The drain of the silicon DMM device was then soldered to the source electrode of the SiC power MOSFET. Four of these assemblies (CM-1, CM-2, CM-3, and CM-4) were created, as shown in Fig. 8.2, to investigate the paralleling operation.

Both the SiC power MOSFETs and the silicon DMM devices were randomly chosen as supplied by the manufacturer. No selection procedure was employed to try to match the devices.

The testing fixture was also created as shown at the top of Fig. 8.2. This allowed all the characterization tests to be performed using a single fixture with either a single composite MOSFET, or with two, three, and four composite MOSFETs working in parallel. This fixture was utilized with the short-circuit test circuit to evaluate the distribution of the current between the devices. The same fixture was used with the double-pulse test circuit to obtain the scaling of the switching capability.

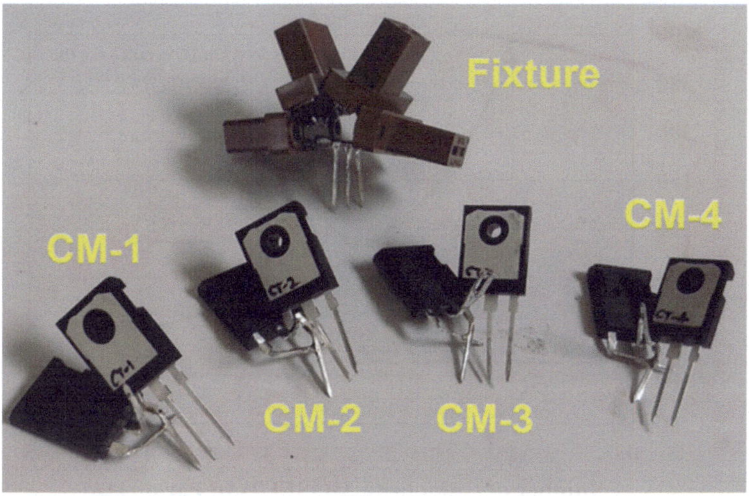

Fig. 8.2 Testing fixture and composite MOSFETs used for investigation of paralleling

The drain of the silicon DMM device could be used as the sense node, as described in Chap. 4, for obtaining the current flowing through each of the composite MOSFETs. This unique feature of the BaSIC topology allowed monitoring the current flowing through each of the devices operating in parallel during both static and dynamic tests.

The same fixture could be used to obtain the static device parameters with a curve tracer. These characteristics included the on-resistance, the transconductance, and the capacitances. These results are described in the subsequent sections.

8.3 Experimental On-Resistance Results

The rated current for the composite MOSFETs in the BaSIC(DMM) topology is expected to scale in proportion to the number of devices connected in parallel. Consequently, the on-resistance was obtained at a current of 2 A for the single composite MOSFET; and at 4 A, 6 A, and 8 A for the cases of two, three, and four composite MOSFETs connected in parallel.

The on-resistance of the four individual SiC MOSFET devices used for this study was found to be 309, 355, 358, and 364 mΩ. This is a variation of 55 mΩ or 19%, which is consistent with the manufacturer's datasheet. The measured on-resistance of the gate-source-shorted Si DMMs was 47, 47, 49, and 50 mΩ. This is a variation of only 3 mΩ or 6%. The measured on-resistance of the composite MOSFETs with a single device was 376 mΩ. It reduced to 203 mΩ, 144 mΩ, and 108 mΩ, when two, three, and four composite MOSFETs were connected in parallel.

A graph of the on-resistance versus number of devices in parallel is shown in Fig. 8.3. The measured on-resistance values are indicated with the red triangles. The

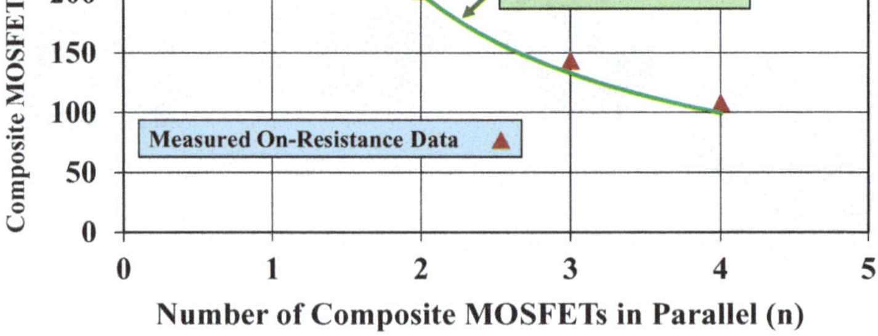

Fig. 8.3 On-resistance of paralleled composite MOSFETs in the BaSIC(DMM) topology

on-resistance decreases with more composite MOSFETs connected in parallel as expected. A green trend line is shown in the figure where the calculated on-resistance is inversely proportional to the number (n) devices connected in parallel. The measured values fall on this trend line demonstrating excellent current sharing between the paralleled composite MOSFET in the BaSIC(DMM) topology.

From the measured on-resistance results, it can be concluded that excellent on-state current scaling can be achieved with the BaSIC(DMM) topology without the need for pre-selection or matching of the devices.

8.4 Experimental Transconductance Results

The transconductance g_M for the composite MOSFETs in the BaSIC(DMM) topology should scale up in proportion to the number of devices connected in parallel. The transconductance was measured at a drain bias of 20 V. The single composite MOSFETs had a g_M of 2–2.5 S at a current of 5 A and a gate bias of 11 V. For the same gate bias of 11 V, two, three, and four composite MOSFETs in parallel had g_M values of 4.2, 6.1, and 7.3 S. The transconductance g_M increased with more devices in parallel as expected.

A graph of the transconductance g_M versus number of devices in parallel is shown in Fig. 8.4. The measured transconductance g_M values are indicated with the blue squares. A blue *linear* trend line is also shown in the figure. The measured values fall on this trend line demonstrating excellent scaling when paralleling composite MOSFET in the BaSIC(DMM) topology.

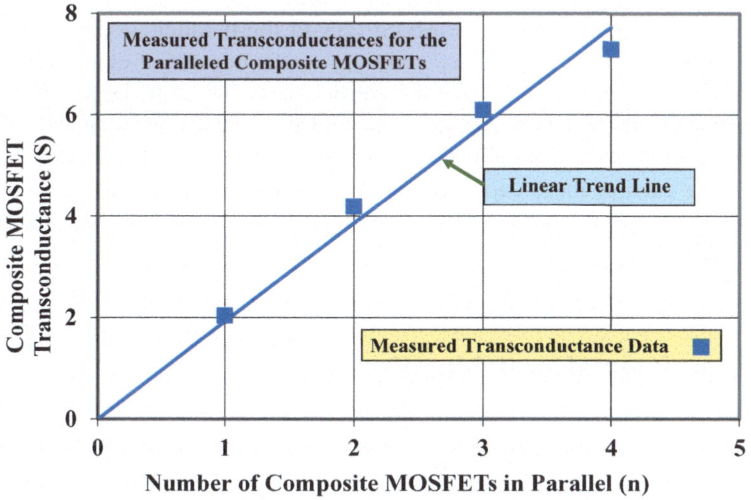

Fig. 8.4 Transconductance of paralleled composite MOSFETs in the BaSIC(DMM) topology

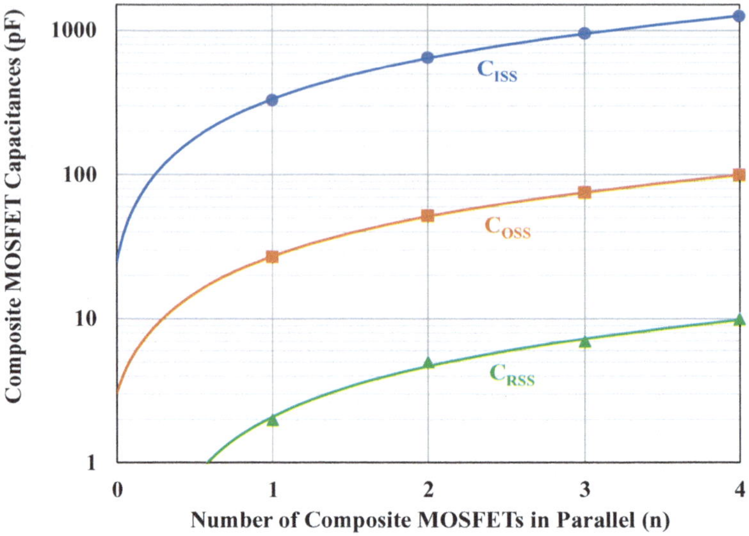

Fig. 8.5 Capacitances of paralleled composite MOSFETs in the BaSIC(DMM) topology

8.5 Experimental Capacitance Results

The capacitances C_{ISS}, C_{OSS}, and C_{RSS} for the composite MOSFETs in the BaSIC(DMM) topology should scale up in proportion to the number of devices connected in parallel. These capacitances are given in the datasheet of the 1.2 kV rated SiC power MOSFET with zero gate bias and a drain bias of 800 V. These test conditions were used for the composite MOSFETs in the paralleled configurations. A graph of the capacitances C_{ISS}, C_{OSS}, and C_{RSS} versus number of devices in parallel is shown in Fig. 8.5 ussing a logarithmic vertical axis due to the large range of values. The measured capacitances C_{ISS}, C_{OSS}, and C_{RSS} values are indicated with the blue, orange, and green data points, respectively. A linear trend line is also shown in the figure for each of these capacitances. The measured values fall on these trend lines demonstrating excellent scaling when paralleling composite MOSFET in the BaSIC(DMM) topology.

8.6 Experimental Switching Results

The turn-on and turn-off switching losses were evaluated using a standard double-pulse test (DPT) setup with a 350 μH load inductor and a Wolfspeed SiC JBS diode as a freewheeling device. Tests were conducted at a drain bias of 800 V with currents of 6 A for the single composite MOSFET. The current for the paralleled composite MOSFETs was scaled to 12, 18, and 24 A for the two, three, and four composite MOSFETs in parallel, respectively. This scaling assumed that each of the paralleled devices would handle 6 A if the paralleling worked perfectly.

The measured turn-on, turn-off, and total switching loss measured for the four cases are plotted in Fig. 8.6 together with the power delivered to the load. The measured values for the turn-off losses E_{OFF} are shown with purple data points. The measured values for the turn-on losses E_{ON} are shown with orange data points. The measured values for the total losses E_T are shown with green data points. A linear trend line is shown using corresponding colors for each case. It can be concluded that all the energy losses are scaling linearly when paralleling the composite MOSFETs in the BaSIC topology.

The total power delivered to the load is also shown in the figure with the blue data points together with a blue linear trend line. It can be seen that the total power scales linearly when paralleling composite MOSFET in the BaSIC(DMM) topology. This allows the BaSIC(DMM) topology to be scaled up to handle large power levels in applications.

One of the advantages of the BaSIC topology is that the current within each of the paralleled devices can be individually monitored using the sense node voltage V_{SENSE}. This was utilized for the case of four composite MOSFETs connected in parallel to observe the current distribution. The total current flowing through the four composite MOSFETs in parallel is shown by the black line in Fig. 8.7 during the first pulse in the DPT test. The current increases to 21 A at 4 μs as shown by the purple circle. The voltages at the sense node for each of the composite MOSFETs are shown in the lower portion of Fig. 8.7. The node voltage values are very close beyond 2 μs showing excellent sharing of the current. The sense node voltage has a value of 0.26 V at 21 A. The current through each of the composite MOSFETs is

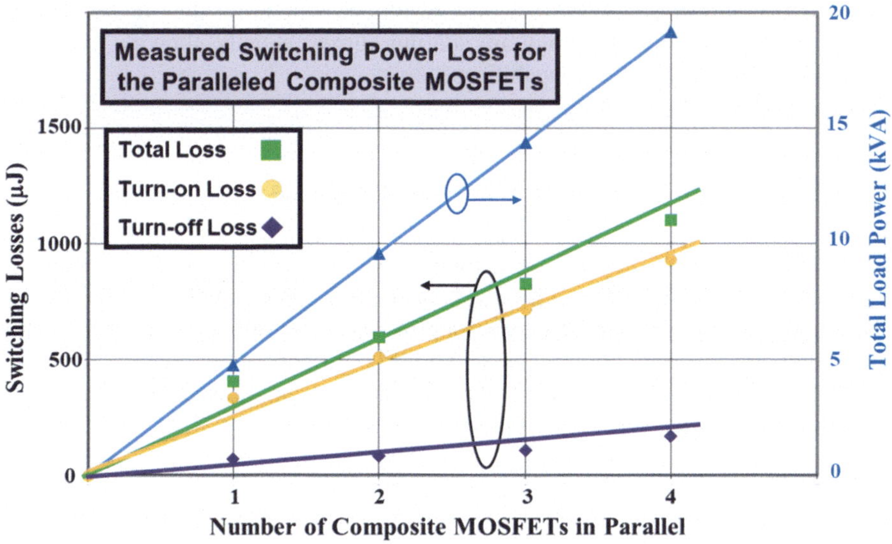

Fig. 8.6 Switching losses of paralleled composite MOSFETs in the BaSIC(DMM) topology

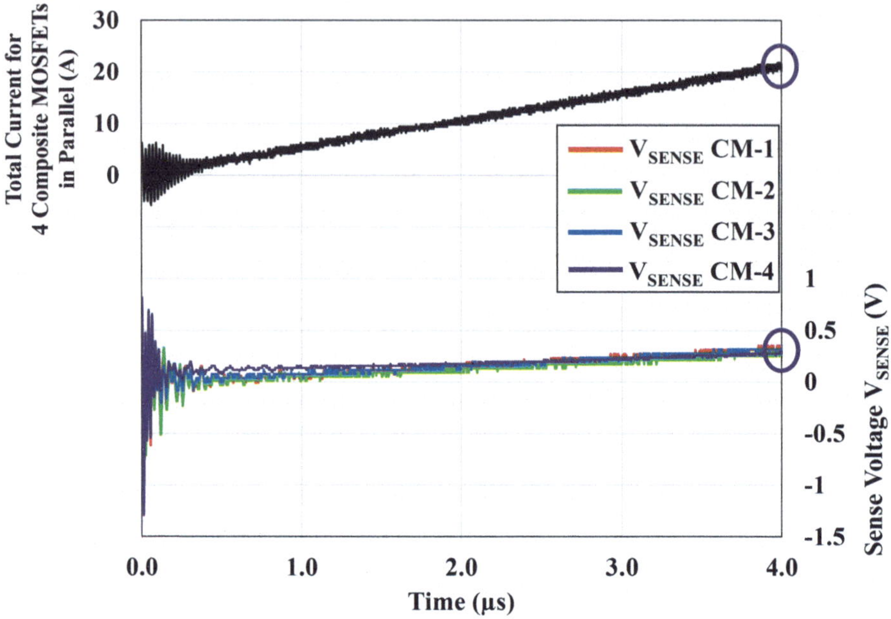

Fig. 8.7 Sense node voltage waveforms for the paralleled composite MOSFETs in the BaSIC(DMM) topology

5.25 A if equal current sharing occurs between the paralleled composite MOSFETs. This would produce a voltage drop of 0.26 V across each of Si DMM devices with measured on-resistances of 47–50 mΩ. This matches the measured sense voltages for the four composite MOSFETs. These results provide proof that the composite MOSFETs in the BaSIC topology are sharing current equally allowing excellent scaling of the power handling capability.

8.7 Experimental Short-Circuit Test Results

Short-circuit tests were first performed with a single composite MOSFET using a drain bias of 800 V. The pulse width was kept at 3 μs to prevent failure of the device. The single composite MOSFET exhibits a peak short-circuit current of 17.5 A. The pulse adiabatically heats the SiC power MOSFET to only about 400 °C, which is well below the melting point of the aluminum metallization used for source contacts, thus preventing catastrophic failure of the device.

The short-circuit tests were repeated with two, three, and four composite MOSFETs in parallel. The peak short-circuit current rose to 35 A, 52 A, and 69 A, respectively. No failure of devices was observed during these tests. These results are graphed in Fig. 8.8 together with a linear trend line. The data points lie along the linear trend line indicating that the composite MOSFETs are sharing the short-circuit current equally. If this did not hold true, any device with the excessive current would fail due to heating above 600 °C.

Confirmation that the composite MOSFETs share current equally during the short-circuit event can be obtained by observing the sense node voltage for each device. In order to illustrate this, the waveforms obtained for the case of four composite MOSFETs operating in parallel are shown in Fig. 8.9. The total current shown by the black line reaches about 70 A which is 4 times the 17.5 A handled by each composite MOSFET. The sense node voltages for each of the composite MOSFETs increase in a very similar manner and have very close values. This demonstrates that the composite MOSFETs are sharing current equally throughout the duration of the short-circuit event.

From these tests, it can be concluded that the BaSIC(DMM) topology with paralleled devices performed in an excellent manner during the short-circuit event. This is important for scaling the power handling capability of this technology. It is worth pointing out that these results were obtained with no effort to screen the SiC power MOSFET or the silicon DMM devices and try to match their characteristics.

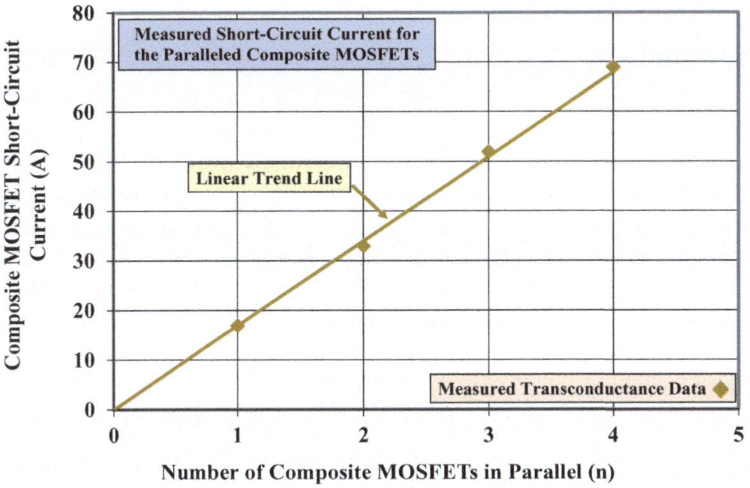

Fig. 8.8 Short-circuit currents observed for paralleled composite MOSFETs in the BaSIC(DMM) topology

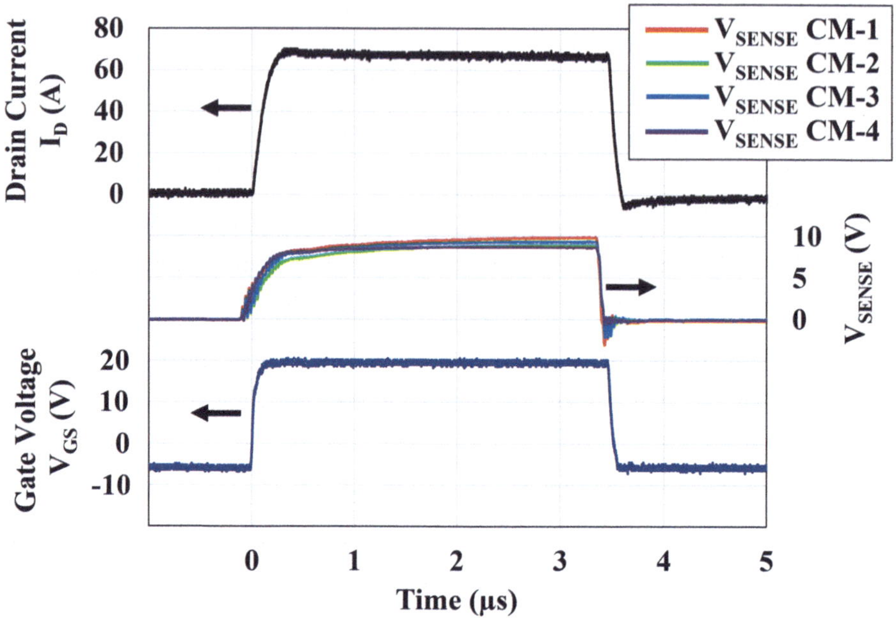

Fig. 8.9 Sense node voltages observed for paralleled composite MOSFETs in the BaSIC(DMM) topology under short-circuit tests

8.8 Summary

The ability to parallel the composite MOSFETs in the BaSIC(DMM) topology has been discussed in this chapter by using commercially available 1.2 kV SiC power MOSFETs and silicon DMM devices. No screening or matching of the device characteristics was used during these tests. Excellent scaling of the on-resistance, transconductance, and capacitances was observed. The power handling capability scaled linearly with more devices connected in parallel with a similar scaling of the power losses. The unique current sensing capability of the BaSIC topology allowed confirmation that the current was equally shared by the composite MOSFETs during the on-state and switching.

Non-destructive short-circuit tests demonstrated the short-circuit current scales in proportion to the number of devices in parallel to the BaSIC(DMM) topology. Equal current sharing between the paralleled devices was confirmed using the sense node voltage for each device. The excellent current sharing during the short-circuit event is important to avoid any device failures.

These test results indicate that the BaSIC(DMM) topology has excellent capability for scaling up the power handling capability to address more applications.

References

1. G. Wang, J. Mookken, J. Rice, M. Schupbach, Dynamic and static behavior of packaged silicon carbide MOSFETs in paralleled applications, in *IEEE Applied Power Electronics Conference and Exposition (APEC)* (2016), pp. 1478–1483
2. H. Li, S. Munk-Nielson, X. Wang, R. Maheshwari, S. Beczcowski, C. Uhrenfeldt, W. Franke, Influences of device and circuit mismatches on paralleling silicon carbide MOSFETs. IEEE Trans. Power Electron. **31**, 621–634 (2016)
3. J. Ke, Z. Zhao, P. Sun, H. Huang, J. Abuogo, X. Cui, New screening method for improving transient current sharing of paralleled SiC MOSFETs, in *Int. Power Electronics Conf.* (2018), pp. 1125–1130
4. P. Sun, Z. Zhao, Y. Cai, Q. Yu, J. Ke, X. Li, J. Li, X. Tang, F. Yang, X. Cui, Influence of chip parameters spread on power loss sharing of paralleled SiC MOSFETs, in *IEEE European Power Electronics Conf.* (2019), pp. 1–10
5. H. Li, S. Zhao, X. Wang, L. Ding, H.A. Mantooth, Parallel connection of silicon carbide MOSFETs—challenges, mechanism, and solutions. IEEE Trans. Power Electron. **38**, 9731–9749 (2023)
6. A. Kanale, B.J. Baliga, Excellent static and dynamic scaling of power handling capability of the BaSIC(DMM) topology with 1.2 kV SiC power MOSFETs, in *IEEE Workshop on Wide Bandgap Power Devices and Applications* (2021), pp. 14–17
7. Wolfspeed product page, https://www.wolfspeed.com/c2m0280120d
8. Littelfuse datasheet, https://m.littelfuse.com/~/media/electronics datasheets/discrete_mosfets/ littelfuse_discrete_mosfets_nchannel_depletion_mode_ixt _16n10_datasheet.pdf

Chapter 9
Optimization of Si DMM

In the previous chapters, it has been demonstrated that the BaSIC topology with a gate-source-shorted silicon depletion mode MOSFET (DMM) in series with the source electrode can significantly improve the short-circuit withstand time of a SiC power MOSFET. These results were obtained by using a commercially available Si DMM device with a blocking voltage rating of 100 V with an actual breakdown voltage of 150 V. However, the maximum voltage imposed on the Si DMM device during all operating conditions was shown to be <10 V. Consequently, the commercially available Si DMM device does not have the most suitable design for this application because its breakdown voltage far exceeds the requirements of the BaSIC(DMM) topology. It is well established that reducing the breakdown voltage of silicon power MOSFETs will lead to much smaller on-resistances.

In this chapter, the characteristics of the Si DMM device are analyzed from the point of view of its use as the non-linear resistance for the BaSIC topology. Based on the analysis, the parameters of the internal device structure are varied to improve the trade-off between reducing the saturation current of the device and its on-resistance. In particular, reducing the gate oxide thickness is shown to significantly improve this trade-off curve. In addition, the U-MOSFET structure is shown to be much superior to the D-MOSFET structure. Significantly lower on-resistances are predicted for an optimized Si DMM device with the saturation current level needed to extend the short-circuit withstand time beyond 10 μs for commercially available 1.2 kV SiC power MOSFETs [1].

A theoretical analysis of the silicon gate-source-shorted depletion mode MOSFET (GSS-DMM) is presented in this chapter. This analysis is not provided in textbooks due to focus on the enhancement mode devices that are usually required for applications. The electrical characteristics of the Si GSS-DMM devices are then generated with various changes to the internal device cell structure. A unique trade-off graph between the drain saturation current and the on-resistance is proposed and used to define the optimum device structure.

B. J. Baliga, A. Kanale, *The BaSIC Topology*,
https://doi.org/10.1007/978-3-031-86630-2_9

9.1 Rationale for Si DMM Optimization

Power semiconductor companies manufacture a large variety of silicon enhancement mode MOSFETs with a wide range of blocking voltages and on-resistances. However, very few silicon depletion mode MOSFETs are manufactured due to their limited applications. The only Si DMM device suitable for the BaSIC(DMM) topology was found to be the Si depletion mode power MOSFET (IXTH16N10D2) with blocking voltage rating of 100 V and on-resistance of 64 mΩ [2]. A short-circuit time (t_{SC}) for a 1.2 kV SiC power MOSFET (C2M0280120D) product was increased by a factor of 1.65× (from 4.8 to 7.9 μs) by utilizing this Si GSS-DMM device. The net on-resistance increased by 17% in this case [3].

The measured breakdown voltage of the commercially available Si DMM device was above 150 V. Since the Si DMM device is not subjected to voltages above 10 V under all operating conditions in the BaSIC topology, the available device is not optimally designed for this application. An optimization of the Si DMM device is desirable to achieve the best trade-off between reducing the saturation current while obtaining the lowest possible on-resistance.

A superior trade-off curve between the short-circuit withstand time (t_{SC}) and the increase in on-resistance (R_{ON}) for the BaSIC(DMM) topology is achieved when the trade-off curve between the saturation current and on-resistance of the Si GSS-DMM is improved. The best trade-off curve for the Si GSS-DMM exhibits both a smaller saturation current and on-resistance simultaneously. This is a unique requirement for the BaSIC topology. Analysis of the Si DMM structural parameters that improve the trade-off curve between the saturation current and on-resistance is therefore required. This analysis has not been previously performed in textbooks and publications because of a focus on enhancement mode devices.

In this chapter, it is:

- Theoretically demonstrated that a much better trade-off between increase in t_{SC} and increase in R_{ON} is achieved by reducing the blocking voltage of the Si GSS-DMM device
- Theoretically demonstrated that the trade-off between increase in t_{SC} and increase in R_{ON} can be further improved by making selective changes to parameters of the Si DMM cell structure
- Theoretically demonstrated that a better trade-off curve between increase in t_{SC} and increase in R_{ON} can be achieved with the trench-gate Si GSS-DMM structure compared with the planar gate Si GSS-DMM structure.

The improvement in the trade-off curve between increase in t_{SC} and increase in R_{ON} achieved by using the new optimized Si GSS-DMM devices in the BaSIC(DMM) topology is theoretically quantified for several commercially available SiC power MOSFETs as examples.

The information provided in this chapter is provided for the power semiconductor industry to encourage the design and manufacturing of better Si depletion mode power MOSFETs for utilization in the BaSIC(DMM) topology to enhance the short-circuit performance of SiC power MOSFETs. A strong growth in the market for the SiC power MOSFET is projected due to their applications in electric vehicles. This creates a large market for the new Si DMM devices that did not exist in the past. Many power semiconductor companies make both Si and SiC power MOSFETs. The information provided in this chapter can be used by them to make a chip set consisting of a Si DMM device and a SiC power MOSFET device that are co-packaged to make the composite MOSFETs for the BaSIC(DMM) topology. These new products can be utilized for motor control applications in electric vehicles and industrial drives due to their superior short-circuit capability.

9.2 Si DMM Structure and Analysis

Commercially available silicon enhancement mode power is fabricated using either the D-MOS or U-MOS structures shown in Fig. 9.1. The peak doping concentration in the P-base region, as determined by the double-diffusion process, is usually chosen to produce a positive threshold voltage for enhancement mode operation [4]. However, both of these structures can operate as depletion mode devices by adjusting the doping concentration of the P-base region to produce a negative threshold

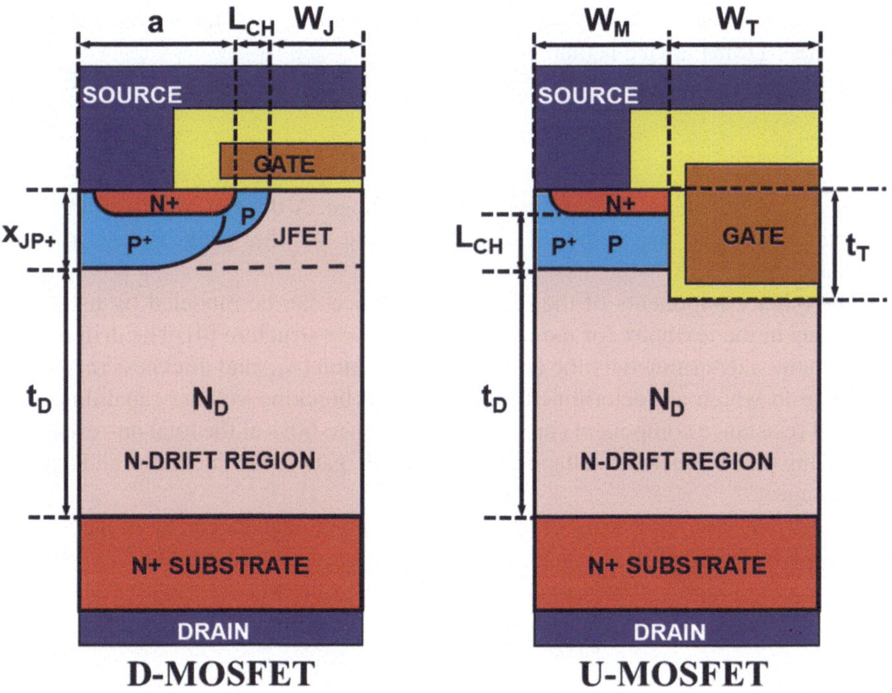

Fig. 9.1 Silicon DMM device structures

voltage. A larger negative threshold voltage is observed when the doping concentration of the P-base region is reduced.

9.2.1 DMM Device Physics and Analysis

The channel resistance for a conventional enhancement mode power MOSFET is given by [4]:

$$R_{CH} = \frac{L_{CH} t_{GOX}}{Z \mu_{ni} \varepsilon_{OX} \left(V_G - V_{TH}\right)} \tag{9.1}$$

where L_{CH} is the channel length, t_{GOX} is the gate oxide thickness, Z is the channel width orthogonal to the cross-section shown in Fig. 9.1, μ_{ni} is the channel inversion layer mobility, ε_{OX} is the gate oxide dielectric constant, V_G is the gate bias voltage, and V_{TH} is the threshold voltage. This expression is applicable to commercially available Si power MOSFETs that are turned on using a positive gate bias with a positive threshold voltage that is necessary for most power electronics applications.

The unusual feature of the Si GSS-DMM device being analyzed here for use in the BaSIC(DMM) topology is that they are operated with zero gate bias (V_G). The threshold voltage (V_{TH}) for the Si DMM device must be designed to be negative to obtain current flow when the gate is shorted to the source. The channel resistance for the Si GSS-DMM device is then given by:

$$R_{CH} = \frac{L_{CH} t_{GOX}}{Z \mu_{ni} \varepsilon_{OX} V_{TH}} \tag{9.2}$$

where V_{TH} is the *magnitude* of the threshold voltage. A desired negative threshold voltage value can be achieved by the choice of the peak P-base doping concentration.

The other components of the on-state resistance can be modeled by using the equations in the textbook for the enhancement mode structure [4]. The drift region component is determined by the doping concentration (N_D) and thickness (t_D) of the drift region which are determined by the desired blocking voltage capability. The channel resistance component can range from 30% to 60% of the total on-resistance depending on the blocking voltage, channel length, gate oxide thickness, and threshold voltage.

The drain saturation current of the conventional enhancement mode power MOSFET is given by [4]:

$$I_{D,SAT} = \frac{Z \mu_{ni} \varepsilon_{OX} \left(V_G - V_{TH}\right)^2}{2 L_{CH} t_{GOX}} \tag{9.3}$$

The drain saturation current is therefore determined by the channel length, gate oxide thickness, and threshold voltage. In the case of the Si GSS-DMM device, the gate bias is zero. The drain saturation is then given by:

$$I_{D,SAT} = \frac{Z\mu_{ni}\varepsilon_{OX}V_{TH}^{2}}{2L_{CH}t_{GOX}} \tag{9.4}$$

There is an interplay between the channel resistance given in Eq. (9.2) and the drain saturation current given in Eq. (9.4) for the Si GSS-DMM device. The saturation current increases and the channel resistance decreases when the magnitude of the negative threshold voltage increases. The threshold voltage can therefore be varied to generate a trade-off curve between the drain saturation current and the on-resistance for the Si DMM device.

9.3 Silicon D-MOSFET Device Analysis

There are two silicon power MOSFETs options manufactured by the industry as shown in Fig. 9.1. Both the D-MOSFET and U-MOSFET structures can be utilized for making an optimum Si DMM device for the BaSIC topology. The analysis of the D-MOSFET is performed in this section.

The commercially available Si DMM device (IXTH16N10D2) can be used to validate the analytical models. Its drift region doping (2.5×10^{15} cm^{-3}) and thickness (8.0 μm) were obtained by assuming that the edge termination limits the blocking voltage to 80% of the ideal parallel plane value. Typical values for the JFET region doping concentration (2×10^{16} cm^{-3}) and width ($W_J = 1.5$ μm) were selected with a P^+ region junction depth (x_{JP+}) of 1.5 μm. The cell parameter "a" in Fig. 9.1 of 3 μm was chosen to allow making the source and base ohmic contacts based on typical alignment tolerances. The gate oxide thickness and channel mobility were assumed to have typical values of 500 Å and 100 cm^2/V-s. The device active area was estimated to be 0.1 cm^2 based on its on-resistance of 64 mΩ and the typical specific on-resistance of Si MOSFETs with blocking voltage of 150 V measured for this device. The channel width (Z), obtained by dividing the active area by the cell width, was found to be 183 cm for this device using the parameters shown in Fig. 9.1 for the D-MOSFET cell. This parameter will be held constant when optimizing the Si DMM device.

9.3.1 Benchmark 100 V Si D-MOSFET Device Analysis

The drain saturation current and on-resistance of the Si D-MOSFET device with the above device parameters were calculated and plotted as shown by the purple line in Fig. 9.2. The on-resistance for the benchmark 100 V rated Si GSS-DMM D-MOSFET

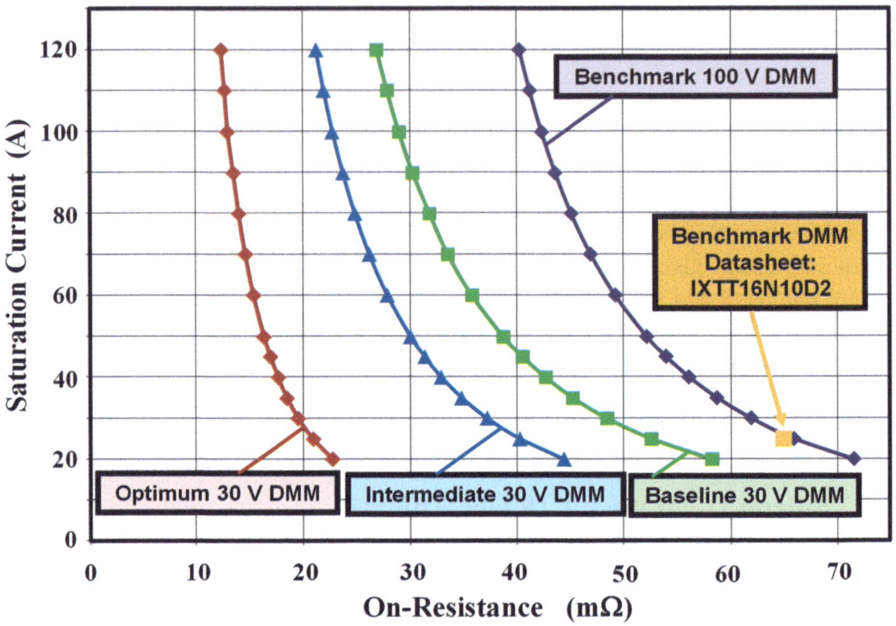

Fig. 9.2 Trade-off curves for the silicon D-MOSFET options

increases from 40 to 72 mΩ when the saturation current is reduced from 120 to 20 A. A similar behavior is observed for all the Si GSS-DMM structures. The operating point of the commercially available 100 V rated IXTH16N10D2 product is marked with the orange data point on the figure. The analytical model precisely matches the measured data for the commercially available 100 V rated GSS-DMM D-MOSFET IXTH16N10D2 product. This provides validation of the analytical models used in this chapter. They can therefore be used to analyze changes made to the device structural parameters.

9.3.2 Baseline 30-V D-MOSFET Structure

The on-resistance of Si enhancement mode power MOSFETs decreases with reduction in the blocking voltage capability [4]. The Si depletion mode MOSFETs will also show this trend although it has not been previously reported. Analysis of the Si GSS-DMM with D-MOSFET structure was performed by increasing the drift region doping to 1.3×10^{16} cm^{-3} and reducing its thickness to 2 μm to achieve a blocking voltage of 30 V. All other device parameters were kept unchanged, including the active area and cell width. The same channel length (1 μm) and gate oxide thickness (500 Å) were used as in the case of the benchmark 100 V D-MOS device. This structure will be referred to as the “Baseline 30 V DMM” in this chapter.

The trade-off curve obtained for this baseline 30 V Si GSS DMM D-MOS device by varying the threshold voltage is shown by the green line in Fig. 9.2. A shift in the trade-off curve to the left-hand side showing improved performance is achieved by reducing the blocking voltage capability to 30 V. For the same saturation current of 50 A, the baseline GSS-DMM D-MOS device has an on-resistance of 39 mΩ compared with 52 mΩ for the 100 V benchmark device.

9.3.3 Intermediate 30-V D-MOSFET Structure

Using the analytical model, an improvement of the trade-off curve was obtained by reducing the channel length (L_{CH}) of the 30 V Si GSS-DMM D-MOS device. High performance 30 V enhancement mode, planar gate, Si power MOSFETs have been reported by Silicon Semiconductor Corporation with very low specific on-resistance and gate charge with a channel length of 0.3 μm [5]. Analysis of the Si GSS-DMM D-MOS device was therefore performed with a reduced channel length of 0.3 μm keeping other parameters unchanged. This structure will be referred to as the "Intermediate 30 V DMM" in this chapter.

The trade-off curve for this intermediate 30 V DMM device is shown in blue in Fig. 9.2. The trade-off curve has shifted to the left-hand side when compared with the baseline 30 V DMM device demonstrating improved performance by reducing the channel length. The same saturation current of 50 A is obtained for the intermediate 30 V DMM device with an on-resistance of 30 mΩ compared with 39 mΩ for the baseline 30 V DMM device.

9.3.4 Optimum 30-V D-MOSFET Structure

Using the analytical models, it was found that even further improvement of the trade-off curve was obtained by reducing the gate oxide thickness of the 30 V Si DMM device. Analysis of the Si GSS-DMM D-MOS was therefore performed using a gate oxide thickness of 100 Å and channel length of 0.3 μm. A gate oxide thickness of 100 Å can be easily grown by reducing the gate oxide growth time. Although unusually small for Si power devices, this value is still larger than oxide thicknesses typically used for modern VLSI CMOS devices. This small gate oxide thickness of 100 Å is acceptable for the Si GSS-DMM device because no gate bias is applied that can stress the gate oxide and it is operated with the gate shorted to the source.

The trade-off curve for this optimum 30 V DMM device is shown in red in Fig. 9.2. The trade-off curve has been shifted to the left-hand side when compared with the intermediate 30 V DMM device demonstrating improved performance by reducing the gate oxide thickness. The same saturation current of 50 A is obtained for the optimum 30 V DMM device with an on-resistance of 16 mΩ compared with 30 mΩ for the intermediate 30 V DMM device.

9.4 Si U-MOSFET Device Analysis

The U-MOSFET structure can also be utilized for making an optimum Si DMM device for the BaSIC topology. The silicon power U-MOSFET structure is shown in Fig. 9.1. The analysis of the U-MOSFET is performed in this section with a focus on the on-resistance and saturation current. No similar analysis of a depletion mode U-MOSFET has been previously published in the literature.

Analytical modeling of the on-resistance for the silicon GSS-DMM U-MOSFET device structure can be performed using the equations in the textbook [4]. These models have been validated using numerical simulations. The saturation current for the U-MOSFET can be obtained using Eq. (9.4). Typical values for the trench width ($W_T = 0.5$ μm) and mesa width ($W_M = 2$ μm) were assumed in the models. The small cell pitch of 2.5 μm for the U-MOSFET structure results in a larger channel width (Z) of 333 cm for the same active area of 0.1 cm^2 as the previous D-MOSFET cases. This reduces the on-resistance and increases the saturated drain current.

9.4.1 Baseline 30-V U-MOSFET Structure

The baseline Si GSS-DMM U-MOSFET device was designed with a channel length of 1 μm and gate oxide thickness of 500 Å to match the baseline D-MOSFET case. The trade-off curve between saturation current and on-resistance for the baseline 30 V GSS-DMM U-MOSFET device was then obtained by adjusting the threshold voltage. This trade-off curve is shown in Fig. 9.3 with the green line. The benchmark 100 V rated GSS-DMM D-MOSFET case (purple line) is included in Fig. 9.3 for comparison. The curve for the baseline 30 V GSS-DMM U-MOSFET lies to the left-hand side of the benchmark structure demonstrating a much superior trade-off curve. The same saturation current of 50 A is obtained for the baseline 30 V GSS-DMM U-MOSFET case with an on-resistance of 13 mΩ compared with 52 mΩ for the 100 V rated benchmark D-MOS device.

9.4.2 Intermediate 30-V U-MOSFET Structure

Using the analytical model, an improvement of the trade-off curve was obtained by reducing the channel length (L_{CH}) of the 30 V Si GSS-DMM U-MOSFET device from 1 to 0.3 μm. This is referred to as the "Intermediate 30 V DMM" in this section.

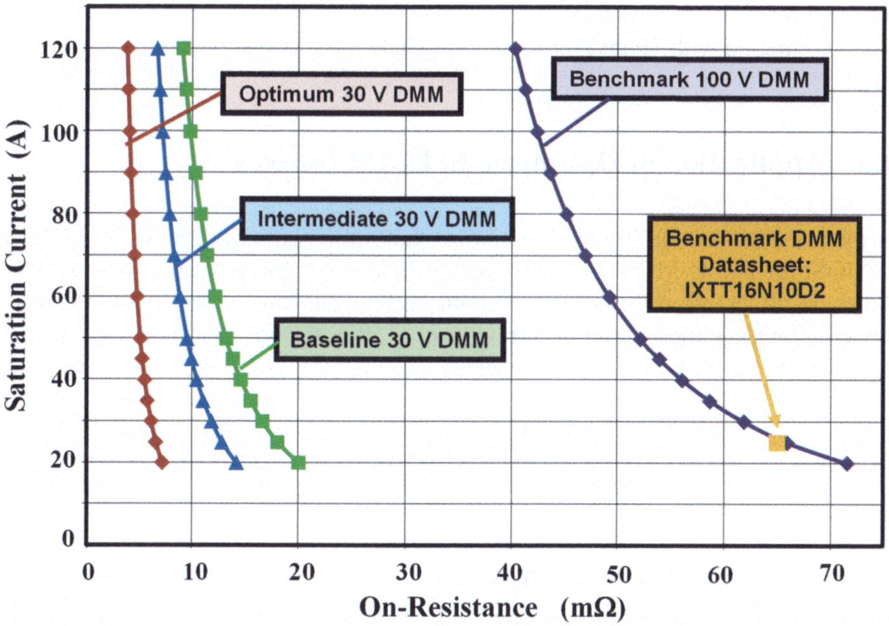

Fig. 9.3 Trade-off curves for the silicon U-MOSFET options

The trade-off curve for this intermediate 30 V DMM device is shown in blue in Fig. 9.3. The trade-off curve has been shifted to the left-hand side when compared with the baseline 30 V DMM U-MOSFET device demonstrating improved performance by reducing the channel length. The same saturation current of 50 A is obtained for the intermediate 30 V DMM U-MOSFET device with an on-resistance of 10 mΩ compared with 13 mΩ for the baseline 30 V DMM U-MOSFET device.

9.4.3 Optimum 30-V U-MOSFET Structure

Using the analytical models, it was found that even further improvement of the trade-off curve was obtained by reducing the gate oxide thickness of the 30 V Si DMM U-MOSFET device. Analysis of the Si GSS-DMM U-MOSFET was therefore performed using a gate oxide thickness of 100 Å and channel length of 0.3 μm. This small gate oxide thickness of 100 Å is acceptable for the Si GSS-DMM device because no gate bias is applied that can stress the gate oxide and it is operated with the gate shorted to the source.

The trade-off curve for this optimum 30 V DMM U-MOSFET device is shown in red in Fig. 9.3. The trade-off curve has been shifted to the left-hand side when compared with the intermediate 30 V DMM device demonstrating improved performance by reducing the gate oxide thickness. The same saturation current of 50 A is obtained for the optimum 30 V DMM U-MOSFET device with an on-resistance of 5 mΩ compared with 10 mΩ for the intermediate 30 V DMM device.

9.5 Application of Optimized Si DMM Devices

In the previous section, it has been shown that optimization of the Si GSS DMM D-MOSFET and U-MOSFET structures leads to very significant improvements in the trade-off between the saturated drain current and the on-resistance of these devices. For the same saturation current level of 50 A, the on-resistance was reduced from 60 mΩ for the commercially available Si D-MOSFET device to only 5 mΩ for the optimized U-MOSFET device. The impact of this on the performance of two commercially available 1.2 kV SiC power MOSFET products is examined in this section.

9.5.1 1.2 kV SiC Power MOSFET C2M0280120D Product

The impact of optimization of the Si DMM devices is first examined for the C2M0280120D product, which has a typical on-resistance of 280 mΩ with a maximum value of 370 mΩ in the datasheet. The short-circuit time and the corresponding drain saturation current were measured for the C2M0280120D product [3, 6] by varying the gate bias. The measured data is plotted in Fig. 9.4 as the blue triangles. A polynomial fit for the trade-off curve between the short-circuit time and the saturation current for this device is shown by the blue line:

$$t_{SC} = 194 \times I_{D,SAT}^{-0.934} \tag{9.5}$$

This polynomial fit can be used to convert the saturation currents obtained by the analysis in the previous section for the various Si GSS-DMM devices into a corresponding short-circuit withstand time for the 1.2 kV SiC power MOSFET C2M0280120D product. These short-circuit withstand times can then be plotted versus the increase in on-resistance to generate useful trade-off curves for the various Si GSS-DMM devices.

9.5.2 Short-Circuit Trade-Off Curve C2M0280120D Product with Si DMM D-MOSFET Devices

A trade-off curve, for the case of a DC drain bias of 800 V and gate bias of 20 V, between the short-circuit time and the increase in on-resistance can be obtained for the 1.2 kV SiC power MOSFET C2M0280120D product by using the information

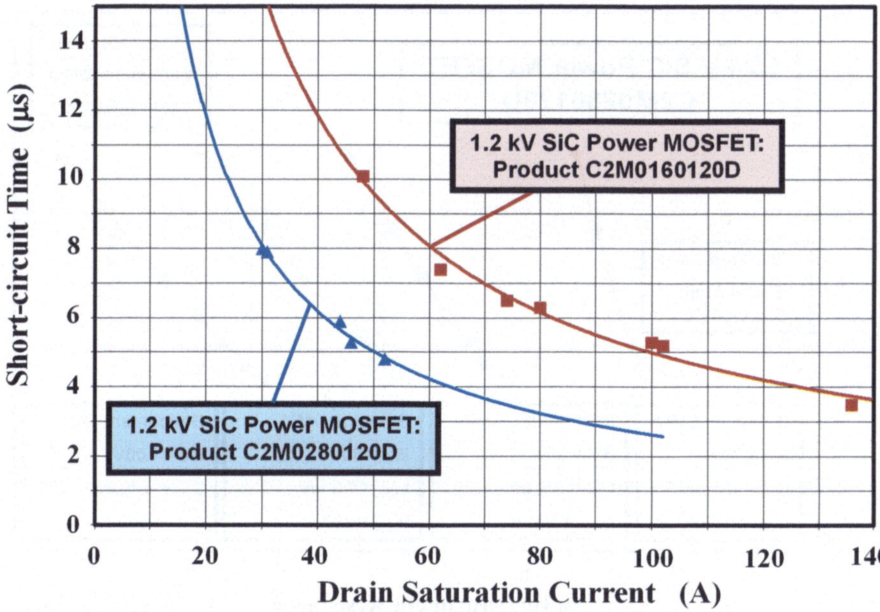

Fig. 9.4 Measured trade-off curve between short-circuit time and drain saturation current for the 1.2 kV SiC power MOSFET C2M0280120D and C2M0160120D products

in Fig. 9.2 for the various Si GSS-DMM D-MOSFET cases. The green line in Fig. 9.5 represents the baseline Si D-MOSFET device, while the blue and red lines represent the intermediate Si D-MOSFET device and the optimum D-MOSFET device, respectively. The purple line in the figure is the trade-off curve obtained for the benchmark commercially available 100 V rated Si D-MOSFET case.

The measured short-circuit time of 4.8 μs [3] for the stand-alone SiC power MOSFET product for a DC supply voltage of 800 V and gate drive voltage of 20 V is also shown in Fig. 9.5 as a reference point with no increase in on-resistance. The orange point in the figure shows the measured short-circuit data point [3] reported for the BaSIC(DMM) topology when using the commercially available 100 V rated Si D-MOSFET IXTT16N10D2 product. A short-circuit time of 7.9 μs was obtained with 17% increase in on-resistance for this case. It can be seen that the measured data point falls close to the purple line providing validation of the modeling described in this section.

The desirable short-circuit time of 10 μs is highlighted in Fig. 9.5 with the horizontal red dashed line. It is achieved for the benchmark Si DMM device with an increase in on-resistance by 18%. It is found that the desirable short-circuit time of 10 μs is achievable with the baseline Si D-MOSFET device with 15% increase in on-resistance. In the case of the intermediate Si D-MOSFET device, the increase in on-resistance reduces to 12% to achieve a short-circuit time of 10 μs. The same desirable short-circuit time of 10 μs is achievable with the optimum Si D-MOSFET device with an increase in on-resistance of only 6%. The optimization of the Si

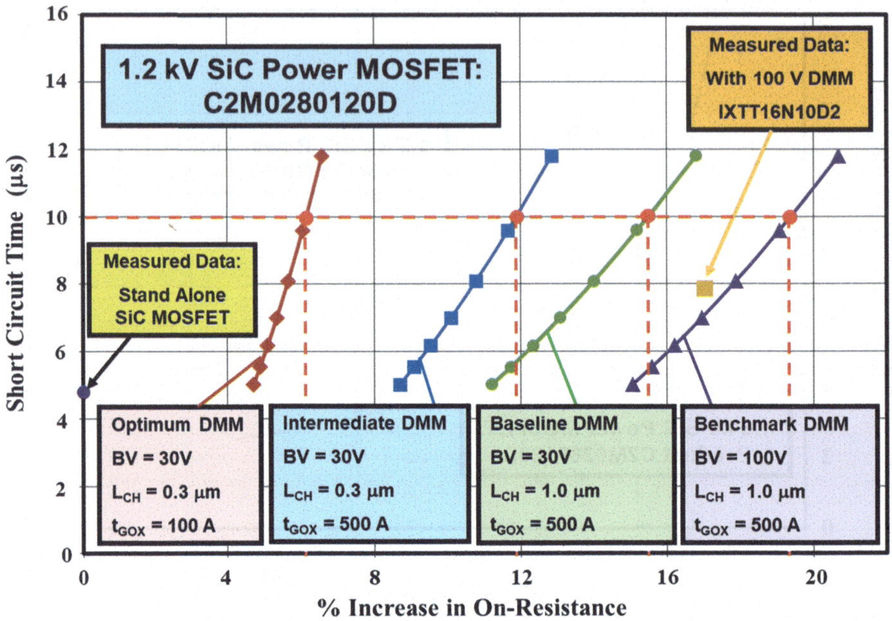

Fig. 9.5 Trade-off curves for the 1.2 kV SiC power MOSFET C2M0280120D product between short-circuit time and increase in on-resistance obtained using the BaSIC topology with various optimized Si GSS DMM D-MOSFET structures

GSS-DMM D-MOSFET device that was discussed in the previous section produces a reduction by a factor 3 times in the % increase in on-resistance.

9.5.3 Short-Circuit Trade-Off Curve C2M0280120D Product with Si DMM U-MOSFET Devices

In the same manner as above, a trade-off curve between the short-circuit time and the increase in on-resistance can be obtained for the 1.2 kV SiC power MOSFET C2M0280120D product by using the information in Fig. 9.3 for the various Si GSS-DMM U-MOSFET cases. The green line in Fig. 9.6 represents the baseline Si U-MOSFET device, while the blue and red lines represent the intermediate Si U-MOSFET device and the optimum U-MOSFET device, respectively. The purple line in the figure is the trade-off curve obtained for the benchmark commercially available 100 V rated Si D-MOSFET case.

The measured short-circuit time of 4.8 μs [3] for the stand-alone SiC power MOSFET product for a DC supply voltage of 800 V and gate drive voltage of 20 V is also shown in Fig. 9.6 as a reference point with no increase in on-resistance. The orange point in the figure shows the measured short-circuit data point [3] reported

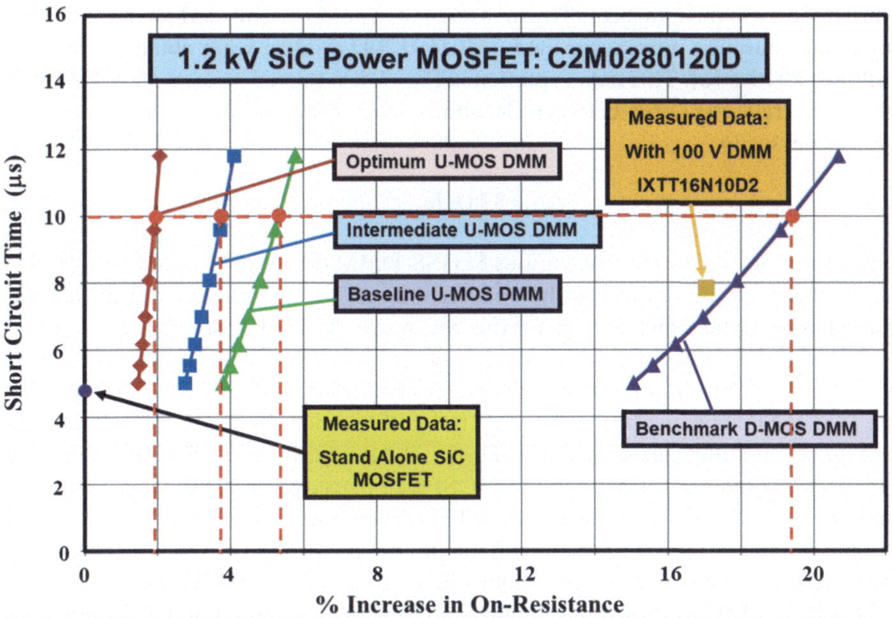

Fig. 9.6 Trade-off curves for the 1.2 kV SiC power MOSFET C2M0280120D product between short-circuit time and increase in on-resistance obtained using the BaSIC topology with various optimized Si GSS DMM U-MOSFET structures

for the BaSIC(DMM) topology when using the commercially available 100 V rated Si D-MOSFET IXTT16N10D2 product.

The desirable short-circuit time of 10 µs is highlighted in Fig. 9.6 with the red horizontal dashed line. It is found that the desirable short-circuit time of 10 µs is achievable with the baseline Si U-MOSFET device with 5.3% increase in on-resistance. In the case of the intermediate Si U-MOSFET device, the increase in on-resistance reduces to 3.8% to achieve a short-circuit time of 10 µs. The same desirable short-circuit time of 10 µs is achievable with the optimum Si U-MOSFET device with an increase in on-resistance of only 2.0%. The optimization of the Si GSS-DMM U-MOSFET device that was discussed in the previous section produces a reduction by a factor 9 times in the % increase in on-resistance.

9.5.4 Short-Circuit Trade-Off Curve C2M0160120D Product with Si DMM D-MOSFET Devices

The same procedure used in the previous sections can be applied to the 1.2 kV SiC power MOSFET C2M0160120D product as another example. This product has a typical on-resistance of 160 mΩ according to its datasheet with a maximum value of 196 mΩ. The procured device samples had a measured value of 168 mΩ, consistent with the datasheet. The measured short-circuit time and the corresponding drain saturation current for the C2M0160120D product were obtained by varying the gate bias [6]. This data is plotted in Fig. 9.4 as the red squares. A polynomial fit for the trade-off curve between the short-circuit time and the saturation current for this device is

$$t_{SC} = 378 \times I_{D,SAT}^{-0.939} \tag{9.6}$$

The saturation currents for the various Si GSS-DMM devices obtained by the analysis in the previous section can be converted by using this polynomial fit into the corresponding short-circuit time for this SiC power MOSFET product.

A trade-off curve, for the case of a DC drain bias of 800 V and gate bias of 20 V, between the short-circuit time and the increase in on-resistance, can be obtained for the 1.2 kV SiC power MOSFET C2M0160120D product by using the information in Fig. 9.2 for the various Si GSS-DMM D-MOSFET cases. The green line in Fig. 9.7 represents the baseline Si D-MOSFET device, while the blue and red lines represent the intermediate Si D-MOSFET device and the optimum D-MOSFET device, respectively. The purple line in the figure is the trade-off curve obtained for the benchmark commercially available 100 V rated Si D-MOSFET case.

The desired 10 μs short-circuit withstand time is highlighted in the figure with a horizontal dashed red line. The benchmark Si 100 V rated GSS-DMM D-MOS

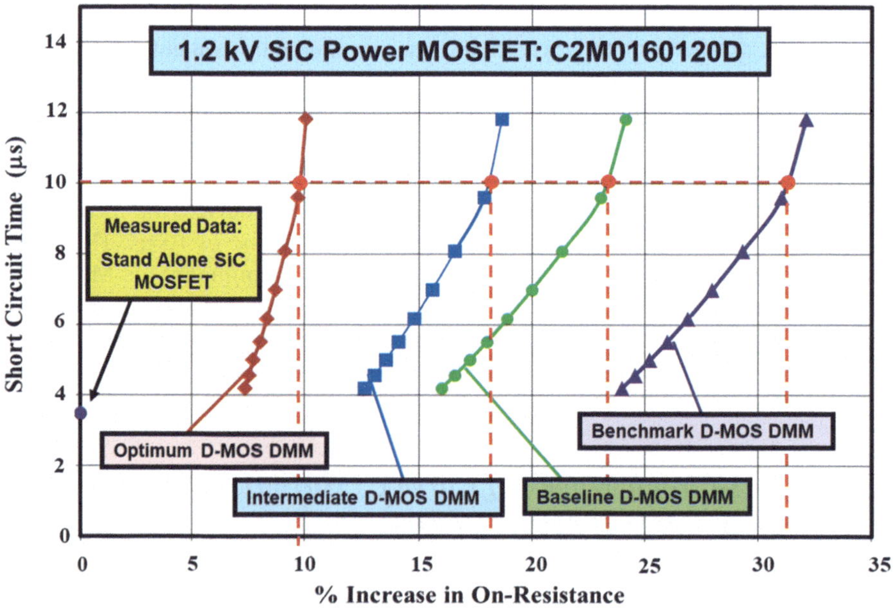

Fig. 9.7 Trade-off curves for the 1.2 kV SiC power MOSFET C2M0160120D product between short-circuit time and increase in on-resistance obtained using the BaSIC topology with various optimized Si GSS DMM D-MOSFET structures

structure produces a large increase in on-resistance of 31.5% to achieve a short-circuit time of 10 μs because of its high blocking voltage rating. Better trade-off curves are obtained by using the improved 30 V GSS-DMM D-MOSFET devices described in the previous section. A short-circuit withstand time of 10 μs is obtained with the baseline Si GSS-DMM D-MOSFET case with 23% increase in on-resistance. The increase in on-resistances reduces to 18% with the intermediate Si GSS-DMM D-MOSFET case and to 10% for the optimum Si GSS-DMM D-MOSFET case.

9.5.5 Short-Circuit Trade-Off Curve C2M0160120D Product with Si DMM U-MOSFET Devices

In the same manner as above, a trade-off curve between the short-circuit time and the increase in on-resistance can be obtained for the 1.2 kV SiC power MOSFET C2M0160120D product by using the information in Fig. 9.3 for the various Si GSS-DMM U-MOSFET cases. The green line in Fig. 9.8 represents the baseline Si U-MOSFET device, while the blue and red lines represent the intermediate Si U-MOSFET device and the optimum U-MOSFET device, respectively. The purple

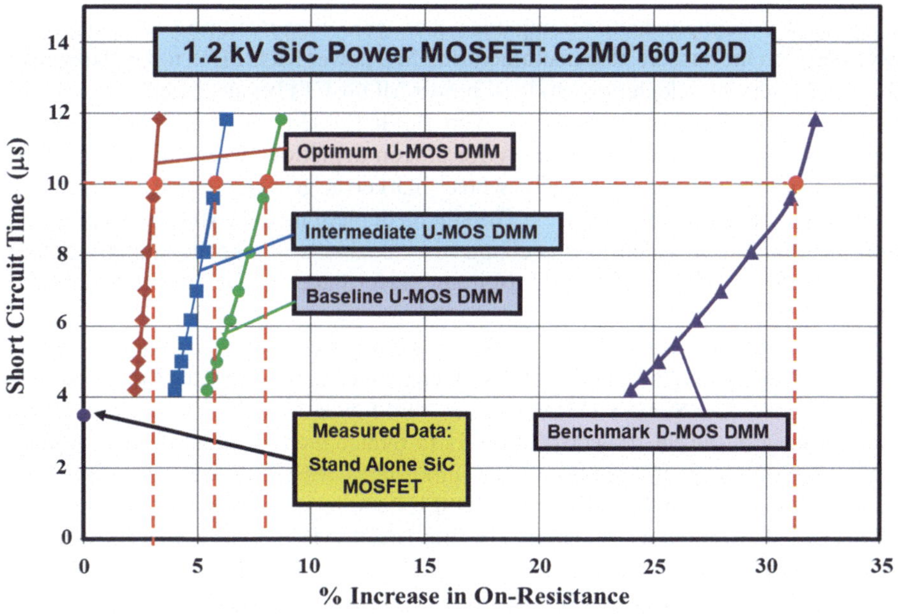

Fig. 9.8 Trade-off curves for the 1.2 kV SiC power MOSFET C2M0160120D product between short-circuit time and increase in on-resistance obtained using the BaSIC topology with various optimized Si GSS DMM U-MOSFET structures

line in the figure is the trade-off curve obtained for the benchmark commercially available 100 V rated Si D-MOSFET case.

The measured short-circuit time of 4.8 μs [3] for the stand-alone SiC power MOSFET product for a DC supply voltage of 800 V and gate drive voltage of 20 V is also shown in Fig. 9.8 as a reference point with no increase in on-resistance. The desirable short-circuit time of 10 μs is highlighted in Fig. 9.8 with the red horizontal dashed line. It is achievable for the benchmark Si DMM D-MOSFET structure with 31.5% increase in on-resistance. It is found that the desirable short-circuit time of 10 μs is achievable with the baseline Si U-MOSFET device with 8% increase in on-resistance. In the case of the intermediate Si U-MOSFET device, the increase in on-resistance reduces to 6% to achieve a short-circuit time of 10 μs. The same desirable short-circuit time of 10 μs is achievable with the optimum Si U-MOSFET device with an increase in on-resistance of only 3.0%. The optimization of the Si GSS-DMM U-MOSFET device that was discussed in the previous section produces a reduction by a factor 10 times in the % increase in on-resistance. These results demonstrate that it is important to manufacture optimized silicon U-MOSFET devices for utilization in the BaSIC(DMM) topology for application to silicon carbide power MOSFETs.

9.6 Summary

Many power semiconductor companies that were manufacturing silicon power MOSFETs in the past are also manufacturing SiC power MOSFETs today. However, they make very few depletion mode Si power MOSFETs because applications usually require normally-off transistors. The BaSIC(DMM) topology creates a new high volume commercial opportunity for silicon depletion mode MOSFETs when paired with SiC power MOSFETs whose market is growing rapidly for electric vehicle and industrial motor drives. The silicon depletion mode MOSFET optimization information provided in this chapter is valuable to the semiconductor power device industry to enhance the short-circuit performance of SiC power MOSFETs in these applications.

Manufacturers can build a co-packaged chip set consisting of the SiC power MOSFET device and the Si GSS-DMM device to make the BaSIC(DMM) composite transistors. The Si GSS DMM devices could be soldered on to the top source metallization of the SiC power MOSFETs to reduce the inductance between them. This creates a compact three-terminal composite MOSFET. A fourth terminal connected to the source of the SiC power MOSFET can serve as a current sense element as discussed in previous chapters. The BaSIC(DMM) topology will provide the desired 10 μs short-circuit capability for SiC power MOSFETs that is needed in motor control applications within electric vehicles and industrial drives.

References

1. A. Kanale, B.J. Baliga, Theoretical optimization of the Si GSS-DMM device in the BaSIC topology for SiC power MOSFET short-circuit capability improvement. IEEE Access **9**, 70039–70047 (2021)
2. Littelfuse datasheet, https://m.littelfuse.com/~/media/electronics datasheets/discrete_mosfets/littelfuse_discrete_mosfets_nchannel_depletion_mode_ixt _16n10_datasheet.pdf
3. A. Kanale, B.J. Baliga, Enhancing short circuit capability of 1.2 kV SiC power MOSFETs using a gate-source-shorted Si depletion-mode MOSFET in series with the source, in *IEEE 13th Int. Conf. Power. Electronics and Drive Systems* (2019), pp. 1–6
4. B.J. Baliga, Power MOSFETs, in *Fundamentals of Power Semiconductor Devices*, 2nd edn. (Springer-Science, 2019), pp. 283–520
5. B.J. Baliga, D.A. Girdhar, Paradigm shift in planar power MOSFET technology. Power Electron. Technol. **29**, 24–32 (2003)
6. A. Kanale, B.J. Baliga, Comparison of current suppression methods to enhance short circuit capability of 1.2 kV SiC power MOSFETs: a new approach using a series-connected, gate-source-shorted Si depletion-mode MOSFET vs use of a series resistance, in *IEEE 7th Workshop on Wide Bandgap Power Devices and Applications* (2019), pp. 53–58

Chapter 10
BaSIC(EMM) for SiC Power MOSFETs

As discussed in Chap. 5, the efficacy of the BaSIC topology was first demonstrated for enhancing the short-circuit capability of silicon carbide power MOSFETs by using a Si depletion mode MOSFET (DMM). A search for silicon depletion mode power MOSFETs produced only one device with the characteristics suitable for the BaSIC topology. It is an IXYS 100 V Si depletion mode MOSFET with on-resistance of 64 mΩ at zero gate bias, i.e., with gate shorted to the source terminal. This device was found to have a measured breakdown voltage of 150 V, well above the 30 V required for the BaSIC topology. It has a relatively large on-resistance with zero gate bias.

In contrast, the power semiconductor industry manufactures a large variety of Si enhancement mode MOSFETs (EMMs) with low on-state resistances due to their widespread applications. It is therefore advantageous to utilize one of these EMM devices for the BaSIC topology applied to the SiC power MOSFET [1].

The BaSIC(EMM) approach is discussed in this chapter using a Wolfspeed 1.2 kV SiC power MOSFET C2M160120D with datasheet on-resistance of 160 mΩ at a drain current of 10 A and gate bias of 20 V. These results were compared with two other short-circuit ruggedness improvement techniques: (a) using reduced gate drive voltage and (b) using a series resistance in subsequent chapters. Trade-off curves between the short-circuit withstand time and the increase in on-resistance and switching energy loss achieved with the BaSIC(EMM) topology are contrasted with these alternate approaches.

10.1 Short-Circuit Configuration: BaSIC(EMM) Topology

The short-circuit tests for the BaSIC(EMM) topology were performed using a single silicon EMM device connected in series with the source electrode of the SiC power MOSFET. This configuration is shown in Fig. 10.1 with the composite

B. J. Baliga, A. Kanale, *The BaSIC Topology*,
https://doi.org/10.1007/978-3-031-86630-2_10

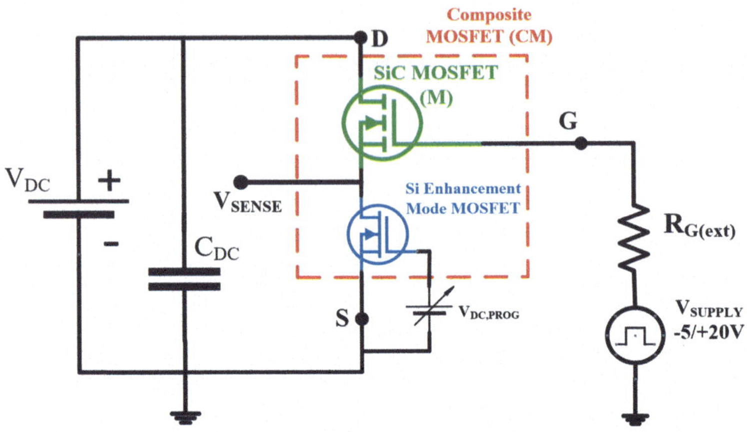

Fig. 10.1 Short-circuit test circuit with BaSIC(EMM) topology

MOSFET labeled as CM. It is worth pointing out that the gate drive voltage is applied to the gate of the SiC power MOSFET as usually done in its power circuit applications. The silicon EMM device only serves as the non-linear resistance Q_{LV} with a DC gate bias $V_{DC,PROG}$ applied to its gate terminal to control its on-state resistance and saturation current level. The DC gate bias $V_{DC,PROG}$ applied to the Si EMM device can be programmed by the power electronics engineer based upon the short-circuit withstand time required for a particular application. This provides a unique flexibility for the power electronics engineer while using a single commercially available SiC power MOSFET.

The short-circuit tests were performed for the BaSIC(EMM) topology by initially using a short duration gate pulse of 1 μs to avoid damaging the SiC power MOSFET. The width of the gate pulse was gradually increased until the device failed while capturing the drain current waveform. This allowed determination of the short-circuit withstand capability of the Composite MOSFET. The same procedure was employed for the stand-alone SiC power MOSFET using the test circuit shown in Fig. 5.1 as an initial benchmark.

10.2 Silicon EMM Device

An Infineon Si OptiMOS enhancement mode MOSFET IPI70N04S-06 with blocking voltage of 40 V was selected for implementation of the BaSIC(EMM) topology [2] due to its low on-state resistance of 6 mΩ at a gate bias of 10 V. Its output characteristics are reproduced from the datasheet at low drain bias voltages in Fig. 10.2

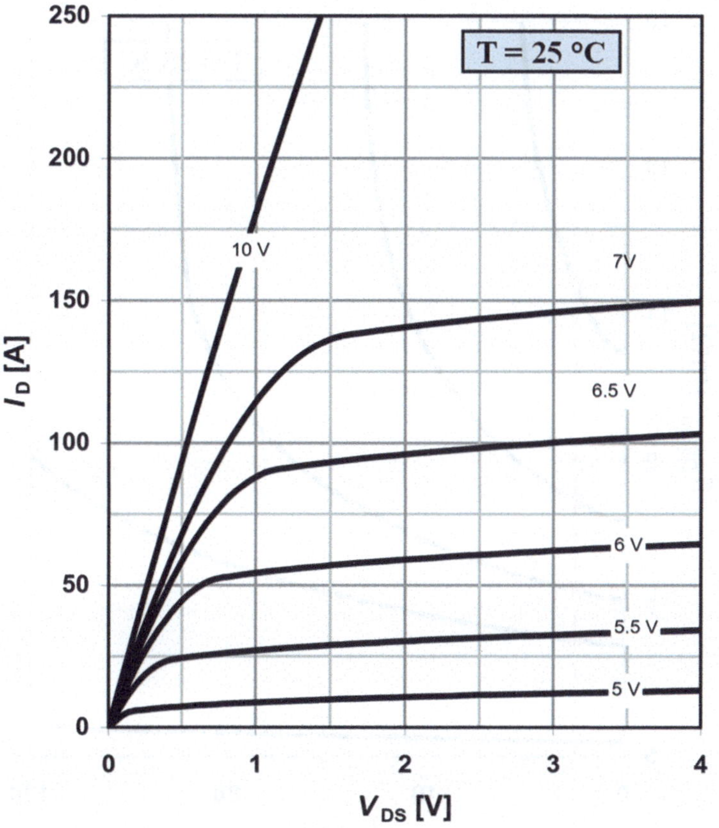

Fig. 10.2 Output characteristics of the silicon EMM device at low drain voltages

for discussion. It can be seen that the device exhibits an on-resistance of 6 mΩ at a gate bias of 10 V with an extremely high drain saturation current level. The drain saturation current level can be brought down to the 25–50 A range needed to enhance the short-circuit withstand time of the SiC power MOSFET when the gate bias voltage is reduced to 5–6 V range.

The on-resistance of the silicon EMM device at 25 °C in its linear regime of operation is provided in its datasheet as a function of the gate bias and drain current [2]. This graph is reproduced here in Fig. 10.3 for discussion. It is appropriate to use the 25 °C data because the silicon EMM does not undergo switching losses under normal circuit operating conditions preventing a significant increase in its junction temperature. The on-state resistances of the silicon EMM device at the SiC power MOSFET rated current of 10 A are highlighted using the red lines and dots. The on-state resistance has a value of 6.5, 7.2, 8.5, and 11.5 mΩ at gate bias values of 7.0, 6.5, 6.0, and 5.5 V, respectively. These resistances are much smaller than that of the Si DMM device used for the BaSIC(DMM) topology in previous chapters.

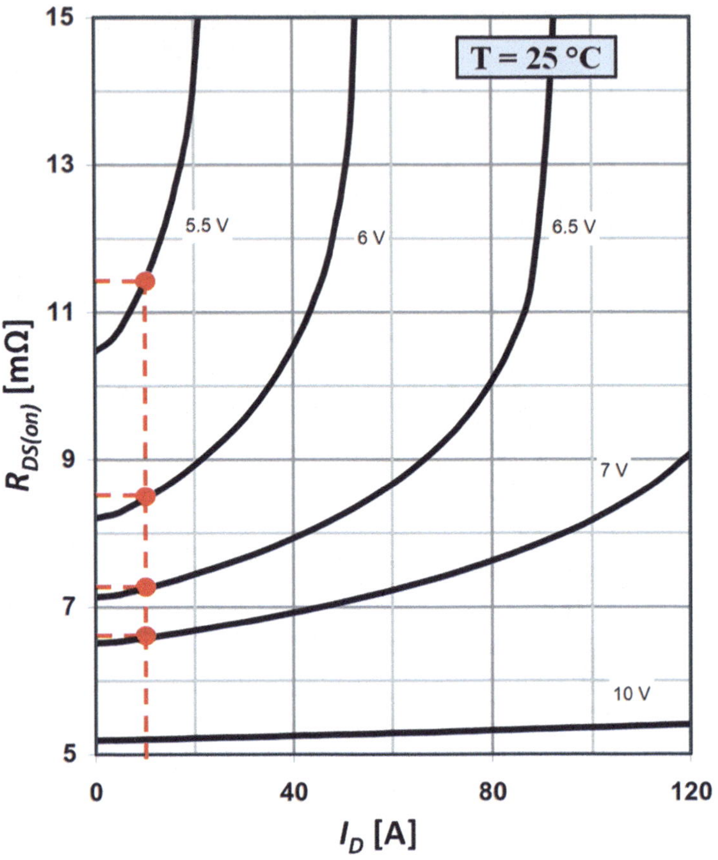

Fig. 10.3 On-resistance of the silicon EMM device

Consequently, the Si EMM device used for the BaSIC(EMM) topology should have smaller impact on the on-resistance and switching losses of the SiC power MOSFET.

The voltage drop across the silicon EMM device (it's drain voltage) increases to about 5 V during the short-circuit event. The effective gate drive voltage for the SiC power MOSFET under short-circuit conditions is therefore reduced, which brings down its saturation current until it automatically matches that for the silicon EMM device. In other words, the voltage $V_{SC}(Q_{LV})$ developed across the silicon EMM under short-circuit conditions automatically adjusts until the SiC power MOSFET saturation current becomes equal to the saturation current of the silicon EMM. In general, the saturation characteristics of both the silicon carbide power MOSFET and silicon EMM device will vary from part to part for the same device due to manufacturing tolerances. The automatic adjustment of the operating point C in Fig. 4.2 ensures seamless operation of the BaSIC(EMM) topology without the need for matching the characteristics of the silicon carbide power MOSFET and silicon EMM device.

10.3 High Voltage Output Characteristics

As explained in Chap. 2, the power transistor operates in its drain current saturation mode while supporting a high voltage equal to the DC supply voltage for the load during the short-circuit event. This operating point is part of the SiC power MOSFETs output characteristics. Knowledge of the high-voltage, high-current output characteristics is therefore invaluable for understanding the short-circuit capability of these power transistors. These characteristics are not provided by the device manufacturer in the datasheets.

A novel method for extracting the high-voltage, high-current output characteristics of power transistors was developed by using the short-circuit test configuration [3]. The short-circuit test configuration shown in Fig. 10.1 was used with a small pulse width of 1.5 μs. During this time, the current increases to a value that is representative of the drain saturation current at the DC supply voltage. The DC supply voltage can be varied in small steps to generate the high- voltage, high-current output characteristic for the chosen gate bias.

The experimentally extracted high-voltage, high-current output characteristic at a gate bias of 20 V for the stand-alone Wolfspeed 1.2 kV SiC power MOSFET C2M160120D is shown in Fig. 10.4 with the red line. The current increases gradually up to 300 V and then saturates a value of 136 A. The experimentally extracted high-voltage, high-current output characteristics for the composite MOSFET CM with DC gate biases of 5.4, 5.7, and 6.0 V applied to the Si EMM device are also shown in Fig. 10.4 with the purple, blue, and green lines, respectively. The current increases gradually up to 400 V and then saturates at values of 62, 74, and 102 A, respectively. These operating points for the four cases are highlighted with the brown circles.

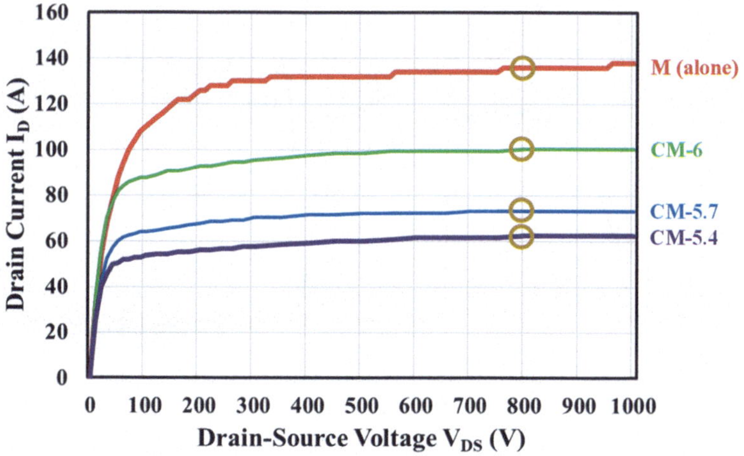

Fig. 10.4 High-voltage, high-current output characteristics: Stand-alone SiC power MOSFET C2M160120D and composite MOSFET CM with DC gate bias voltages of 5.4, 5.7, and 6.0 V applied to the Si EMM device

10.4 Short-Circuit Test Results

The short-circuit withstand time for the composite MOSFET CM can be obtained by using the test configuration shown in Fig. 10.1 with various values programmed for the DC gate bias applied to the Si EMM device. The short-circuit pulse width is gradually increased while capturing the current waveform until the SiC power MOSFET fails. The DC supply voltage is monitored as shown by the black line in Fig. 10.5. It is important to make sure that the capacitor C_{DC} is adequate to maintain the voltage at the desired value, 800 V in this case.

The measured short-circuit waveforms are shown in Fig. 10.5. The red line was obtained for the stand-alone SiC power MOSFET C2M160120D with a peak short-circuit current of 136 A. The device failed when the pulse width was increased to 3.5 μs. The green line labeled Case 2a was obtained for the composite MOSFET M with a DC gate bias of 6.0 V applied to the Si EMM device with a peak short-circuit current of 98 A. The SiC power MOSFET C2M160120D within Case 2a failed when the pulse width was increased to 5.2 μs. The blue line labeled Case 2b was obtained for the composite MOSFET M with a DC gate bias of 5.7 V applied to the Si EMM device with a peak short-circuit current of 74 A. The SiC power MOSFET C2M160120D within Case 2b failed when the pulse width was increased to 6.5 μs. The ochre line labeled Case 2c was obtained for the composite MOSFET M with a DC gate bias of 5.4 V applied to the Si EMM device with a peak short-circuit

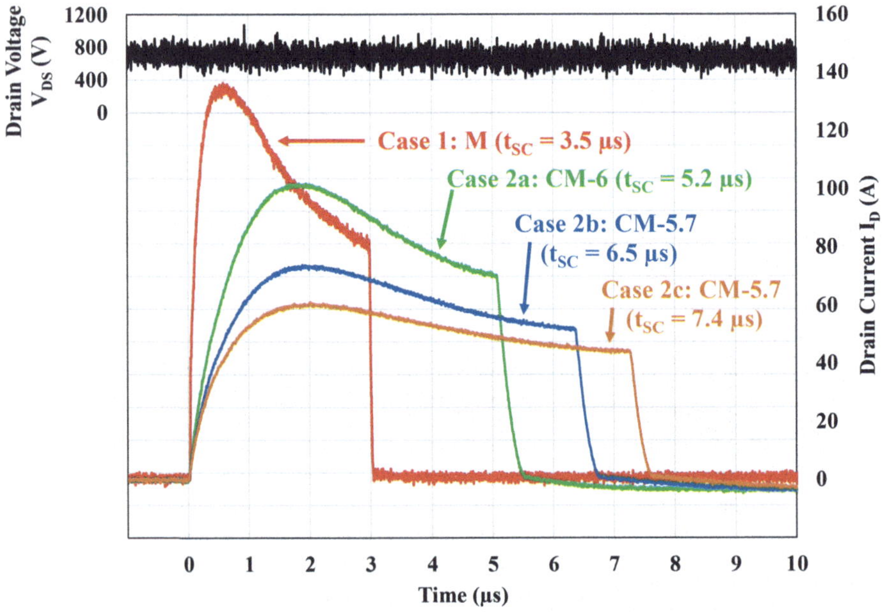

Fig. 10.5 Measured short-circuit waveforms: Stand-alone SiC power MOSFET C2M160120D and composite MOSFET CM with DC gate bias voltages of 5.4, 5.7, and 6.0 V applied to the Si EMM device

current of 62 A. The SiC power MOSFET C2M160120D within Case 2c failed when the pulse width was increased to 7.4 μs.

The short-circuit withstand time increases in inverse proportion to the peak short-circuit current level as expected from the theory provided in Chap. 2. These test provide validation that the short-circuit time can be greatly increased by using the BaSIC(EMM) topology. This topology was able to reduce the short-circuit current level and significantly increase the short-circuit withstand time for the SiC power MOSFET. The tests demonstrate that the BaSIC(EMM) topology is able to increase the short-circuit withstand time for a commercially available SiC power MOSFET by a factor of 2.1 times.

10.5 Impact on Normal Circuit Operation

The impact of the resistance of the Si EMM device on the on-state resistance and switching behavior of the composite MOSFET produced in the BaSIC(EMM) topology with a SiC power MOSFET must be quantified. Trade-off curves between enhancing the short-circuit time and an increase in on-resistance and switching losses can then be developed to fully comprehend the value of the BaSIC(EMM) topology.

10.5.1 Impact on On-Resistance

The on-state characteristics of the four configurations discussed above were obtained at lower drain voltages by using a curve tracer with a gate drive voltage of 20 V applied to the SiC power MOSFET M (product C2M160120D). The results for stand-alone SiC power MOSFET M are shown in Fig. 10.6 by the red line. The on-state operating point for all the devices was obtained its rated current of 10 A, highlighted with the red horizontal line. The on-resistance at this operating point is 168 mΩ for the stand-alone SiC power MOSFET M. The typical value cited in the datasheet for this transistor is 160 mΩ with a maximum value of 196 mΩ. The measured values fall within the datasheet parametric variation of 17% above the typical value.

The measured on-state characteristics for the composite MOSFET with three DC gate bias values applied to the Si EMM device are also shown in Fig. 10.6. The on-state operating point for each composite MOSFET can be evaluated at the rated current of 10 A for the SiC power MOSFET. The on-resistance at this operating point is 196 mΩ for CM-5.4 with DC gate bias of 5.4 V applied to its Si EMM device. The on-resistance at this operating point is 175 mΩ for CM-5.7 with DC gate bias of 5.7 V applied to its Si EMM device. The on-resistance at this operating point is 170 mΩ for CM-6.0 with DC gate bias of 6.0 V applied to its Si EMM device. These changes are within the 17% device-to-device variation when supplied

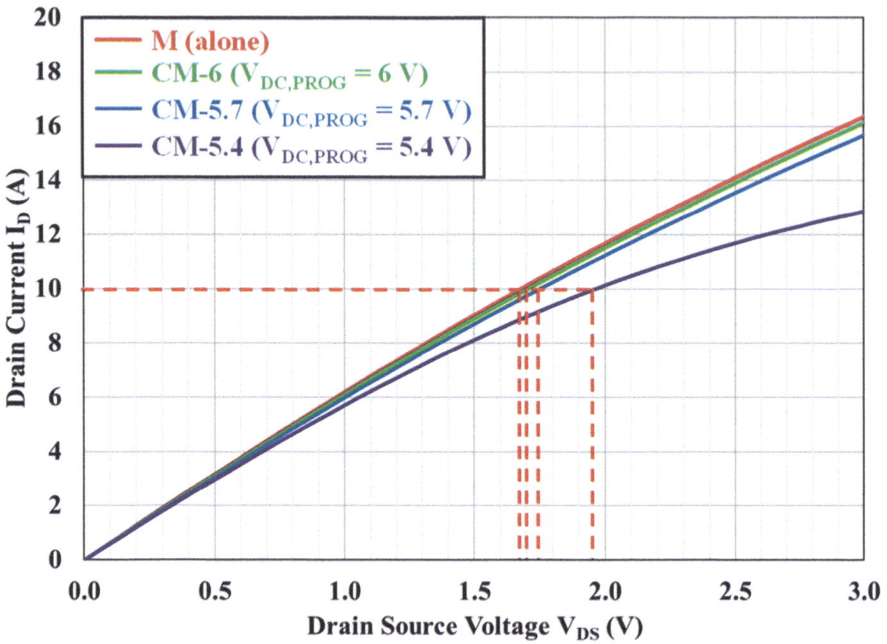

Fig. 10.6 Measured on-state characteristics of the stand-alone SiC power MOSFET M and the composite MOSFETs with DC-bias of 5.4, 5.7, and 6.0 V applied to the Si EMM device

by the manufacturer. The Si EMM device is therefore contributing only a small additional resistance to the SiC power MOSFET product C2M160120D in the BaSIC(EMM) topology.

In order to understand the on-resistance values obtained for the composite MOSFETs, it is necessary to examine the output characteristics of the actual Si EMM devices used for the BaSIC(EMM) topology with DC gate bias values in the range of interest to suppress the short-circuit current. The results of this measurement are shown in Fig. 10.7 with gate bias values ranging from 3.0 to 7.5 V. Curves with gate bias values of 5.4, 5.7, and 6.0 V are pertinent to the composite MOSFETs discussed in the previous section. The on-resistances were extracted at a drain current of 10 A, shown by the red dashed line in Fig. 10.7, to be relevant to the composite MOSFETs. The Si EMM exhibits an on-resistance of 10 mΩ at a gate bias of 6.0 V. The on-resistance of the Si EMM increases to 15 mΩ at a gate bias of 5.7 V. For lower gate bias values, the Si EMM deviated from the linear regime and exhibits quasi-saturation. Consequently, the on-resistance of the Si EMM increases to 35 mΩ at a gate bias of 5.4 V. These measured resistances for the Si EMM device are consistent with the observed trend in the measured on-resistances of the composite MOSFETs.

The parameters of the Si EMM device that are of importance for the BaSIC topology are its on-resistance and saturated drain current at each gate bias. It is useful to create a trade-off curve between these parameters with the gate bias as a

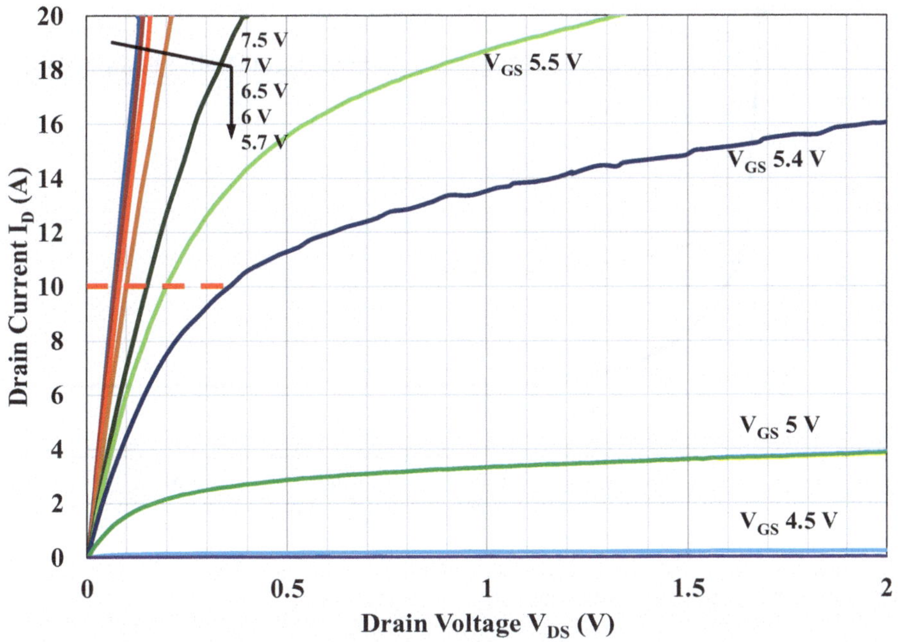

Fig. 10.7 Measured output characteristics of the Si EMM device (Infineon Si OptiMOS enhancement mode MOSFET IPI70N04S-06) at various gate bias voltages

parametric variable. This trade-off curve can be generated as shown in Fig. 10.8 by the blue points and line by using the room temperature values extracted from the Si EMM manufacturer's datasheet. Based on Fig. 10.8, it can be concluded that the chosen Si EMM for the BaSIC(EMM) topology is satisfactory for reducing the $I_{D,SAT(EMM)}$ to 65 A with a gate bias of 6 V while obtaining a low on-resistance of 10 mΩ. It is worth noting that reducing the $I_{D,SAT(EMM)}$ below 60 A with smaller gate bias significantly increases the on-resistance for the selected Si EMM device. Extrapolating the short-circuit data in Fig. 10.5, it is found that the desired short-circuit withstand time of 10 μs can be achieved if the peak short-circuit current is reduced to 50 A. From the blue trade-off line in Fig. 10.8, it can be seen that a $I_{D,SAT(EMM)}$ of 50 A is obtained with a gate bias of 5.7 V, allowing achieving the desired short-circuit withstand time of 10 μs. However, this is inconsistent with the short-circuit test results shown in Fig. 10.5, where the peak short-circuit current reaches 74 A, thereby reducing the short-circuit withstand time to only 6.5 μs. The same comments are applicable to the Si EMM operated with a gate bias of 5.4 V whose room temperature saturation current is 23 A in the datasheet, while its peak short circuit is 62 A.

The observed inconsistency occurs because the Si EMM heats up during the SC event. Its temperature rise, computed by using the adiabatic model used for short-circuit analysis of a power device [1] with $I_D = 50$ A, $V_{DS} = 5$ V, is about 200 °C after 2 μs. The rise in temperature reduces the threshold voltage from 3 to 1.5 V

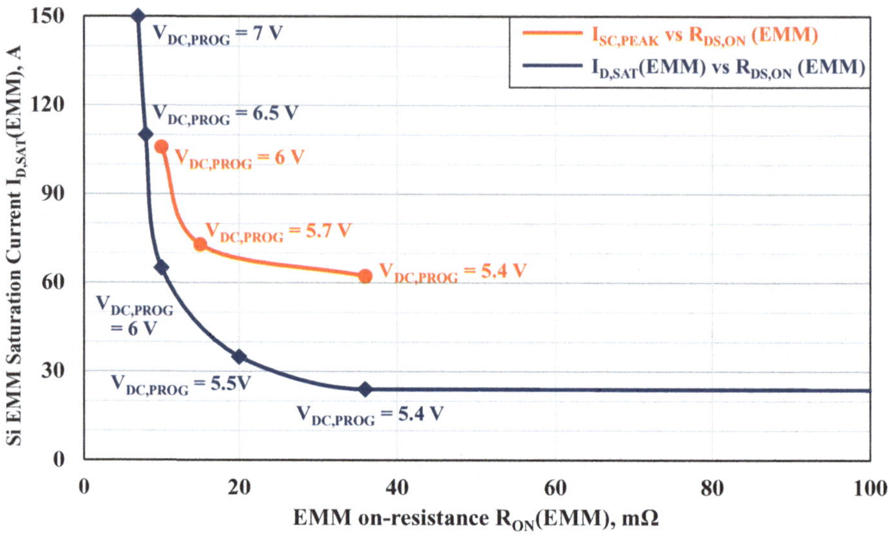

Fig. 10.8 Measured trade-off characteristics of the Si EMM device (Infineon Si OptiMOS IPI70N04S-06) between drain saturation current and on-resistance

according to the Si EMM's datasheet. The saturation current for V_{GS} = 5.4 V will then increase by a factor of 2.6×, in proportion with the change in (V_{GS}–V_{TH}) [2], leading to a value of 62 A as observed in Fig. 10.5.

A revised trade-off curve for the Si EMM device can be prepared by using the measured peak short-circuit currents from Fig. 10.5 as shown by the red points and line in Fig. 10.8. It can be observed that the heating of the Si EMM during the short-circuit event degrades the trade-off curve between I_{SC} and R_{on}. Based on this degraded trade-off curve, it can be concluded that the desired short-circuit withstand time of 10 μs cannot be achieved by using the Si EMM (Infineon Si OptiMOS IPI70N04S-06) selected for this work because it's peak SC current cannot be reduced to 50 A without a very large increase in the on-resistance. A more suitable Si EMM device can be found using the selection methodology described in the next chapter.

10.5.2 Impact on Capacitances

The capacitances of a power MOSFET determine its switching performance [4]. The input, output, and reverse transfer capacitances of all the devices were therefore measured as a function of the drain voltage. These measured capacitances are displayed in Fig. 10.9. The red lines apply for the stand-alone SiC power MOSFET product C2M160120D. The measured values for the C_{ISS}, C_{OSS}, and C_{RSS} are consistent with values provided in the datasheet for the device.

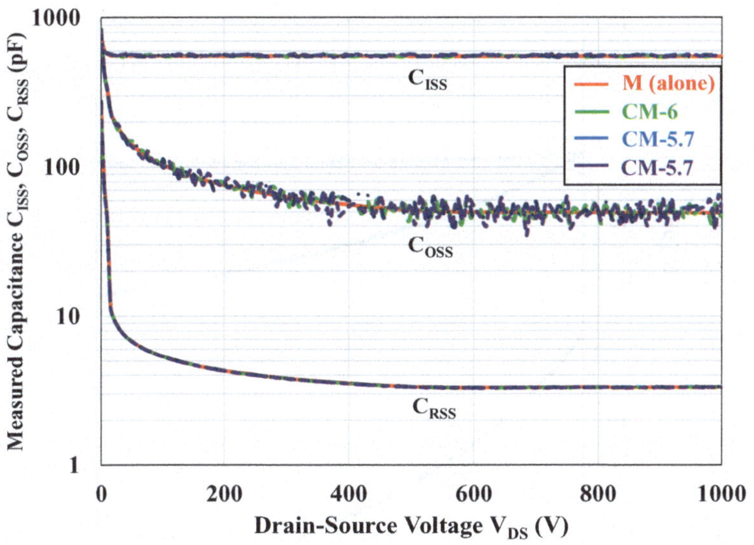

Fig. 10.9 Measured capacitances for the stand-alone SiC power MOSFET M and the composite MOSFETs with DC-bias of 5.4, 5.7, and 6.0 V applied to the Si EMM device

The purple, green, and blue lines in the figure apply to the composite MOSFET M with the three DC gate bias voltages applied to the Si EMM device. They overlap with the lines for the stand-alone SiC power MOSFET. This demonstrates that the BaSIC(EMM) topology does not alter the capacitances. The explanation for this behavior is the same as that given in Chap. 5 for the BaSIC(DMM) topology. Equations for the equivalent capacitances were provided there. The capacitances of the Si EMM are reproduced from its datasheet in Fig. 10.10. The output capacitance C_{OSS} of the Si EMM device at small drain bias is large (2000 pf). Consequently, the capacitance of the composite MOSFETs becomes close to that of the stand-alone SiC power MOSFET.

10.5.3 Impact on Transconductance

The transconductance g_M of the power transistor plays a role in the switching transient for the drain voltage and current during operation with an inductive load [4]. Its value can be ascertained from the transfer characteristics of the device. The measured transfer characteristics for all the cases at a drain bias of 10 V are shown in Fig. 10.11. The transconductance, extracted by taking a derivative of the drain current with respect to the gate voltage, is also shown in the figure. The value of the transconductance can be compared for the four cases at the same drain current level of 10 A. It is given by the values at the points indicated on the transconductance curve.

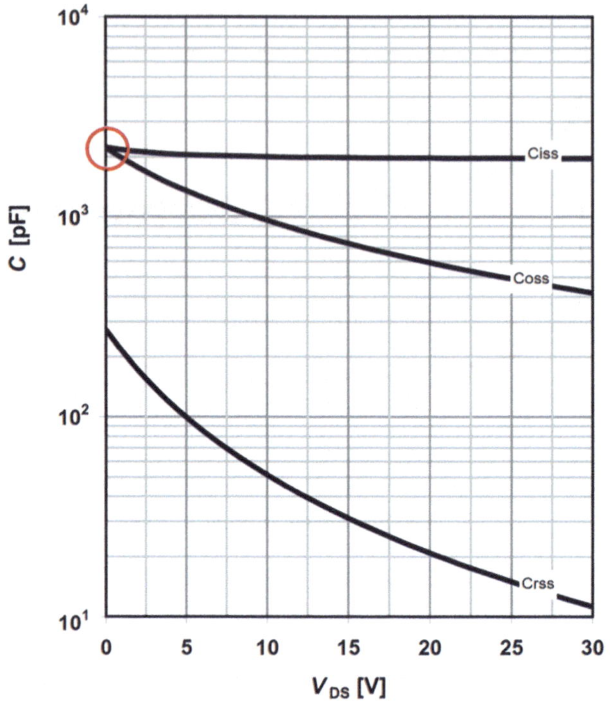

Fig. 10.10 Capacitances for the Si EMM device from its datasheet

The stand-alone SiC power MOSFET product has a transconductance of 4.3 S, which is close to the value in its datasheet. The transconductances of the composite MOSFETs are 4.2, 4.1, and 3.9 S for DC gate bias of 6.0, 5.7, and 5.4 V, respectively. The reduction of the transconductance due to a resistance added in series with the source of the SiC power MOSFET is given in Eq. (5.3). The reduction of the transconductance of the composite MOSFETs is small because of the low on-state resistance of the Si EMM device in the case of all three DC gate bias values.

10.5.4 Impact on Switching Losses

The double-pulse test, whose circuit diagram was shown in Fig. 5.15, was used to evaluate the switching power losses of all the power transistors. The double-pulse testing of all the power transistors was performed using SiC JBS rectifiers as free-wheeling diodes. The test were performed at a drain current of 10 A and drain supply voltage of 800 V. The rate of change of voltage for a power MOSFET slows down when the transconductance g_M becomes smaller. A smaller dV_D/dt produces greater switching losses. It can therefore be expected that the switching losses for the composite MOSFET CM will be larger than for the stand-alone SiC power

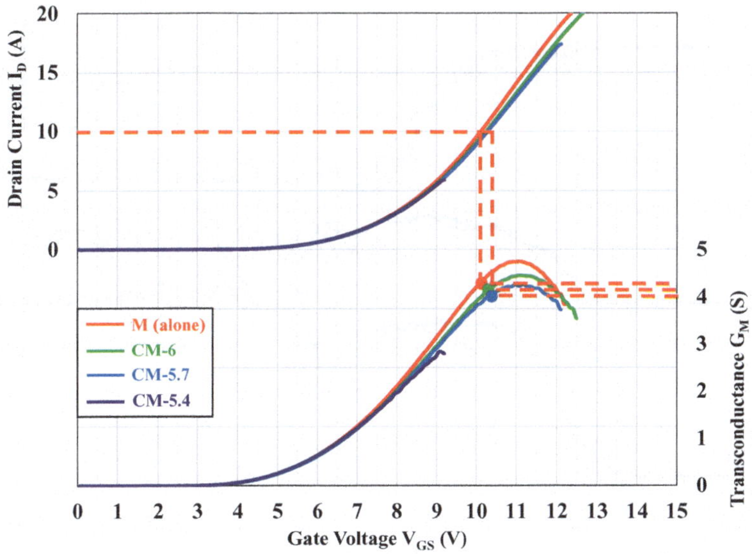

Fig. 10.11 Measured transconductances for the stand-alone SiC power MOSFET M and the composite MOSFETs with DC-bias of 5.4, 5.7, and 6.0 V applied to the Si EMM device

MOSFET because they have slightly smaller transconductance values as discussed in Sect. 10.5.3.

The measured turn-on transients for all the transistor configurations are shown in Fig. 10.12. It can be observed that the composite MOSFETs have a slightly smaller the dV_D/dt compared with the stand-alone SiC power MOSFET. The energy loss per cycle was extracted by integration of the instantaneous drain current and voltage. The turn-on loss for the stand-alone SiC power MOSFET was found to be 314 mJ. The composite MOSFETs had turn-on losses of 322, 368, and 374 μJ for the case of DC gate bias of 6.0, 5.7, and 5.4 V applied to the Si EMM device. The composite MOSFETs exhibit progressively larger turn-on losses when the resistance of the Si EMM increases with smaller DC gate bias.

The measured turn-off transients for the three transistor configurations are shown in Fig. 10.13. Very little difference between all the cases is observed. The turn-off loss for the stand-alone SiC power MOSFET was 68 μJ, while that for the composite MOSFETs was found to be 65, 64, and 63 μJ for the case of DC gate bias of 6.0, 5.7, and 5.4 V applied to the Si EMM device.

The total energy loss per cycle for the stand-alone SiC power MOSFET was 382 μJ, while that for the composite MOSFETs was found to be 387, 432, and 437 μJ for the case of DC gate bias of 6.0, 5.7, and 5.4 V applied to the Si EMM device. The energy loss per cycle for the composite MOSFETs are larger than that for the stand-alone SiC power MOSFET by 1, 13, and 14% for the case of DC gate bias of 6.0, 5.7, and 5.4 V applied to the Si EMM device.

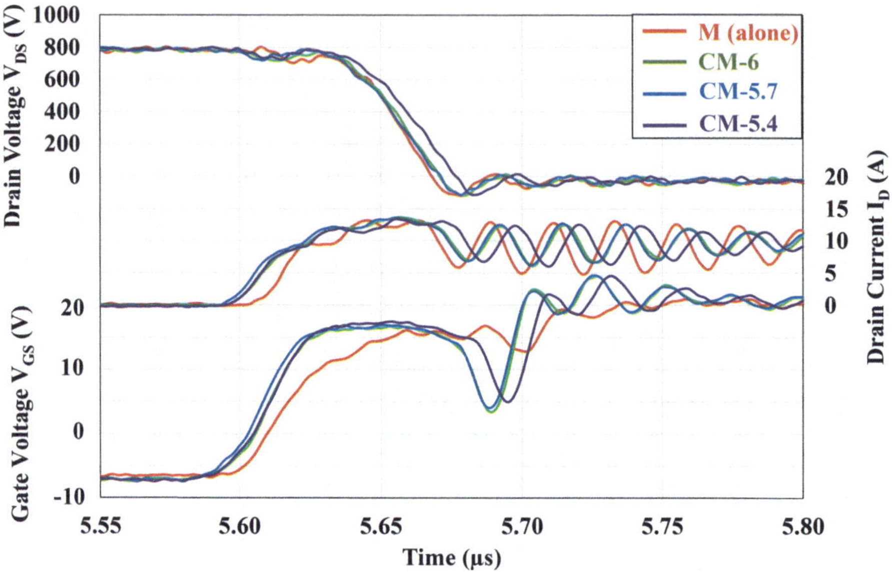

Fig. 10.12 Measured turn-on transients for the stand-alone SiC power MOSFET M; and the composite MOSFETs with DC-bias of 5.4, 5.7, and 6.0 V applied to the Si EMM device

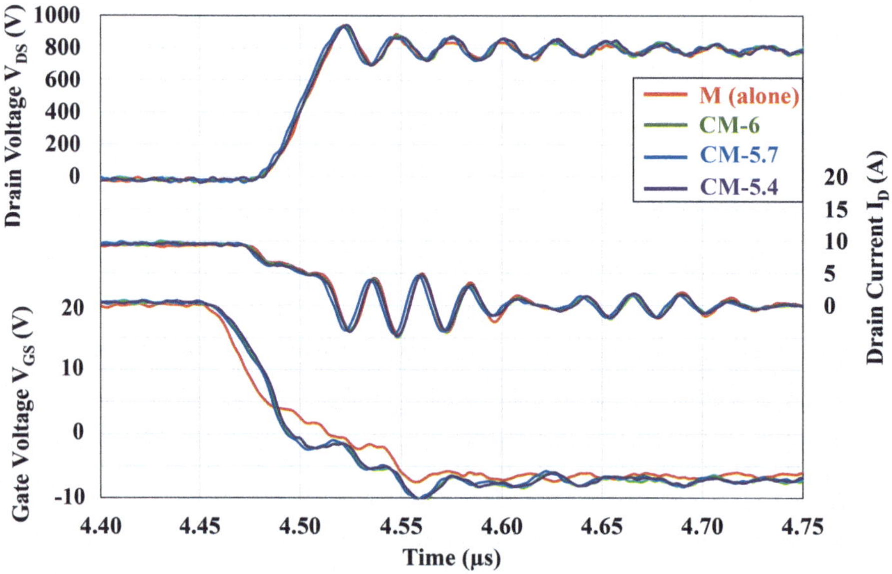

Fig. 10.13 Measured turn-off transients for the stand-alone SiC power MOSFET M and the composite MOSFETs with DC-bias of 5.4, 5.7, and 6.0 V applied to the Si EMM device

In conclusion, the double-pulse tests demonstrate that the composite MOSFETs have larger switching losses than the stand-alone SiC power MOSFET. However, the increase in the switching losses is small.

10.6 Sense Node Voltage

It was pointed out in Chap. 4 that the voltage across the non-linear resistance Q_{LV} can be used for monitoring the current level in the SiC power MOSFET Q_{HV}. This is possible for the BaSIC(EMM) topology for the SiC power MOSFET if the Si EMM device operates in its linear regime at the on-state current level of the SiC power MOSFET. The sense node voltage V_{SENSE} is therefore proportional to the current flowing through the SiC power MOSFET. This is a very beneficial feature of the BaSIC topology allowing power electronics engineers to make sure that the various SiC power MOSFETs within the H-bridge or other circuits do not exceed their reliable operating limits.

Validation of using the sense node voltage to monitor the current flowing through the SiC power MOSFET was experimentally evaluated with the silicon EMM device operating at a DC gate bias of 5.7 V. The double-pulse test circuit waveforms are shown in Fig. 10.14. The current ramps up to 10 A at 4.5 μs after application of

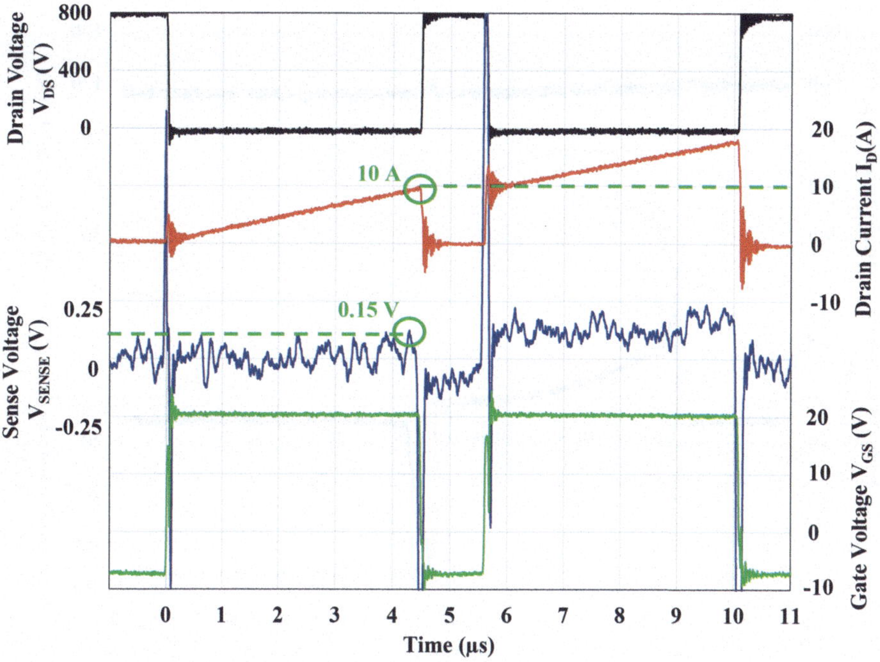

Fig. 10.14 Sense node voltage for the composite MOSFET CM-5.7 during normal operation

the first gate drive pulse as shown by the red line. The sense node voltage V_{SENSE} is shown by the blue line. There is considerable noise in the waveform because the circuit was implemented using discrete devices connected together with wires. The sense node voltage V_{SENSE} has a value of about 0.15 V when the current has a value of 10 A. The equivalent sense resistance is therefore 15 mΩ as expected from the measured characteristics of the silicon EMM device in Fig. 10.7.

The sense node voltage V_{SENSE} can be utilized to detect the short-circuit event as well. The Si EMM enters its current saturation regime during short-circuit conditions. The measured waveform of the sense node voltage during a short-circuit pulse is shown in Fig. 10.15 for the composite MOSFET CM-6. The sense voltage V_{SENSE} spikes to 13 V after initiating the short-circuit event and then reduces with time. The effective gate bias for the SiC power MOSFET then increases with time to allow the larger current flow shown in the figure up to 2 μs. The sense voltage V_{SENSE} reaches a value of 5 V (above its value before the SC event as pointed out on the figure) at 1 μs. The large change in value for V_{SENSE} can be detected using level crossing logic for fast detection of the SC event and shutting down the gate drive before device failure. This is discussed in more detail in a later chapter.

A relatively short blanking time of 1 μs is sufficient for the BaSIC(EMM) topology as indicated in Fig. 10.15 by the brown arrow. This time interval is much smaller than the typical 3 μs blanking time needed for the conventional IGBT DESAT gate control circuits. The difference is due to the sense node voltage V_{SENSE} operating at

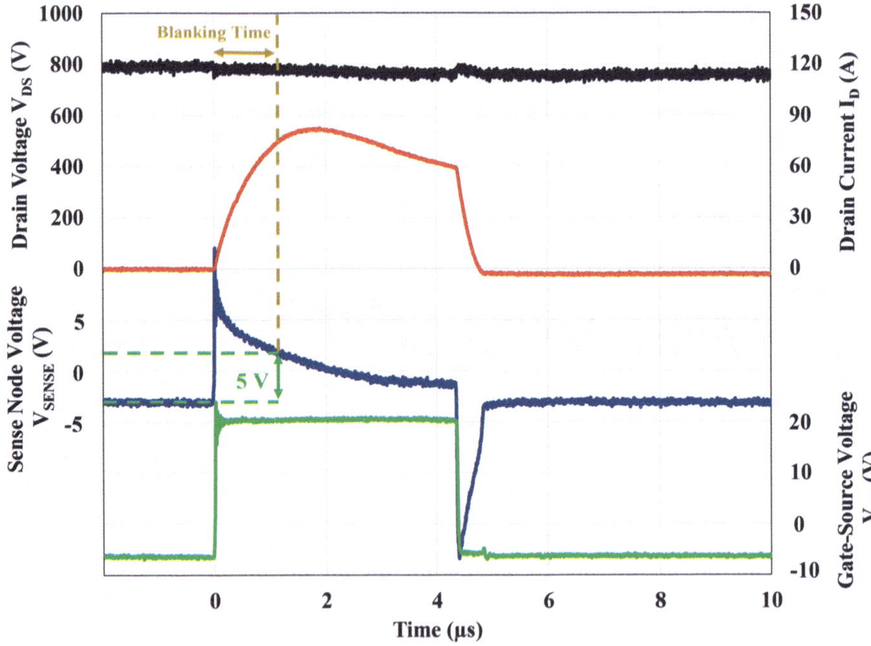

Fig. 10.15 Sense node voltage for the composite MOSFET CM-6 during short-circuit operation

low voltages at all times compared with the DESAT circuit where the drain terminal voltage of the power transistor is being monitored. The spike in the sense node voltage V_{SENSE} can be reduced if the SiC power MOSFET and silicon EMM are co-packaged in a 4-terminal module to minimize the stray inductance.

10.7 Summary

The benefits of the BaSIC topology with a silicon enhancement mode MOSFET (EMM) as the non-linear resistance are demonstrated in this chapter for the case of a silicon carbide power MOSFET. A wide variety of Si EMMs are manufactured by the industry making this approach practical. As an example, an increase in the short-circuit withstand time by 210% was obtained for the BaSIC(EMM) topology with an increase in the on-state resistance by 17% and the switching losses by 14%. This trade-off between improving the short-circuit withstand time and degradation of normal circuit operation is much superior to previous reported methods. Even superior trade-off results can be achieved by using the Si EMM selection methodology described in the later chapter.

References

1. A. Kanale, B.J. Baliga, A new user-configurable method to improve short-circuit ruggedness of 1.2-kV SiC power MOSFETs. IEEE Trans. Power Electron. **36**, 2059–2067 (2021)
2. https://www.infineon.com/cms/en/product/power/mosfet/12v-300v-n-channel-power-mosfet/irl60hs118/
3. A. Kanale, K. Han, B.J. Baliga, S. Bhattacharya, Superior short circuit performance of 1.2kV SiC JBSFETs compared to 1.2kV SiC MOSFETs, in *12th European Conference on Silicon Carbide and Related Materials*, Paper WE.P.MO9 (2018)
4. B.J. Baliga, *Fundamentals of Power Semiconductor Devices*, 2nd edn. (Springer Science, New York, 2019)

Chapter 11
BaSIC(EMM) Topology Versus Gate Drive Voltage Reduction

The BaSIC(EMM) approach was discussed in the previous chapter using a Wolfspeed 1.2 kV SiC power MOSFET C2M160120D with datasheet on-resistance of 160 mΩ at a drain current of 10 A and gate bias of 20 V. These results are compared with short-circuit ruggedness improvement achieved by using reduced gate drive voltage in this chapter [1]. Trade-off curves between the short-circuit withstand time and the increase in on-resistance and switching energy loss achieved with the BaSIC(EMM) topology are contrasted with the gate drive voltage reduction approach.

11.1 Gate Voltage Reduction Method

The saturation current of the SiC power MOSFET can be reduced to enhance the short-circuit withstand time by decreasing the gate drive voltage V_{GS}. This approach also impacts the on-resistance and switching losses of the power transistor. A comparison between reducing the gate drive voltage approach and the BaSIC(EMM) topology is provided in this section to evaluate the benefits of using this method.

11.2 High Voltage Output Characteristics

In order to make a good comparison between the two approaches, it is appropriate to match the drain saturation current for the SiC power MOSFET for both cases. The gate drive voltages were, therefore, chosen to obtain approximately the same drain saturation current as observed for the BaSIC(EMM) cases with DC gate drive voltages of 5.4, 5.7, and 6.0 V applied to the Si EMM device. The high-voltage, high-current output characteristics were measured using the short-circuit test

B. J. Baliga, A. Kanale, *The BaSIC Topology*,
https://doi.org/10.1007/978-3-031-86630-2_11

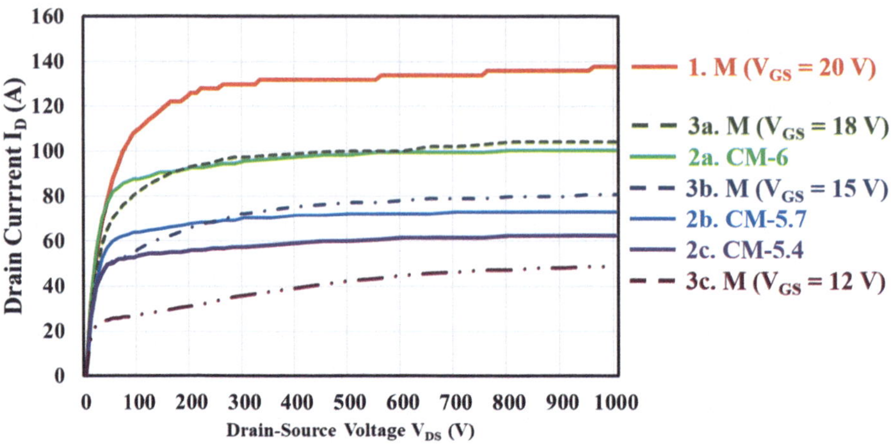

Fig. 11.1 High-voltage, high-current output characteristics: Stand-alone SiC power MOSFET C2M160120D with 12, 15, 18, and 20 V gate drive voltages. The characteristics of the composite MOSFET CM with DC gate bias voltages of 5.4, 5.7, and 6.0 V applied to the Si EMM device are included for comparison

configuration with a small pulse width of 1.5 μs. The DC supply voltage was varied in small steps to generate the high-voltage, high-current output characteristic for each chosen gate drive voltage.

The experimentally extracted high-voltage, high-current output characteristic at gate bias values 12, 15, 18, and 20 V for the stand-alone Wolfspeed 1.2 kV SiC power MOSFET C2M160120D are shown in Fig. 11.1 with the purple, blue, and green dashed lines and the solid red line, respectively. The current increases gradually up to 300 V and then saturates a value of 136 A at a drain voltage of 800 V for the gate bias of 20 V. The experimentally extracted high-voltage, high-current output characteristics for the composite MOSFET CM with gate biases of 5.4, 5.7, and 6.0 V applied to the Si EMM device are also shown in Fig. 11.1 with the purple, blue, and green solid lines, respectively, for comparison. The saturation current at a drain bias of 800 V obtained with a gate drive voltage of 18 V closely matches the CM-6 BaSIC(EMM) case. The saturation current at a drain bias of 800 V obtained with a gate drive voltage of 15 V matches the CM-5.7 BaSIC(EMM) case. The saturation current at a drain bias of 800 V obtained with a gate drive voltage of 12 V is 30% smaller than the CM-5.4 BaSIC(EMM) case.

11.3 Short-Circuit Test Results

The short-circuit withstand time for the SiC power MOSFET C2M160120D was obtained by using the test configuration shown in Fig. 5.1. The short-circuit pulse width is gradually increased while capturing the current waveform until the SiC power MOSFET fails. The DC supply voltage is monitored as shown by the black line in Fig. 11.2. It is important to make sure that the capacitor C_{DC} is adequate to maintain the voltage at the desired value, 800 V in this case.

The measured short-circuit waveforms are shown in Fig. 11.2. The red line was obtained for the stand-alone SiC power MOSFET C2M160120D with gate bias of 20 V. A peak short-circuit current of 136 A was observed for this case. The device failed when the pulse width was increased to 3.5 μs. The green line was obtained for the stand-alone SiC power MOSFET C2M160120D with gate bias of 18 V. A peak short-circuit current of 100 A was observed for this case. The device failed when the pulse width was increased to 5.3 μs. The blue line was obtained for the stand-alone SiC power MOSFET C2M160120D with gate bias of 15 V. A peak short-circuit current of 78 A was observed for this case. The device failed when the pulse width was increased to 6.3 μs. The purple line was obtained for the stand-alone SiC power MOSFET C2M160120D with gate bias of 12 V. A peak short-circuit current of 48 A was observed for this case. The device failed when the pulse width was increased to 10.1 μs.

The measured short-circuit withstand time increases in inverse proportion to the peak short-circuit current level as shown in Fig. 11.3 using the data on the stand-alone SiC power MOSFET with various gate drive voltages and data taken with the

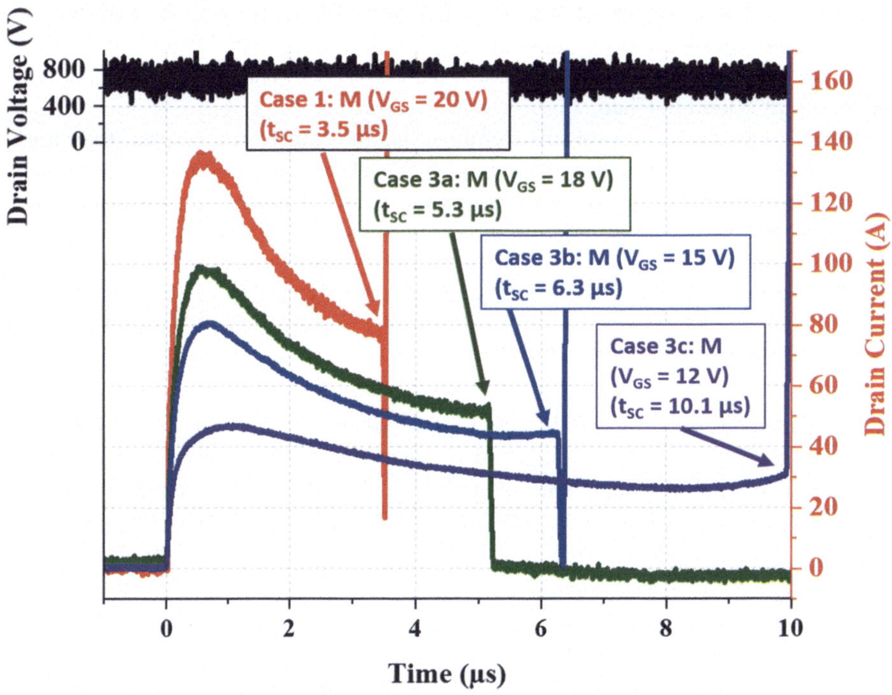

Fig. 11.2 Measured short-circuit waveforms: Stand-alone SiC power MOSFET C2M160120D with various gate drive voltages

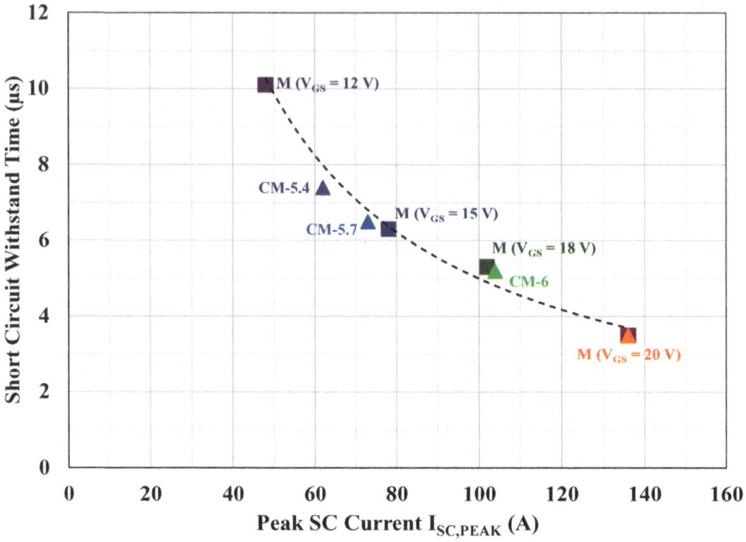

Fig. 11.3 Measured trade-off curve between short-circuit withstand time and peak short-circuit current

BaSIC(EMM) topology using composite MOSFETs CM. A single dashed line can be drawn through the data points demonstrating that the method used to reduce the peak short-circuit current does not matter. The observed behavior is consistent with the theory provided in Chap. 2.

These short-circuit test results demonstrate the ability to increase the short-circuit withstand time for the SiC power MOSFET by reducing its gate drive voltage. The impact of this approach on the normal circuit operating performance of the SiC power MOSFET is discussed in the next section.

11.4 Impact on On-Resistance

The on-state characteristics of the stand-alone SiC power MOSFET C2M160120D were obtained at lower drain voltages by using a curve tracer with various gate drive voltages. The measured data is plotted in Fig. 11.4 with a dashed redline highlighting the on-state current level of 10 A. The on-resistances at this current level are 168, 181, 220, and 380 mΩ for the case of gate drive voltages of 20, 18, 15, and 12 V, respectively. The typical value cited in the datasheet for this transistor is 160 mΩ at a gate drive voltage of 20 V and drain current of 10 A, with a maximum value of 196 mΩ. The measured value falls within the datasheet parametric variation of 17% above the typical value. The on-resistance increases substantially by 226% when the gate drive voltage is reduced to 12 V. This is detrimental to on-state power losses.

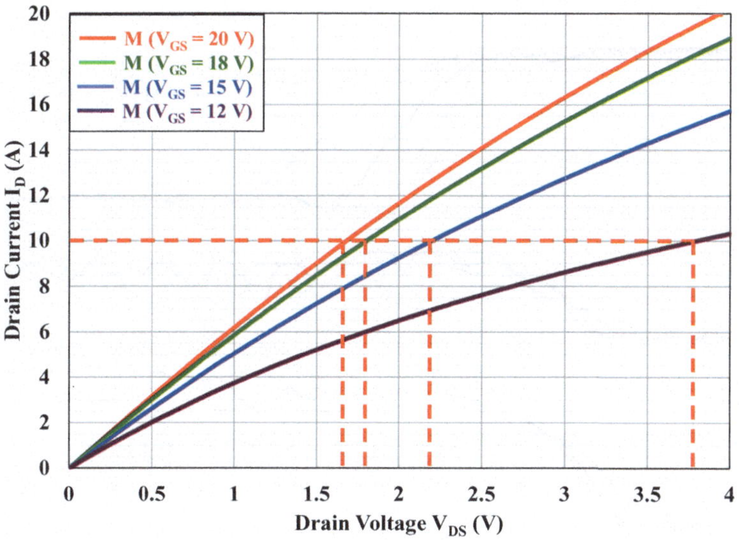

Fig. 11.4 Measured on-state characteristics of the stand-alone SiC power MOSFET M with various gate drive voltages

11.5 Impact on Switching Losses

The double-pulse test, whose circuit diagram is shown in Fig. 5.15, was used to evaluate the switching power losses for the stand-alone SiC power MOSFET C2M160120D with various gate drive voltages. The double-pulse testing of all cases was performed using SiC JBS rectifiers as freewheeling diodes. The test was performed at a drain current of 10 A and drain supply voltage of 800 V.

The rate of change of voltage for a power MOSFET slows down when the gate drive voltage is reduced according to Eq. (5.4). The smaller dV_D/dt produces greater switching losses. It can therefore be expected that the switching losses for the stand-alone SiC power MOSFET will increase when the gate drive voltage is reduced.

The measured turn-on transients for the stand-alone SiC power MOSFET C2M160120D with gate drive voltages of 20, 18, 15, and 12 V are shown in Fig. 11.5. It can be seen that the dV_D/dt becomes progressively smaller with reduction of the gate drive voltage consistent with Eq. (5.4). The energy loss per cycle was extracted by integration of the instantaneous drain current and voltage. The turn-on loss for the stand-alone SiC power MOSFET increased from 314 mJ to 346, 432, and 575 μJ when the gate drive voltage was reduced to 18, 15, and 12 V, respectively.

The measured turn-off transients for the stand-alone SiC power MOSFET C2M160120D with gate drive voltages of 20, 18, 15, and 12 V are shown in Fig. 11.6. The turn-off occurs at an earlier time when the gate drive voltage is reduced because it decays to the plateau voltage in a shorter time due to the lower

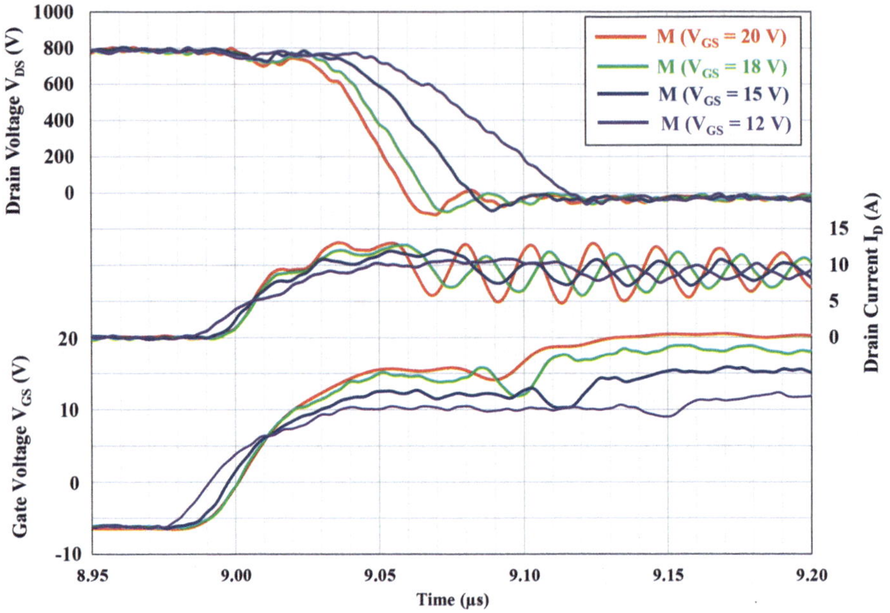

Fig. 11.5 Measured turn-on transients for the stand-alone SiC power MOSFET M with various gate drive voltages

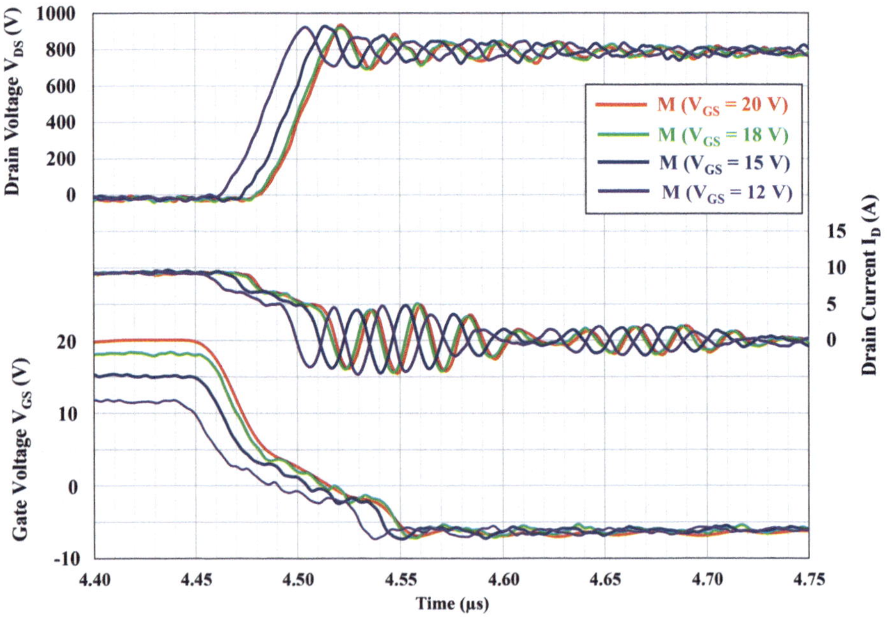

Fig. 11.6 Measured turn-off transients for the stand-alone SiC power MOSFET M with various gate drive voltages

initial voltage. However, very little difference in the transition times is observed between all the cases. The turn-off loss for the stand-alone SiC power MOSFET was found to be 68, 66, 68, and 68 μJ for the case of gate drive voltages of 20, 18, 15, and 12 V.

The total energy loss per cycle for the stand-alone SiC power MOSFET was found to change from 382 μJ, to 412, to 500, and to 643 μJ when the gate drive voltage was changed from of 20 to 18 to 15 and to 12 V. The energy loss per cycle for the stand-alone SiC power MOSFET increased by 68% when the gate drive voltage was reduced to 12 V. In conclusion, the double-pulse tests demonstrate that the switching losses for the stand-alone SiC power MOSFET increase substantially when the gate drive voltage is reduced.

11.6 Trade-Off Curves

Comparison of the performance of the BaSIC(EMM) topology with the approach of reducing the gate drive voltage can be achieved by making a trade-off curve between the increase in short-circuit withstand time and the increase in the on-resistance. The data obtained for the two approaches are shown in Fig. 11.7 with trend lines. The trade-off curve for the BaSIC(EMM) topology lies to the left-hand side of that

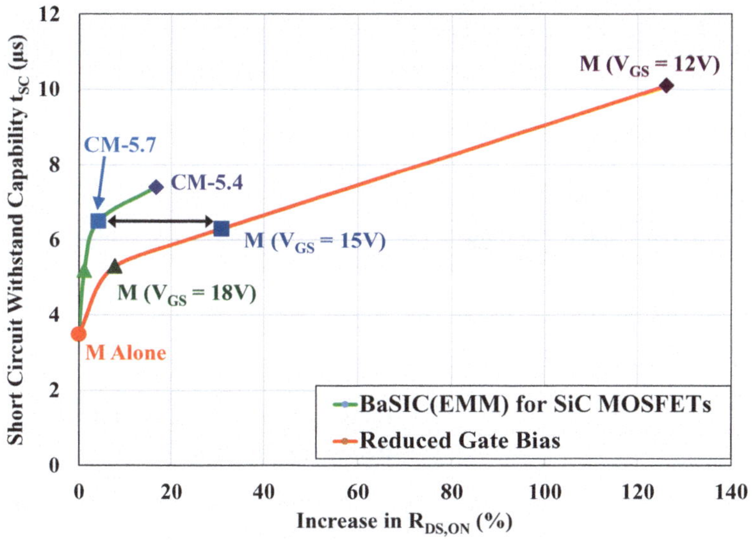

Fig. 11.7 Trade-off curves for the stand-alone SiC power MOSFET M with gate drive voltage reduction versus the BaSIC(EMM) topology: on-resistance comparison

for the approach using reduction of the gate drive voltage demonstrating that it has superior performance. A quantitative evaluation of the difference in performance of the two approaches can be made by choosing a short-circuit withstand time of 6.5 μs, an increase by 86% over the 3.5 μs observed for the stand-alone SiC power MOSFET with gate drive voltage of 20 V. This is highlighted using the black double-headed arrow in Fig. 11.7. The increase in on-resistance for the BaSIC(EMM) topology is only 4% compared with 31% increase when the gate drive voltage is reduced. This is a factor 8 times better performance with the BaSIC(EMM) topology. An even bigger difference between the approaches is achievable by using the selection methodology for the silicon EMM device discussed in the next chapter.

Similarly, a trade-off curve prepared between short-circuit withstand time and switching energy loss can also be used to compare the two approaches. The data for the two approaches are plotted in Fig. 11.8 with trend lines. A quantitative difference between the two approaches can be made by choosing a short-circuit withstand time of 6.5 μs, an increase by 86% over the 3.5 μs observed for the stand-alone SiC power MOSFET with gate drive voltage of 20 V. This is highlighted using the black double-headed arrow in Fig. 11.8. The increase in total switching energy loss for the BaSIC(EMM) topology is 13% compared with 31% with reduction of the gate drive voltage. This is a factor 2.4 times better performance achieved using the BaSIC(EMM) topology. An even bigger difference between the approaches is achievable by using the selection methodology for the silicon EMM device discussed in the next chapter.

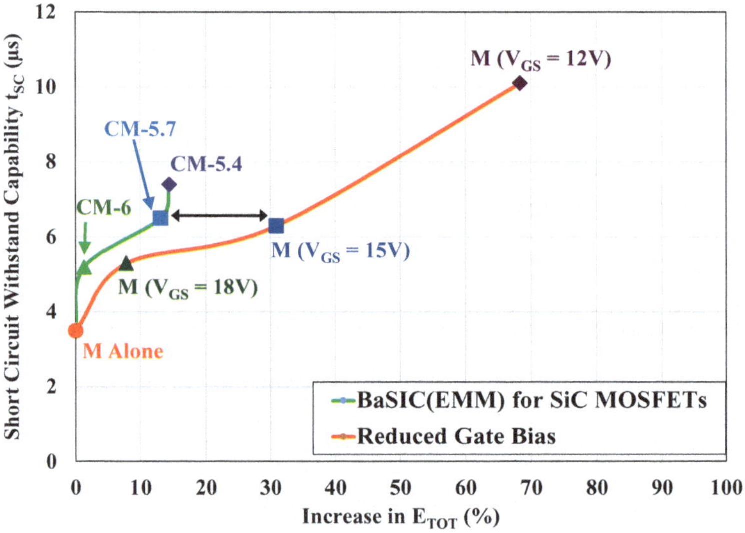

Fig. 11.8 Trade-off curves for the stand-alone SiC power MOSFET M with gate drive voltage reduction versus the BaSIC(EMM) topology: switching loss comparison

11.7 Summary

The benefits of using the BaSIC(EMM) topology are compared with gate drive voltage reduction approach in this chapter for the case of a silicon carbide power MOSFET. A short-circuit withstand time of 6.5 μs, an increase by 86% over the 3.5 μs observed for the stand-alone SiC power MOSFET with gate drive voltage of 20 V, was achieved with the BaSIC(EMM) topology with an increase in the on-state resistance by 4% and the switching losses by 11%. In contrast, the gate drive reduction approach produced the same short-circuit withstand time of 6.5 μs with an increase in the on-state resistance by 31% and the switching losses by 31%. The trade-off performance of the BaSIC(EMM) topology is therefore much superior to gate drive voltage reduction approach.

Reference

1. A. Kanale, B.J. Baliga, A new user-configurable method to improve short-circuit ruggedness of 1.2-kV SiC power MOSFETs. IEEE Trans. Power Electron. **36**, 2059–2067 (2021)

Chapter 12
BaSIC(EMM) Topology Versus Ballast Resistance

The BaSIC(EMM) approach was discussed in Chap. 11 using a Wolfspeed 1.2 kV SiC power MOSFET product C2M160120D with datasheet on-resistance of 160 mΩ at a drain current of 10 A and gate bias of 20 V. These results were compared with short-circuit ruggedness improvement achieved by reducing the gate drive voltage in Chap. 11.

In this chapter, both of these approaches are compared with short-circuit ruggedness improvement achieved by using a series ballast resistance connected to the source terminal of the SiC power MOSFET [1]. Trade-off curves between the short-circuit withstand time and the increase in on-resistance and switching energy loss achieved with the BaSIC(EMM) topology are contrasted with the series ballast resistance approach as well as the gate drive voltage reduction approach.

12.1 Series Ballast Resistance Method

The series ballast resistance method was described in Chap. 7 for comparison with the BaSIC(DMM) topology. The saturation current of the SiC power MOSFET can be reduced to enhance the short-circuit withstand time by the addition of the series ballast resistance. The voltage drop across the series ballast resistance under short-circuit conditions opposes the gate drive voltage for the SiC power MOSFET. This brings down its drain saturation current level leading to a longer short-circuit withstand time. This approach obviously adds to the on-resistance, and it increases the switching losses of the power transistor due to a reduction of the transconductance.

The performance of the Wolfspeed 1.2 kV SiC power MOSFET C2M160120D product with series ballast resistances is evaluated in this chapter for comparison with the BaSIC(EMM) topology performance described for the same device in Chap. 10. Trade-off curves of short-circuit withstand time against the increase in

B. J. Baliga, A. Kanale, *The BaSIC Topology*,
https://doi.org/10.1007/978-3-031-86630-2_12

on-resistance and switching losses are used to demonstrate the differences between this approach and the BaSIC(EMM) case. The results described in Chap. 11 for the gate drive voltage reduction approach are included to compare all three approaches.

12.2 High Voltage Output Characteristics

In order to make a good comparison between the three approaches, it is appropriate to match the drain saturation current for the SiC power MOSFET for all the cases at high drain voltages. The gate drive voltages were, therefore, chosen to obtain approximately the same drain saturation current as observed for the BaSIC(EMM) cases with DC gate drive voltages of 5.4, 5.7, and 6.0 V applied to the Si EMM device as discussed in Chap. 11. Series ballast resistance values of 165, 65, and 15 mΩ were chosen to achieve a similar match of the drain saturation current. The high-voltage, high-current output characteristics were measured using the short-circuit test configuration with a small pulse width of 1.5 μs. The DC supply voltage was varied in small steps to generate the high-voltage, high-current output characteristic for each chosen gate drive voltage.

The experimentally extracted high-voltage, high-current output characteristics at gate bias values 12, 15, 18, and 20 V for the stand-alone Wolfspeed 1.2 kV SiC power MOSFET C2M160120D are shown in Fig. 12.1 with the brown, blue, green, and the red solid lines, respectively. The current increases gradually up to 300 V and then saturates a value of 136 A for a gate bias of 20 V. The experimentally extracted high-voltage, high-current output characteristics for the composite MOSFET CM

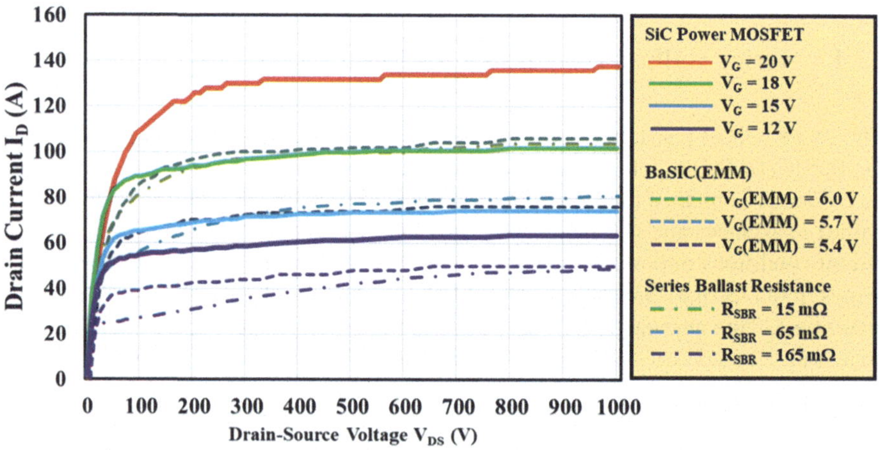

Fig. 12.1 High-voltage, high-current output characteristics: Stand-alone SiC power MOSFET C2M160120D with 12, 15, 18, and 20 V gate drive voltages (solid lines). The characteristics of the composite MOSFET CM with DC gate bias voltages of 5.4, 5.7, and 6.0 V applied to the Si EMM device (dashed lines). The characteristics of the composite MOSFET CM with series ballast resistances of 15, 65, and 165 mΩ (dot-dashed lines)

with DC gate biases of 5.4, 5.7, and 6.0 V applied to the Si EMM device are shown in Fig. 12.1 with the purple, blue, and green dashed lines, respectively, for comparison. The experimentally extracted high-voltage, high-current output characteristics for the composite MOSFET CM with series ballast resistances of 165, 65, and 15 mΩ are shown in Fig. 12.1 with the purple, blue, and green dot-dashed lines, respectively. The saturation current at a drain bias of 800 V obtained with a gate drive voltage of 18 V closely matches the CM-6 BaSIC(EMM) case and the series ballast resistance of 15 mΩ. The saturation current at a drain bias of 800 V obtained with a gate drive voltage of 15 V closely matches the CM-5.7 BaSIC(EMM) case and the series ballast resistance of 65 mΩ. The saturation current at a drain bias of 800 V obtained with a gate drive voltage of 12 V matches the series ballast resistance of 165 mΩ.

12.3 Short-Circuit Test Results

The short-circuit withstand time for the SiC power MOSFET C2M160120D was obtained by using the test configuration shown in Fig. 5.1. The short-circuit pulse width is gradually increased while capturing the current waveform until the SiC power MOSFET fails. The DC supply voltage is monitored as shown by the black line in Fig. 12.2. It is important to make sure that the capacitor C_{DC} is adequate to maintain the voltage at the desired value, 800 V in this case.

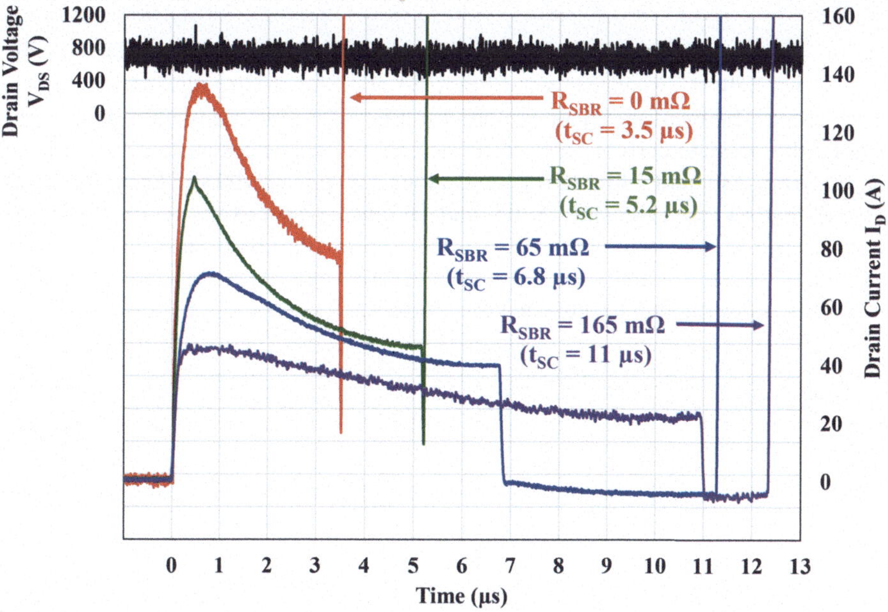

Fig. 12.2 Measured short-circuit Waveforms: SiC power MOSFET C2M160120D with various series ballast resistances

The measured short-circuit waveforms are shown in Fig. 12.2 for the three series ballast resistance cases together with that for the stand-alone SiC power MOSFET for comparison. The red line was obtained for the stand-alone SiC power MOSFET C2M160120D with gate bias of 20 V. A peak short-circuit current of 136 A was observed for this case. The device failed when the pulse width was increased to 3.5 μs. The green line was obtained for the SiC power MOSFET with a series ballast resistance of 15 mΩ. A peak short-circuit current of 106 A was observed for this case. The device failed when the pulse width was increased to 5.2 μs. The blue line was obtained for the SiC power MOSFET with a series ballast resistance of 65 mΩ. A peak short-circuit current of 70 A was observed for this case. The device failed when the pulse width was increased to 6.8 μs. The purple line was obtained for the SiC power MOSFET with a series ballast resistance of 165 mΩ. A peak short-circuit current of 47 A was observed for this case. The device failed when the pulse width was increased to 11 μs. The measured short-circuit withstand time increases in inverse proportion to the peak short-circuit current level as shown in Fig. 11.3, which is consistent with the theory provided in Chap. 2.

These short-circuit test results demonstrate the ability to increase the short-circuit withstand time for the SiC power MOSFET by the addition of the series ballast resistance. The impact of this approach on the normal circuit operating performance of the SiC power MOSFET is discussed in the next section.

12.4 Impact on On-Resistance

The on-state characteristics of the SiC power MOSFET C2M160120D with the three series ballast resistances were obtained at lower drain voltages by using a curve tracer. The measured data is plotted in Fig. 12.3 with a dashed redline highlighting the on-state current level of 10 A. The on-resistances at this current level are 168, 185, 230, and 340 mΩ for stand-alone SiC power MOSFET and the cases of series ballast resistances of 15, 65, and 165 mΩ, respectively. As expected, the on-resistance is increased by the value of the series ballast resistance. The on-resistance increases substantially by 200% when a series ballast resistance of 165 mΩ is used. This is detrimental to on-state power losses.

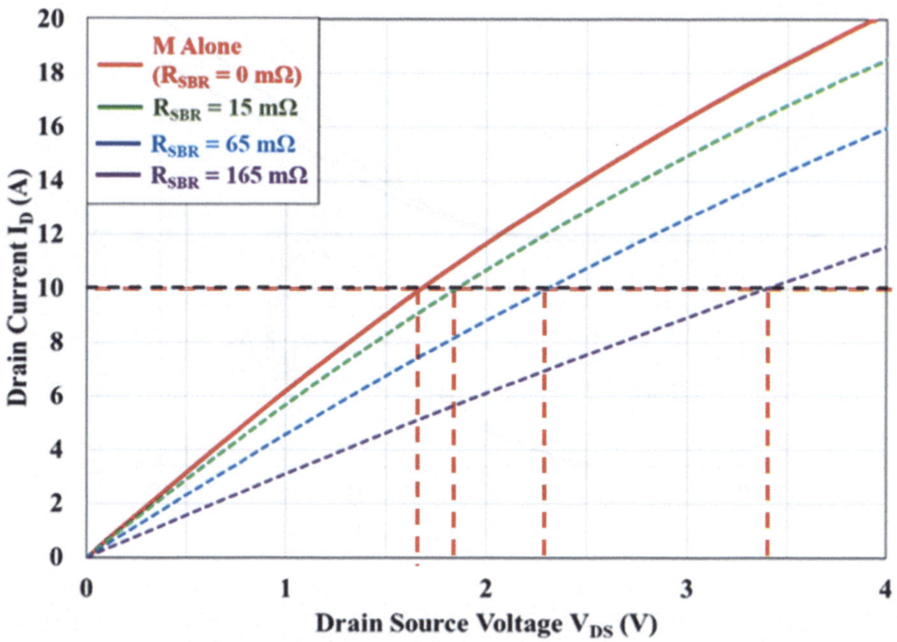

Fig. 12.3 Measured on-state characteristics of the stand-alone SiC power MOSFET M with various series ballast resistances

12.5 Impact on Transconductance

The transconductance g_M of the power MOSFET plays a role in the switching transient for the drain voltage during operation with an inductive load. Its value can be extracted from the transfer characteristics of the device. The transfer characteristics for the SiC power MOSFET C2M160120D with the three series ballast resistances were measured at a drain bias of 20 V. They are compared with the transfer curve obtained for the stand-alone SiC power MOSFET C2M160120D in Fig. 12.4. The transconductance, extracted by taking a derivative of the drain current with respect to the gate voltage, is shown in the lower portion of the figure. The value of the transconductance can be compared for the three cases at the same drain current level of 10 A indicated by the red dashed line. It is given by the values at the four points indicated on the transconductance curves.

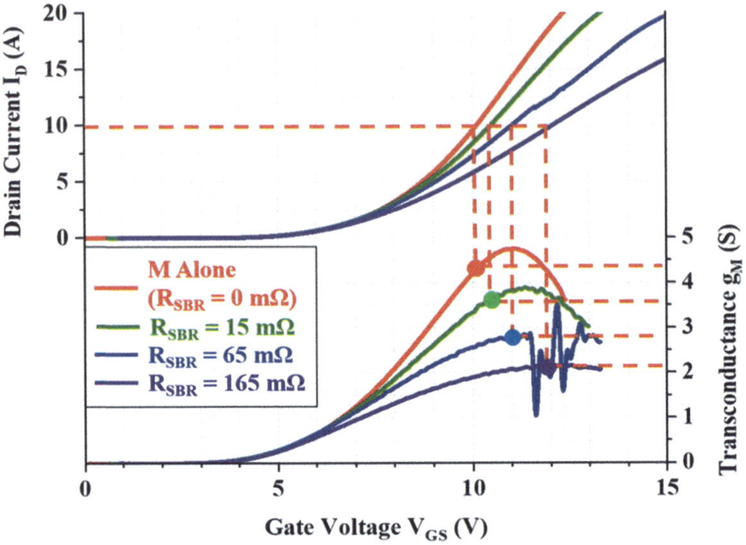

Fig. 12.4 Transconductances for the SiC power MOSFET with the series ballast resistances

The stand-alone SiC power MOSFET has a transconductance of 4.4 S, which is close to the value in its datasheet. The transconductances of the MOSFETs with series resistances of 15, 65, and 165 mΩ are 3.7, 2.9, and 2.1 S. The reduction in the transconductance is associated with the resistance of the ballast resistance as given in Eq. (5.3). The transconductance value predicted by this equation is 3.9 S for the case of a resistance R_S of 15 mΩ which matches the measured value for the 15 mΩ series ballast resistance. The transconductance value predicted by this equation is 3.3 S for the case of a resistance R_S of 65 mΩ which matches the measured value for the 65 mΩ series ballast resistance case. The transconductance value predicted by this equation is 2.4 S for the case of a resistance R_S of 165 mΩ which matches the measured value for the 165 mΩ series ballast resistance case. The impact of the reduction of the transconductance on the switching losses is discussed in the next section.

12.6 Impact on Switching Losses

The double-pulse test, whose circuit diagram was shown in Fig. 5.15, was used to evaluate the switching power losses for the SiC power MOSFET C2M160120D with various series ballast resistances. The double-pulse testing of all cases was performed using SiC JBS rectifiers as freewheeling diodes. The test was performed at a drain current of 10 A and drain supply voltage of 800 V.

The rate of change of the drain voltage for a SiC power MOSFET slows down when a resistance is added in series with the source electrode because of a reduction of the transconductance according to Eq. (5.4). The smaller dV_D/dt produces greater switching losses. It can therefore be expected that the switching losses for the SiC power MOSFET will increase in the presence of a series ballast resistance.

The measured turn-on transients for the stand-alone SiC power MOSFET C2M160120D with various series ballast resistances are shown in Fig. 12.5. The dV_D/dt becomes progressively smaller with increasing values for the series ballast resistance. The energy loss per cycle was extracted by integration of the instantaneous drain current and voltage. The turn-on loss for the SiC power MOSFET increased from 314 mJ for the stand-alone device to 351, 419, and 433 μJ for series ballast resistance values of 15, 65, and 165 mΩ, respectively.

The measured turn-off transients for the SiC power MOSFET C2M160120D did not change with the addition of the series ballast resistances and are therefore not shown here. The turn-off losses were 62, 64, and 64 μJ for series ballast resistance values of 15, 65, and 165 mΩ, respectively, compared with 68 μJ for the stand-alone device. The total energy loss per cycle for the stand-alone SiC power MOSFET was found change from 382 μJ to 413, to 483, and to 497 μJ with series ballast resistance values of 15, 65, and 165 mΩ, respectively. The energy loss per cycle for the stand-alone SiC power MOSFET increased by 30% with series ballast resistance values of 165 mΩ. In conclusion, the double-pulse tests demonstrate that the switching losses for the stand-alone SiC power MOSFET increase substantially when a series ballast resistance is added.

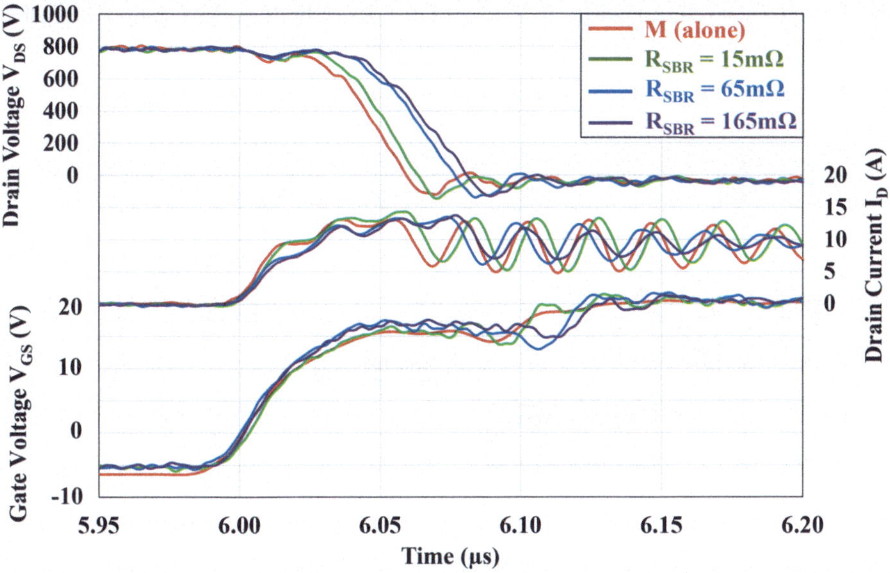

Fig. 12.5 Measured turn-on transients for the SiC power MOSFET with various series ballast resistances

12.7 Trade-Off Curves

The performance of the BaSIC(EMM) topology can be compared with the approach of adding a series ballast resistance by making a trade-off curve between the increase in short-circuit withstand time and the increase in the on-resistance. The data obtained for the two approaches are shown in Fig. 12.6 with red and blue trend lines. The trade-off curve for the BaSIC(EMM) topology lies to the left-hand side of that for the approach using series ballast resistances indicating its superior performance. The trade-off curve for the gate drive voltage reduction approach is also included in Fig. 12.6 with a green trend line. It can be seen that the series ballast resistance method and the gate drive reduction method produce similar results.

A quantitative evaluation of the improvement in performance with the BaSIC(EMM) topology can be made by choosing a short-circuit withstand time of 6.5 μs, which is 86% larger than the 3.5 μs observed for the stand-alone SiC power MOSFET with gate drive voltage of 20 V. The increase in on-resistance for the BaSIC(EMM) topology is only 4% compared with more than 30% increase with the other approaches. This is a factor 8 times better performance with the BaSIC(EMM) topology. This is highlighted using the black double-headed arrow. An even bigger difference between the approaches is achievable by using the selection methodology for the silicon EMM device discussed in the next chapter.

Similarly, a trade-off curve prepared between short-circuit withstand time and switching energy loss can be used to compare the three approaches, as shown with the data plotted in Fig. 12.7 with trend lines. A quantitative difference between the BaSIC(EMM) topology with the series ballast resistance approach can be made by choosing a short-circuit withstand time of 6.5 μs, which is 86% larger than the

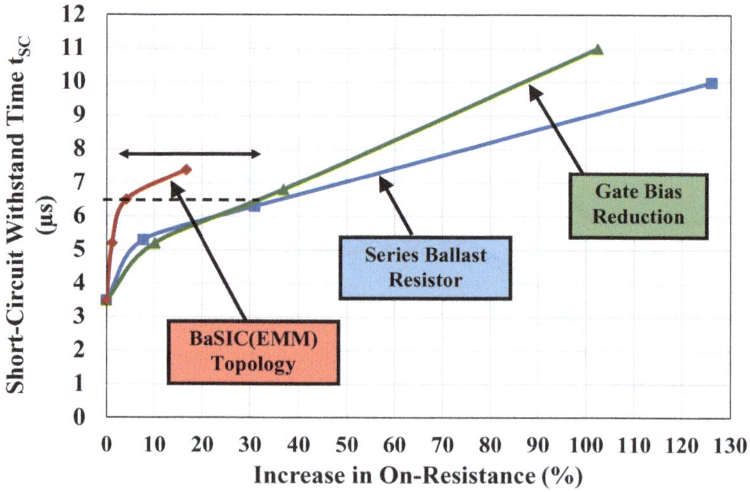

Fig. 12.6 Trade-off curves for the SiC power MOSFET with series ballast resistance versus with gate drive voltage reduction and the BaSIC(EMM) topology—on-resistance comparison

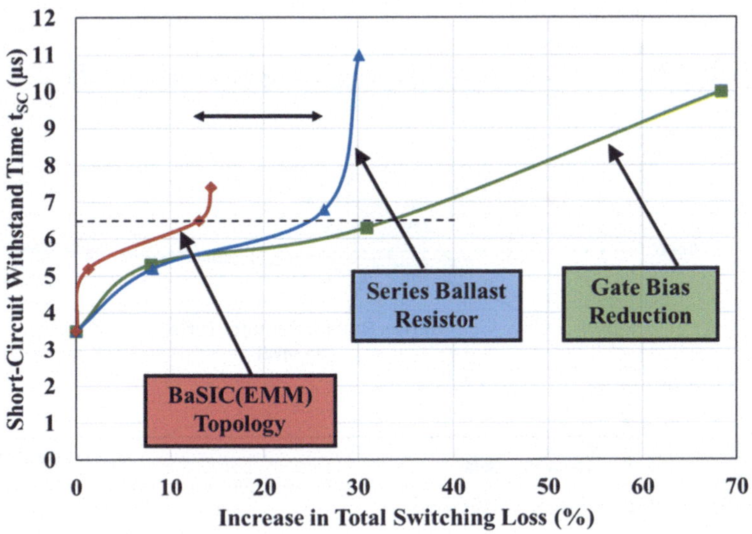

Fig. 12.7 Trade-off curves for the SiC power MOSFET with series ballast resistance versus with gate drive voltage reduction and the BaSIC(EMM) topology: switching loss comparison

3.5 µs observed for the stand-alone SiC power MOSFET with gate drive voltage of 20 V. The increase in total switching energy loss for the BaSIC(EMM) topology is 13% compared with 26% with the series ballast resistance. A factor 2.0 times better performance is therefore achieved using the BaSIC(EMM) topology. This is highlighted using the black double-headed arrow. An even bigger difference between the approaches is achievable by using the selection methodology for the silicon EMM device discussed in the next chapter.

12.8 Summary

A comparison of the performance obtained by using the BaSIC(EMM) topology and the series ballast resistance approach has been provided in this chapter. In the case of the SiC power MOSFET C2M160120D product, a short-circuit withstand time of 6.5 µs, which is 86% larger than the 3.5 µs observed for the stand-alone SiC power MOSFET with gate drive voltage of 20 V, was achieved with the BaSIC(EMM) topology with an increase in the on-state resistance by 4% and the switching losses by 11%. In contrast, the series ballast resistance approach produced the same short-circuit withstand time of 6.5 µs with an increase in the on-state resistance by 31% and the switching losses by 26%. The trade-off performance of the BaSIC(EMM) topology is therefore much superior to series ballast resistance approach.

A pictorial contrast between the three approaches for improving the short-circuit withstand capability is provided in Fig. 12.8 by depicting the output characteristics

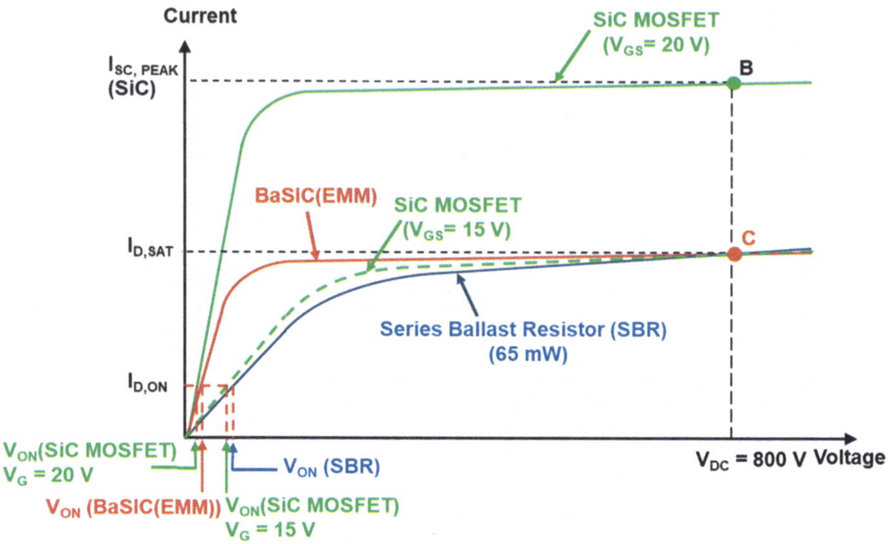

Fig. 12.8 Comparison of the output characteristics of the stand-alone SiC power MOSFET with gate drive voltages of 20 and 15 V with those for the series ballast resistance and the BaSIC(EMM) topology

of the composite MOSFETs. The output characteristic for the stand-alone SiC power MOSFET with gate bias of 20 V is shown by the solid green line. It has an operating point B during short-circuit conditions with very high current. The output characteristic for the stand-alone SiC power MOSFET with gate bias of 15 V is shown by the dashed green line with a reduced saturation current level. The output characteristics for the BaSIC(EMM) topology are shown by the red line. The output characteristics for the SiC power MOSFET with series ballast resistance are shown by the blue line. The same short-circuit withstand time of 6.5 μs was obtained for the BaSIC(EMM) topology, the SiC MOSFET case with gate drive voltage of 15 V, and the series ballast resistance of 65 mΩ due to a common operating point C under short-circuit conditions. These conditions can therefore be used for the comparison. It can be seen that the composite MOSFET in the BaSIC(EMM) topology has a much better output characteristics (red line) with lower on-state resistance and on-state voltage drop (V_{ON}) to achieve the same drain saturation current level as the other two approaches.

Reference

1. A. Kanale, B.J. Baliga, Comparison of three methods [gate bias reduction, series ballast resistor, and BaSIC(EMM)] to improve short-circuit capability of 1.2 kV SiC power MOSFETs, in *PCIM Europe* (2021), pp. 306–315

Chapter 13
Selection Methodology for Si EMM Device

The performance of the BaSIC(EMM) topology using a Wolfspeed 1.2 kV SiC power MOSFET C2M160120D was compared in Chap. 12 with short-circuit ruggedness improvement achieved by reducing the gate drive voltage or by adding a series ballast resistance. These results were obtained by a using an Infineon Si OptiMOS enhancement mode MOSFET IPI70N04S-06. It is not obvious that this device has the best characteristics for the BaSIC(EMM) topology among the large number of available Si power MOSFETs produced by various manufacturers.

A selection methodology for the Si EMM device used for the BaSIC(EMM) topology is needed based on the information provided by manufacturers in their datasheets. This chapter describes a procedure for methodically selecting the Si EMM device to optimize and enhance the performance of the BaSIC(EMM) topology [1]. The characteristics of the Si EMM devices provided in their datasheets that the most appropriate for this application are defined here. The key device parameters are (1) the on-resistance at 25 °C for the Si EMM devices at various gate bias voltages and (2) the drain saturation currents at these gate bias voltages at 150–175 °C. The choice of these temperatures is justified by the thermal analysis of the Si EMM devices during normal circuit operation and during the short-circuit event.

Many types of Si EMM devices are commercially available from numerous vendors with a wide range of breakdown voltages and on-resistances. Silicon power MOSFETs have been manufactured with blocking voltages ranging from 20 to 1200 V. The industry produces devices with the planar-gate or D-MOSFET structure and the trench-gate or U-MOSFET structure.

It is shown that a short-circuit time to 10 μs can be achieved for the 1.2 kV SiC 160 mΩ power MOSFET C2M160120D product with an increase in on-resistance of only 3% by using the selection methodology described in this chapter. This performance has been experimentally validated.

B. J. Baliga, A. Kanale, *The BaSIC Topology*,
https://doi.org/10.1007/978-3-031-86630-2_13

13.1 Trade-Off Curve

The first step in the selection methodology is to ascertain the peak short-circuit current level required to achieve the desired short-circuit withstand time of 10 μs for the SiC power MOSFET. This is achieved by using the short-circuit test circuit in Fig. 5.1 to perform measurements on the SiC power MOSFET with various gate drive voltages. The peak short-circuit current can be reduced by decreasing the gate drive voltage as discussed in Chap. 11. The analytical model for short-circuit conditions based on adiabatic heating that was discussed in Chap. 2 indicates that the short-circuit withstand time (t_{SC}) is inversely proportional to the peak short-circuit current level $I_{P,SC}$. As an example, this behavior is observed for the 1.2 kV 160 mΩ SiC power MOSFET product C2M0160120D as shown in Fig. 13.1. The blue dotted line in Fig. 13.1 is a polynomial fit to the data points with t_{SC} proportional to $I_{P,SC}$ to the power of −0.99 which is very close to the −1.00 predicted by the adiabatic theory with constant short-circuit current versus time. From these results, it can be concluded that a short-circuit withstand time of 10 μs is achievable for this device when the peak short-circuit current is reduced to 50 A as indicated by the red dashed lines in the figure. This is therefore the *primary (and first) criterion* that must be met when selecting the Si EMM device for the BaSIC(EMM) topology for the case of the 1.2 kV 160 mΩ SiC power MOSFET C2M0160120D product.

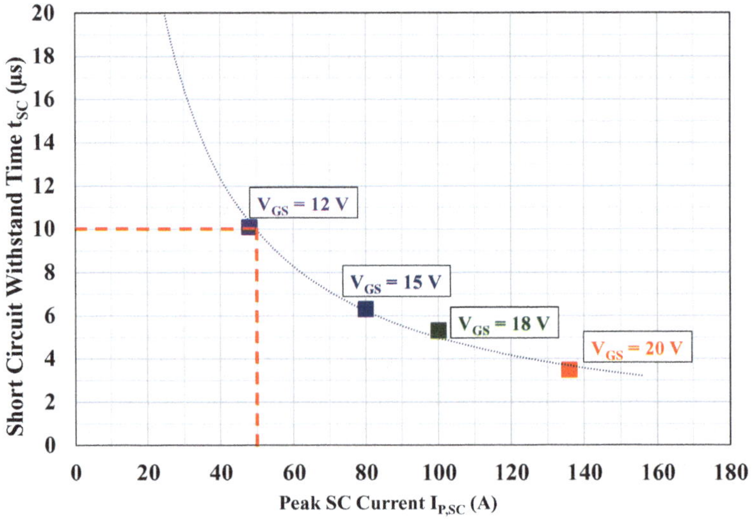

Fig. 13.1 Trade-off curve between short-circuit withstand time and peak short-circuit current for the stand-alone SiC power MOSFET C2M160120D using gate bias as the variable

13.2 Si EMM On-State Resistance

As discussed in Chap. 10, the BaSIC(EMM) topology operates with the Si EMM in its on-state with a low on-resistance during the normal on-state and switching conditions in power circuits for the SiC power MOSFET. The Si EMM device does not incur any switching losses during normal circuit operation because a constant (DC) bias is applied to the gate of the Si EMM. The power dissipation in the Si EMM is consequently small. As an example, the power dissipation in a 5 mΩ Si EMM device is 0.25 W for at a duty cycle of 50% for the case of the 1.2 kV SiC power MOSFET operating at its rated on-state current of 10 A. This will produce a temperature rise of less than 5 °C based on a typical thermal resistance of 12.5 °C/W for the IRF6722 Si power MOSFET product. It can be concluded that the 25 °C on-resistances provided in Si EMM datasheets should therefore be used during the selection process. This provides the *second criterion* when selecting the Si EMM devices for the BaSIC(EMM) topology.

13.3 Si EMM Drain Saturation Current

The current increases until the Si EMM enters its saturation region during the short-circuit event with the BaSIC(EMM) topology, allowing it to limit the current level. As shown in Chap. 10, the voltage across the Si EMM increases to about 8 V during the short-circuit event. The gate-source voltage across the SiC power MOSFET is reduced from 20 V by this value to 12 V until its saturation current automatically equals that of the Si EMM. This allows the SiC MOSFET to support the DC supply voltage (typically 800 V) while sustaining the saturation current of 50 A as dictated by the Si EMM. The power dissipation in the Si EMM is then 400 W assuming a constant short-circuit current level of 50 A.

The temperature of the Si EMM device increases during the SC event due to the above power dissipation. Using the adiabatic model for short-circuit events described in Chap. 2, the temperature rise in the Si EMM is calculated to be 125–150 °C in 5–10 μs for a device with a typical active area of 0.05 cm^2. A depletion width of 1 μm was used with a volumetric specific heat of 1.66 J/cm^3 K for silicon for this calculation. Consequently, the peak short-circuit current is determined by the drain saturation current of the Si EMM at either 150 or 175 °C that is provided in Si EMM datasheets, rather than values at 25 °C. This is the *third selection criterion* for the Si EMM device used in the BaSIC(EMM) topology.

13.4 Si EMM Trade-Off Curves

Based on the discussion in the previous sections, the selection of the Si EMM requires knowledge of its 25 °C on-resistance as a function of DC gate bias and its saturation current at 150 °C for the same DC gate bias values. Unfortunately, most Si EMM vendors do not provide high current output characteristics at either 25 or 150 °C in their datasheets. In general, they would have to be measured after acquiring devices of different voltage ratings from various vendors. However, this type of data is available in the datasheets for International Rectifier Corporation (IRF) Si MOSFETs that is published on the Infineon website [2].

The information in the IRF datasheets is therefore utilized in this chapter to demonstrate the Si EMM selection process for achieving improved performance of the BaSIC(EMM) topology. Using these data, it is possible to generate a trade-off curve between drain saturation current at 150 °C and the on-resistance at 25 °C for each product by using the DC gate bias applied to the Si EMM device as a parametric variable. The trade-off curves for various Si EMM products can then be compared to determine which device provides the lowest on-resistance at the drain saturation current level required to get the desired short-circuit time of 10 μs.

The enhancement mode Si power MOSFETs are available from IRF with blocking voltages from 20 V to over 200 V. It is well known that the specific on-resistance (on-resistance for 1 cm^2 active area) for a Si power MOSFET increases with blocking voltage capability due to thicker and higher resistivity drift regions required to support the larger voltage [3]. Since the Si EMM is required to support less than 10 V in the BaSIC(EMM) topology under all circuit operating conditions (including the short-circuit event), any products with blocking voltage of above 20 V are satisfactory candidates for this topology. The pertinent characteristics taken from the datasheets of some typical IRF products are discussed in this chapter for the BaSIC(EMM) application.

13.5 30 V Si EMM Devices

Many IRF power MOSFET products are available with a blocking voltage of 30 V due to their applications in power management circuits. In this section, 30 V devices with various on-resistance values are compared to determine the best trade-off curve.

13.5.1 IRF7313 Product

This 30 V rated device has an on-resistance of 23 mΩ at a gate bias $V_{GS} = 10$ V. Its output characteristics provided in the datasheet at 25 and 150 °C are shown in Fig. 13.2. The on-resistance at 25 °C can be obtained at various gate bias voltages

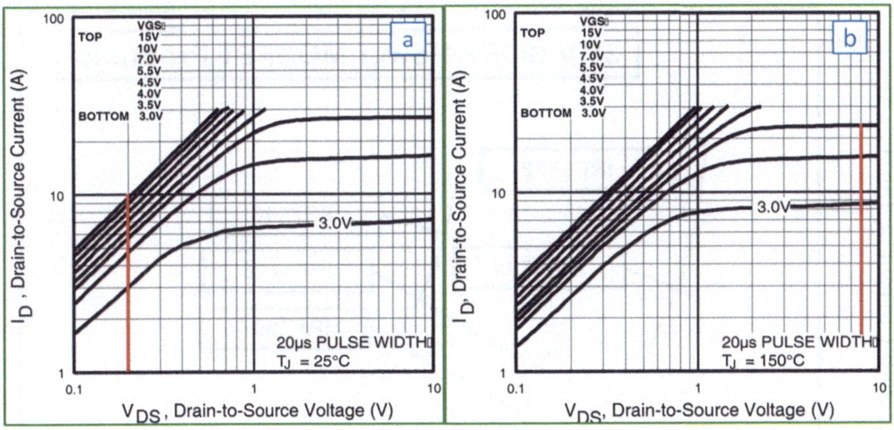

Fig. 13.2 Output characteristics for the 30 V 23 mΩ Si power MOSFET product IRF7313 with different gate bias voltages: (**a**) 25 °C and (**b**) 150 °C

by extracting the drain current at $V_{DS} = 0.2$ V along the red line in Fig. 13.2a. The saturated drain current at 150 °C can be extracted at $V_{DS} = 8$ V for various gate bias voltages along the red line in Fig. 13.2b. Alternately, the transfer characteristics at 150 °C in the datasheets can be used to determine the drain saturation currents for all the Si EMM devices.

It is worth pointing out that for this device, the drain saturation current decreases from 28 to 23 A, a change of 13%, for a fixed gate bias of 4.0 V when the temperature increases from 25 to 150 °C. This is a desirable behavior for the Si EMM in the BaSIC(EMM) topology because the short-circuit current will be reduced when the Si EMM heats up during the short-circuit event. It is noteworthy that not all Si EMM devices exhibit a decrease in the saturation current with increase in temperature. A reduction of saturation current with temperature generates a more favorable trade-off curve for the BaSIC(EMM) device in the proposed methodology. This behavior therefore becomes a *fourth selection criterion* for the Si EMM devices.

The data extracted from the datasheet can be combined to generate the trade-off curve for the IRF7313 Si EMM product as shown in Fig. 13.3 by the purple line. From the trade-off curve, it is found that the on-resistance for this device is 28 mΩ to obtain a drain saturation current of 50 A, indicated by the green dashed line in Fig. 13.3. The net on-resistance for the BaSIC(EMM) topology will then increase by 17% over the 168 mΩ value measured for the 1.2 kV SiC power MOSFET product C2M0160120D at a gate bias of 20 V.

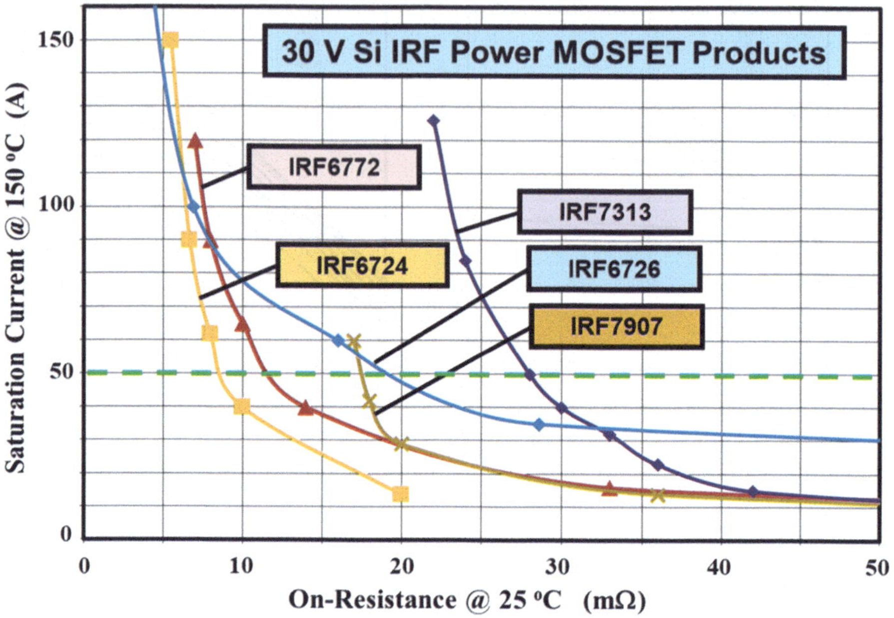

Fig. 13.3 Trade-off curves for 30 V IRF Si power MOSFET products with different on-resistance values at 25 °C using saturation current at 150 °C in their datasheets

13.5.2 IRF7907 Product

This 30 V rated device has an on-resistance of 13.7 mΩ at a gate bias $V_{GS} = 10$ V. Its output characteristics provided in the datasheet at 25 and 150 °C are shown in Fig. 13.4. The on-resistance at 25 °C can be obtained at various gate bias voltages by extracting the drain current at $V_{DS} = 0.2$ V along the red line in Fig. 13.4a. The saturated drain current at 150 °C can be extracted at $V_{DS} = 8$ V for various gate bias voltages along the red line in Fig. 13.4b. For this device, the drain saturation current at $V_{GS} = 3.5$ V does not change when the device heats up from 25 to 150 °C. This is acceptable behavior for the Si EMM device in the BaSIC(EMM) topology although a reduction of the saturation current with temperature is preferable.

The extracted data can be combined to generate the trade-off curve for this device as shown in Fig. 13.3 by the brown line. From the trade-off curve, it is determined that the on-resistance for this device is 17.5 mΩ to obtain a drain saturation current of 50 A. The net on-resistance for the BaSIC(EMM) topology will then increase by 10%. This Si EMM device is therefore superior to the IRF7313 product for the BaSIC(EMM) application.

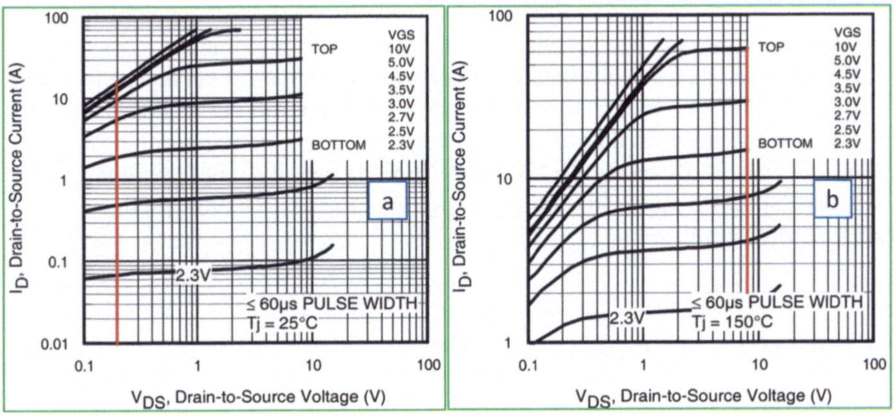

Fig. 13.4 Output characteristics for the 30 V 13.7 mΩ Si power MOSFET product IRF7907 with different gate bias voltages: (**a**) 25 °C and (**b**) 150 °C

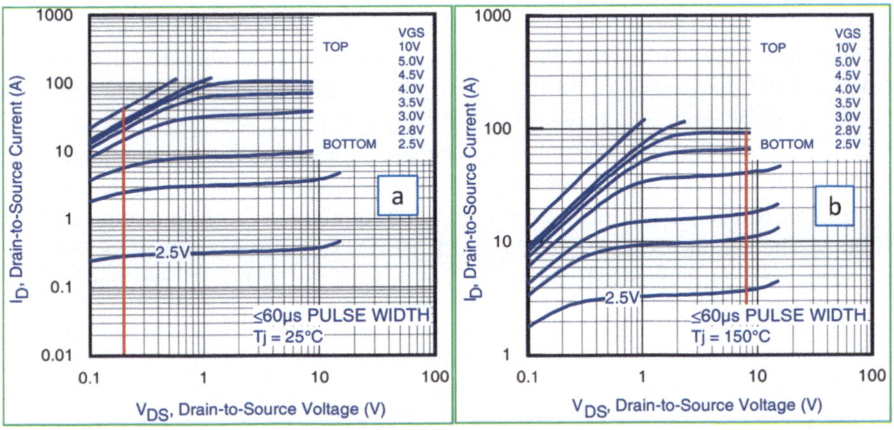

Fig. 13.5 Output characteristics for the 30 V 4.7 mΩ Si power MOSFET product IRF6722 with different gate bias voltages: (**a**) 25 °C and (**b**) 150 °C

13.5.3 IRF6722 Product

This 30 V rated device has an on-resistance of 4.7 mΩ at a gate bias $V_{GS} = 10$ V. Its output characteristics provided in the datasheet at 25 and 150 °C are shown in Fig. 13.5. The on-resistance at 25 °C can be obtained at various gate bias voltages by extracting the drain current at $V_{DS} = 0.2$ V along the red line in Fig. 13.5a. The saturated drain current at 150 °C can be extracted at $V_{DS} = 8$ V for various gate bias voltages along the red line in Fig. 13.5b. For this device, the drain saturation current at $V_{GS} = 4.0$ V does not change when the device heats up from 25 to 150 °C. This is acceptable behavior for the Si EMM device in the BaSIC(EMM) topology although a reduction of the saturation current with temperature is preferable.

The extracted data can be combined to generate the trade-off curve for this device as shown in Fig. 13.3 by the maroon line. From the trade-off curve, it is determined that the on-resistance for this device is 11 mΩ to obtain a drain saturation current of 50 A. The net on-resistance for the BaSIC(EMM) topology will then increase by 7%. This Si EMM device is therefore superior to the IRF7907 product for the BaSIC(EMM) application.

13.5.4 IRF6724 Product

This 30 V rated device has an on-resistance of 1.9 mΩ at a gate bias V_{GS} = 10 V. Its output characteristics provided in the datasheet at 25 and 150 °C are shown in Fig. 13.6. The on-resistance at 25 °C can be obtained at various gate bias voltages by extracting the drain current at V_{DS} = 0.2 V along the red line in Fig. 13.6a. The saturated drain current at 150 °C can be extracted at V_{DS} = 8 V for various gate bias voltages along the red line in Fig. 13.6b. For this device, the drain saturation current at V_{GS} = 3.5 V increases by 15% from 25 to 150 °C. This is an undesirable behavior for the Si EMM device in the BaSIC(EMM) topology.

The extracted data can be combined to generate the trade-off curve for this device as shown in Fig. 13.3 by the ochre line. From the trade-off curve, it is determined that the on-resistance for this device is 8.5 mΩ to obtain a drain saturation current of 50 A. The net on-resistance for the BaSIC(EMM) topology will then increase by 5%. This Si EMM device is therefore superior to the IRF6722 product for the BaSIC(EMM) application.

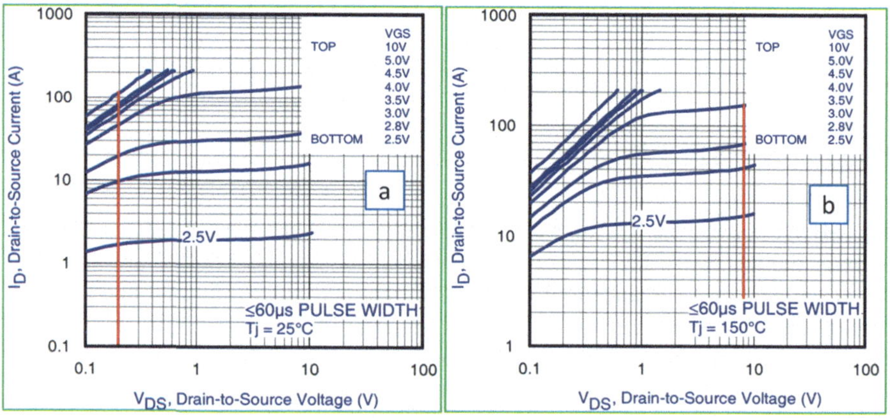

Fig. 13.6 Output characteristics for the 30 V 1.9 mΩ Si power MOSFET product IRF6724 with different gate bias voltages: (**a**) 25 °C and (**b**) 150 °C

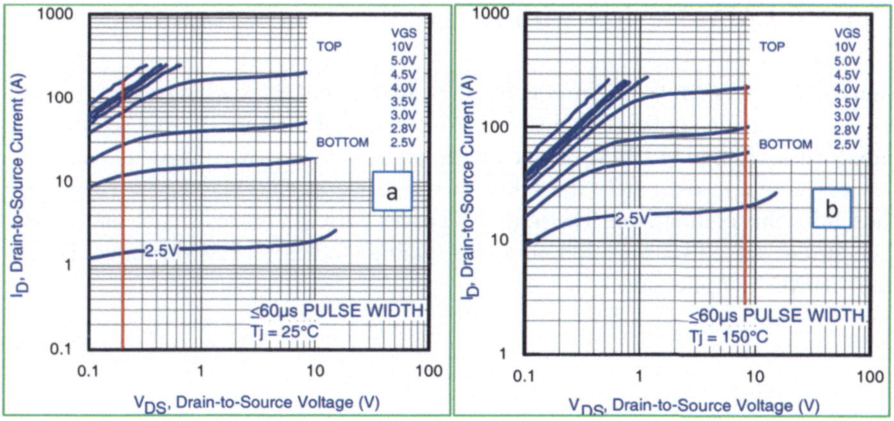

Fig. 13.7 Output characteristics for the 30 V 1.3 mΩ Si power MOSFET product IRF6726 with different gate bias voltages: (**a**) 25 °C and (**b**) 150 °C

13.5.5 IRF6726 Product

This 30 V rated device has an on-resistance of 1.3 mΩ at a V_{GS} = 10 V. Its output characteristics provided in the datasheet at 25 and 150 °C are shown in Fig. 13.7. The on-resistance at 25 °C can be obtained at various gate bias voltages by extracting the drain current at V_{DS} = 0.2 V along the red line in Fig. 13.7a. The saturated drain current at 150 °C can be extracted at V_{DS} = 8 V for various gate bias voltages along the red line in Fig. 13.7b. For this device, the drain saturation current at V_{GS} = 3.5 V does not change from 25 to 150 °C. This is an acceptable behavior for the Si EMM device in the BaSIC(EMM) topology.

The extracted data can be combined to generate the trade-off curve for this device as shown in Fig. 13.3 by the blue line. From the trade-off curve, it is determined that the on-resistance for this device is 19 mΩ to obtain a drain saturation current of 50 A. The net on-resistance for the BaSIC(EMM) topology will then increase by 12%. This Si EMM device is therefore inferior to the IRF6724 product for the BaSIC(EMM) application. This example demonstrates that selecting a Si EMM with the lowest on-resistance at V_{GS} = 10 V on the datasheet does not necessarily provide the best performance for the BaSIC(EMM) topology. The proposed selection methodology using a trade-off curve between on-resistance and saturation current is essential to identify the best Si EMM device.

13.6 Si EMM Devices with Larger Blocking Voltages

Many IRF power MOSFET products are available with a blocking voltage above 30 V due to their applications in power management circuits. In this section, the impact of selecting Si EMM devices with larger breakdown voltages is analyzed. Devices with approximately the same on-resistance at gate bias V_{GS} = 10 V were chosen for this comparison. The trade-off curves between on-resistance and saturation current serve as a selection guide for finding the best Si EMM product for the BaSIC(EMM) application.

13.6.1 IRF135B Product

This 135 V rated device has an on-resistance of 6.7 mΩ at a gate bias V_{GS} = 10 V. Its output characteristics provided in the datasheet at 25 and 150 °C are shown in Fig. 13.8. The on-resistance at 25 °C can be obtained at various gate bias voltages by extracting the drain current at V_{DS} = 0.2 V along the red line in Fig. 13.8a. The saturated drain current at 175 °C can be extracted at V_{DS} = 8 V for various gate bias voltages along the red line in Fig. 13.8b. Data at 150 °C was not provided by the manufacturer. For this device, the drain saturation current at V_{GS} = 4.0 V increases from 25 to 175 °C by a factor of 8 times. This severely degrades its performance in the BaSIC(EMM) topology.

The extracted data can be combined to generate the trade-off curve for this device as shown in Fig. 13.9 by the red line. From the trade-off curve, it is determined that the on-resistance for this device is 90 mΩ to obtain a drain saturation current of 50 A. The high value increases the net on-resistance for the BaSIC(EMM) topology by

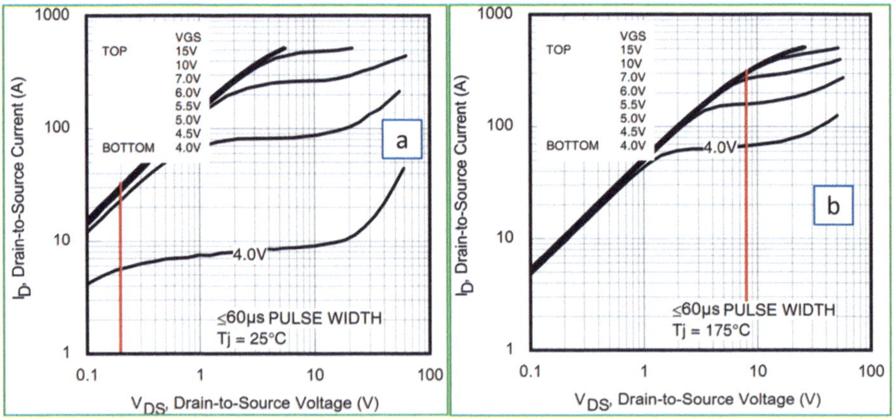

Fig. 13.8 Output characteristics for the 135 V 6.7 mΩ Si power MOSFET product IRF135B with different gate bias voltages: (**a**) 25 °C and (**b**) 175 °C

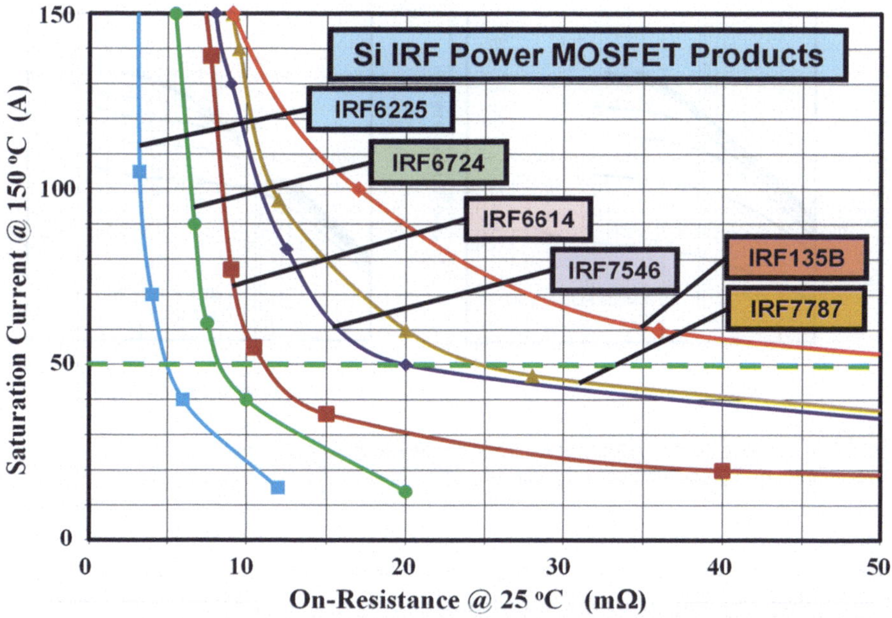

Fig. 13.9 Trade-off curves for IRF Si power MOSFET products with different blocking voltage ratings: on-resistance values at 25 °C versus saturation current at 150 °C in their datasheets

56%. It is obvious that increasing the blocking voltage of the Si EMM device unnecessarily to 135 V has severely degraded the performance of the BaSIC(EMM) topology.

13.6.2 IRF7787 Product

This 75 V rated device has an on-resistance of 6.7 mΩ at a gate bias $V_{GS} = 10$ V. Its output characteristics provided in the datasheet at 25 and 175 °C are shown in Fig. 13.10. The on-resistance at 25 °C can be obtained at various gate bias voltages by extracting the drain current at $V_{DS} = 0.2$ V along the red line in Fig. 13.10a. The saturated drain current at 175 °C can be extracted at $V_{DS} = 8$ V for various gate bias voltages along the red line in Fig. 13.10b. Data at 150 °C was not provided by the manufacturer. For this device, the drain saturation current at $V_{GS} = 5.5$ V increases from 25 to 175 °C by a factor of 1.4 times. This degrades its performance in the BaSIC(EMM) topology.

The extracted data can be combined to generate the trade-off curve for this device as shown in Fig. 13.9 by the brown line. From the trade-off curve, it is determined that the on-resistance for this device is 25 mΩ to obtain a drain saturation current of 50 A. A reduction of the Si EMM device on-resistance is observed by reducing the

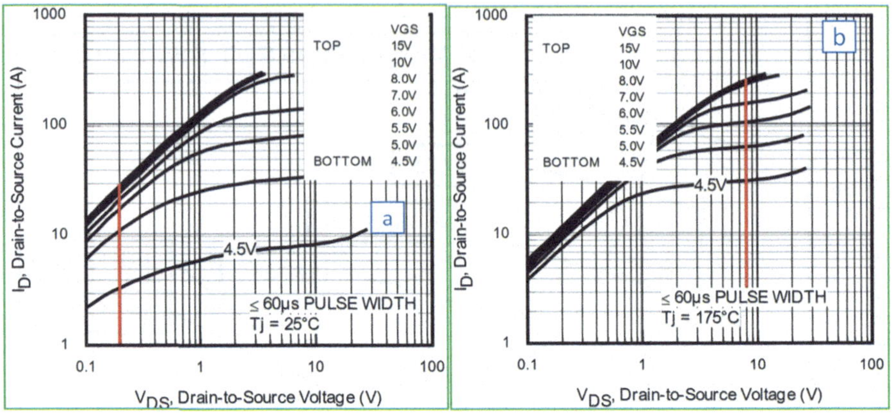

Fig. 13.10 Output characteristics for the 75 V 6.9 mΩ Si power MOSFET product IRF7787 with different gate bias voltages: (**a**) 25 °C and (**b**) 175 °C

blocking voltage is half. The net on-resistance for the BaSIC(EMM) topology increases by 16% by using this 75 V rated Si EMM device. It is obvious that reducing the blocking voltage of the Si EMM device from 135 to 75 V greatly improves the performance of the BaSIC(EMM) topology.

13.6.3 IRF7546 Product

This 60 V rated device has an on-resistance of 6.6 mΩ at a gate bias $V_{GS} = 10$ V. Its output characteristics provided in the datasheet at 25 and 175 °C are shown in Fig. 13.11. The on-resistance at 25 °C can be obtained at various gate bias voltages by extracting the drain current at $V_{DS} = 0.2$ V along the red line in Fig. 13.11a. The saturated drain current at 175 °C can be extracted at $V_{DS} = 8$ V for various gate bias voltages along the red line in Fig. 13.11b. Data at 150 °C was not provided by the manufacturer. For this device, the drain saturation current at $V_{GS} = 5.5$ V increases from 25 to 175 °C by a factor of 1.5 times. This degrades its performance in the BaSIC(EMM) topology.

The extracted data can be combined to generate the trade-off curve for this device as shown in Fig. 13.9 by the purple line. From the trade-off curve, it is determined that the on-resistance for this device is 20 mΩ to obtain a drain saturation current of 50 A. A reduction of the on-resistance is observed by reducing the blocking voltage of the Si EMM device from 75 to 60 V. The net on-resistance for the BaSIC(EMM) topology increases by 12.5% with this 60 V rated Si EMM product. It is obvious that reducing the blocking voltage of the Si EMM device improves the performance of the BaSIC(EMM) topology.

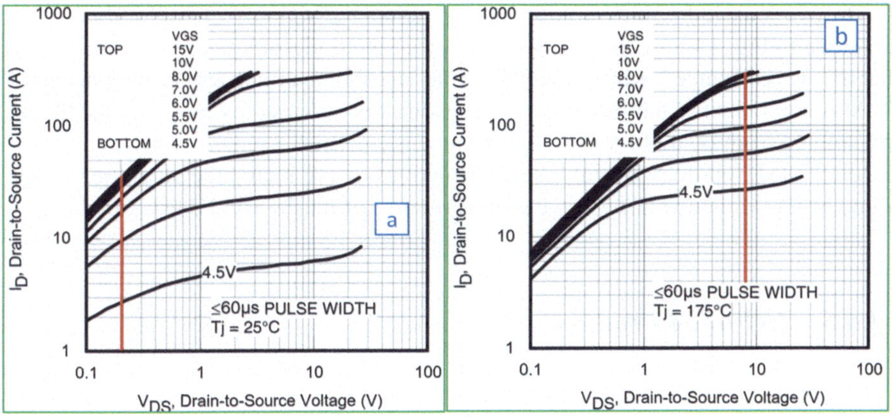

Fig. 13.11 Output characteristics for the 60 V 6.9 mΩ Si power MOSFET product IRF7546 with different gate bias voltages: (**a**) 25 °C and (**b**) 175 °C

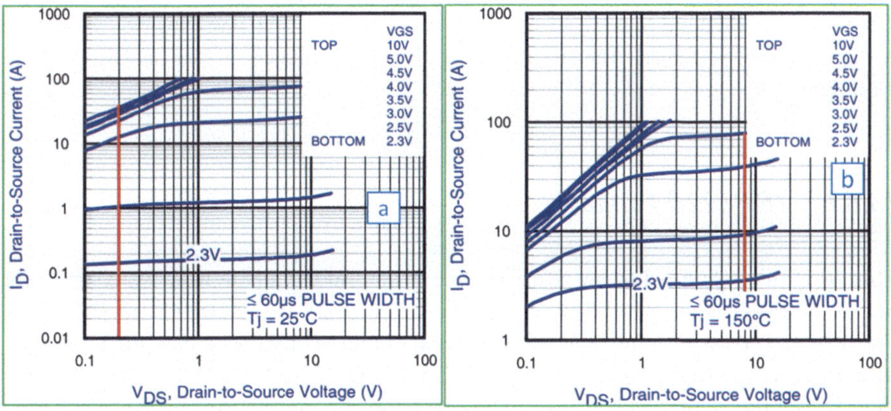

Fig. 13.12 Output characteristics for the 40 V 6.9 mΩ Si power MOSFET product IRF6614 with different gate bias voltages: (**a**) 25 °C and (**b**) 175 °C

13.6.4 IRF6614 Product

This 40 V rated device has an on-resistance of 5.9 mΩ at a gate bias $V_{GS} = 10$ V. Its output characteristics provided in the datasheet at 25 and 175 °C are shown in Fig. 13.12. The on-resistance at 25 °C can be obtained at various gate bias voltages by extracting the drain current at $V_{DS} = 0.2$ V along the red line in Fig. 13.12a. The saturated drain current at 150 °C can be extracted at $V_{DS} = 8$ V for various gate bias voltages along the red line in Fig. 13.12b. For this device, the drain saturation current at $V_{GS} = 3.5$ V does not change from 25 to 150 °C. This is satisfactory for good performance in the BaSIC(EMM) topology.

The extracted data can be combined to generate the trade-off curve for this device as shown in Fig. 13.9 by the maroon line. From the trade-off curve, it is determined that the on-resistance for this device is 11 mΩ to obtain a drain saturation current of 50 A. A significant reduction of the on-resistance is observed by reducing the blocking voltage from 60 to 40 V. The lower value increases the net on-resistance for the BaSIC(EMM) topology by only 7%. It is obvious that reducing the blocking voltage of the Si EMM device improves the performance of the BaSIC(EMM) topology.

13.6.5 IRF6724 Product

This 30 V rated device has an on-resistance of 4.7 mΩ at a gate bias V_{GS} = 10 V. Its characteristics were discussed in the previous section in conjunction with Fig. 13.3. The trade-off curve for this device is shown by the green line in Fig. 13.9. From the trade-off curve, it is determined that the on-resistance for this device is 8 mΩ to obtain a drain saturation current of 50 A. The net on-resistance for the BaSIC(EMM) topology will then increase by only 5%. It is obvious that reducing the blocking voltage of the Si EMM device improves the performance of the BaSIC(EMM) topology.

13.6.6 IRF6225 Product

This 20 V rated device has an on-resistance of 3.2 mΩ at a gate bias V_{GS} = 10 V. Its output characteristics provided in the datasheet at 25 and 150 °C are shown in Fig. 13.13. The on-resistance at 25 °C can be obtained at various gate bias voltages by extracting the drain current at V_{DS} = 0.2 V along the red line in Fig. 13.13a. The saturated drain current at 150 °C can be extracted at V_{DS} = 8 V for various gate bias voltages along the red line in Fig. 13.13b. For this device, the drain saturation current at V_{GS} = 2.0 V reduces by a factor of 2 times from 25 to 150 °C. This is very beneficial for good performance in the BaSIC(EMM) topology.

The extracted data can be combined to generate the trade-off curve for this device as shown in Fig. 13.9 by the blue line. From the trade-off curve, it is determined that the on-resistance for this device is 5 mΩ to obtain a drain saturation current of 50 A. A further reduction of the on-resistance has been achieved by reducing the blocking voltage from 30 to 20 V. This lower value increases the net on-resistance for the BaSIC(EMM) topology by only 3%. It is obvious that reducing the blocking voltage of the Si EMM device greatly improves the performance of the BaSIC(EMM) topology.

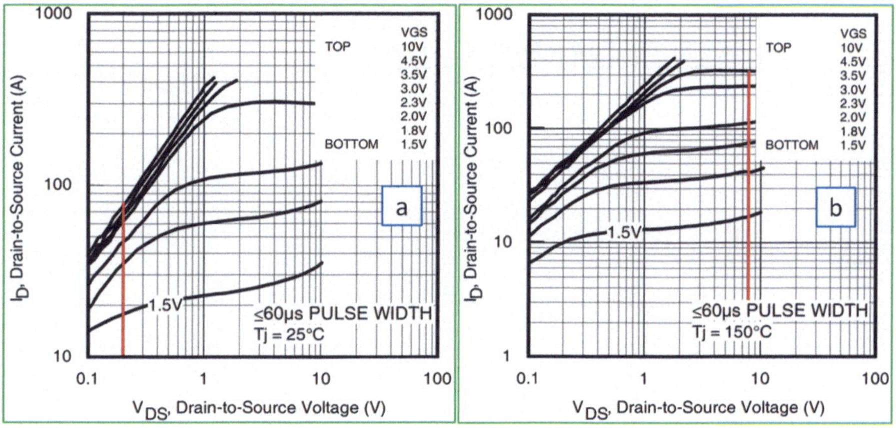

Fig. 13.13 Output characteristics for the 20 V 6.9 mΩ Si power MOSFET product IRF6225 with different gate bias voltages: (**a**) 25 °C and (**b**) 150 °C

13.7 Comparison of Si EMM Products

In the previous sections, it was demonstrated that Si EMM devices can be selected using the datasheets from manufacturers websites to achieve a peak short-circuit current of 50 A, which is required to obtain a short-circuit withstand time of 10 μs for the CREE/Wolfspeed 1.2 kV 160 mΩ SiC power MOSFET product C2M0160120 chosen for this study. The on-resistance for the Si EMM devices when operating at the DC gate bias required to obtain a saturation current of 50 A is substantially larger than when operated at a gate bias of 10 V used in datasheets. This is illustrated in Fig. 13.14 for the Si EMM power MOSFET products with a blocking voltage of 30 V. The red line in the figure gives the on-resistance at the gate bias required to obtain a saturation current of 50 A while the blue line shows it's value at V_{GS} = 10 V. As an example, the on-resistance increases from 1.9 to 15 mΩ, an increase by a factor of 7.9× for the lowest on-resistance case.

It can be seen from Fig. 13.9 that the on-resistance of Si EMM devices at the gate bias required to obtain a saturation current for 50 A became larger with increasing blocking voltage capability. These on-resistance values are plotted in Fig. 13.15 to show the very strong trend of increasing on-resistance with blocking voltage for the IRF products. The on-resistance increased by a factor of 18 times when the blocking voltage was raised from 20 and 135 V. Based on this plot, it is clear that Si EMM devices with lowest acceptable blocking voltage capability of 20 V should be selected for the BaSIC(EMM) topology.

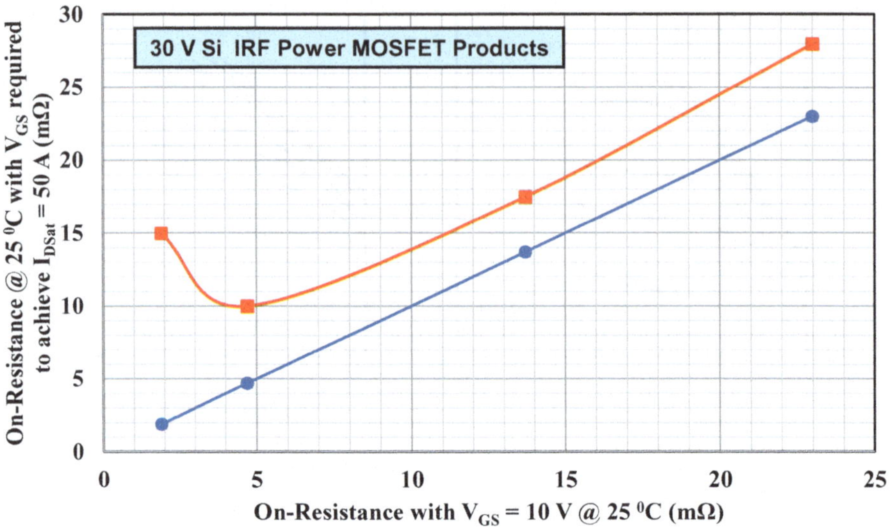

Fig. 13.14 On-resistance of 30 V Si power MOSFET products at V_{GS} required to achieve a saturation current of 50 A vs on-resistance at $V_{GS} = 10$ V

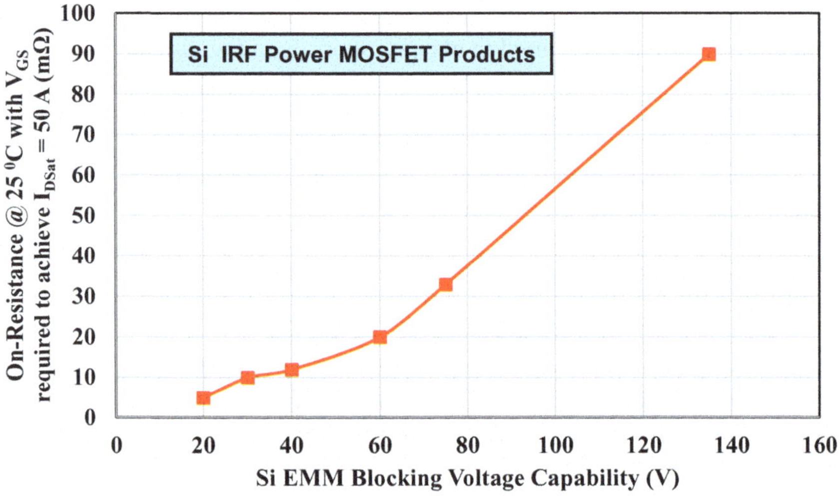

Fig. 13.15 On-resistance of Si power MOSFET products with various blocking voltages at V_{GS} required to achieve a saturation current of 50 A

13.8 Improved Trade-Off Curves

In Chap. 10, an Infineon Si power MOSFET IPi70N04S4 with blocking capability of 40 V was used to improve the short-circuit time for the Wolfspeed 1.2 kV 160 mΩ SiC power MOSFET product C2M0160120. The experimentally measured trade-off curve for this case is reproduced in Fig. 13.16 as the maroon line. This maroon trade-off curve was obtained my measuring the short-circuit time at a DC supply voltage of 800 V with 20 V gate drive applied to the composite transistor. The gate bias $V_{DC,PROG}$ applied to the Si EMM was varied to change the short-circuit current and generate the maroon trade-off curve. It can be seen that a short-circuit withstand time of up to 7.4 μs was obtained with an increase in on-resistance of 17%. The ideally desired short-circuit time of 10 μs is not achievable with this Si EMM device because its on-resistance becomes prohibitively large (>300% increase) as can be inferred from the figure.

Based on the Si EMM selection methodology discussed in this chapter, it was found that the Si EMM IRLR6225 with blocking capability of 20 V is a better choice for the BaSIC(EMM) topology. The saturation current and on-resistance data for this device were shown in Fig. 13.12. The saturation current for each gate bias ($V_{DC,PROG}$) can be converted to the corresponding short-circuit time using the curve-fit polynomial blue line in Fig. 13.1. This information is combined to generate the trade-off graph represented by the blue points and the blue trend line in Fig. 13.16. It can be seen (red dashed lines) that a short-circuit withstand time of 10 μs is achievable by using Si EMM IRLR6225 device with only 3% increase in on-resistance. This demonstrates that the selection methodology proposed and discussed in this chapter can be used to achieve a short-circuit time of over 10 μs for the CREE/Wolfspeed 1.2 kV 160 mΩ SiC power MOSFET product C2M0160120 with very small increase in on-resistance.

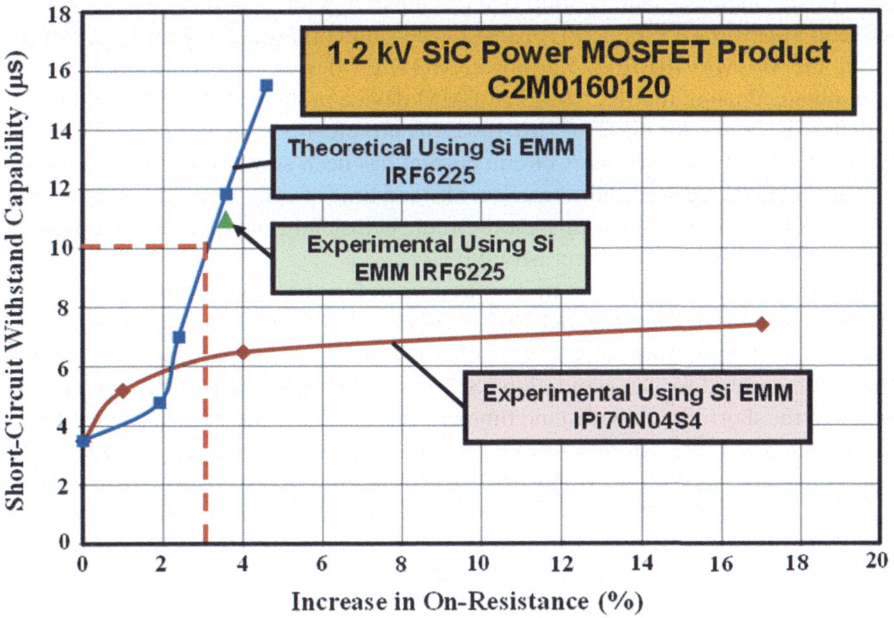

Fig. 13.16 Trade-off curve for the BaSIC(EMM) topology: on-resistance

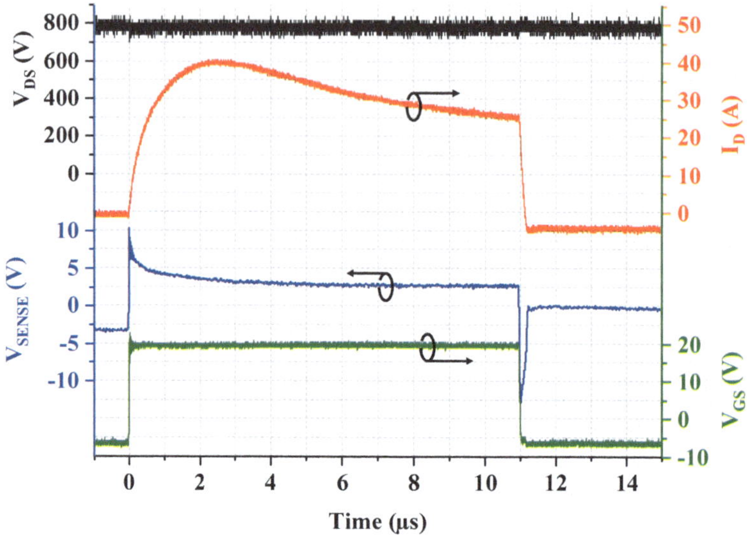

Fig. 13.17 Short-circuit test waveforms observed for the BaSIC(EMM) topology with the Si IRF6225 device in series with the source of the 1.2 kV SiC power MOSFET C2M0160120 product

The validation of the selection methodology discussed in this chapter requires experimental demonstration that a short-circuit time of over 10 μs can be achieved using the BaSIC(EMM) topology, and that the trade-off curve between t_{SC} and percentage R_{on} increase can be improved compared with that observed using the Infineon Si power MOSFET IPi70N04S4. The validation was accomplished by procuring the Si EMM IRLR6225 Si power MOSFET device. The short-circuit waveform obtained using the IRLR6225 Si EMM device in the BaSIC(EMM) topology with the SiC power MOSFET C2M0160120D product is shown in Fig. 13.17. It can be observed that the peak short-circuit current has been successfully reduced to 40 A resulting in the desired short-circuit withstand time of over 10 μs (actually 11 μs). It is worth pointing out that this short-circuit withstand time of 11 μs is observed *without failure* of the SiC power MOSFET.

It can be seen from the blue trade-off curve in Fig. 13.9 that a saturation current of 40 A is achieved for the IRLR6225 Si power MOSFET device with an on-resistance of 6 mΩ. This value is an increase of 3.6% in the on-resistance over the 168 mΩ measured for the stand-alone SiC power MOSFET. The measured combination of the short-circuit withstand time of 11 μs and 3.6% increase in on-resistance is shown in the trade-off graph in Fig. 13.16 by the green triangle. The measured data point falls very close to the predicted trade-off curve for the BaSIC(EMM) topology with the IRLR6225 Si power MOSFET device. This provides validation that a superior trade-off curve between the short-circuit time and percentage increase in on-resistance can be achieved using the selection methodology discussed in the paper.

13.9 Summary

Using the criteria established in this chapter, the Si EMM selection methodology consists of the following steps:

(a) Determine the peak short-circuit current ($I_{P,SC}$) for the SiC power MOSFET that allows it to withstand a short-circuit time of 10 μs by adjusting its gate drive voltage.
(b) Determine the values for the on-resistance for Si EMM devices at 25 °C from their datasheets at various gate bias voltages.
(c) Determine the values for the drain saturation currents for Si EMM devices at 150 or 175 °C from their datasheets at various gate bias voltages.
(d) Prepare a trade-off curve of the drain saturation current at 150 or 175 °C versus the on-resistance at 25 °C for each Si EMM device using the gate bias as a parametric variable.
(e) Determine the on-resistance for each Si EMM at 25 °C corresponding to a saturation current that matches the $I_{P,SC}$ determined in step (a).
(f) Select the Si EMM that produces a drain saturation current at 150 or 175 °C of $I_{P,SC}$ that was determined in step (a) with the smallest on-resistance at 25 °C.
(g) Determine the DC gate bias for the selected Si EMM corresponding to the above operating point.

It was found that the lowest Si EMM on-resistance, at the V_{GS} required to obtain a saturation current of $I_{P,SC}$, is not always achieved with devices having the smallest on-resistance at $V_{GS} = 10$ V as reported in the datasheets. Consequently, it is important to utilize the trade-off curve proposed in this chapter when selecting the best Si EMM product.

The methodology proposed and described in this chapter can be applied for the short-circuit protection of SiC power MOSFETs over a broad range of blocking voltages and on-resistances by using the BaSIC(EMM) topology. The Si EMM IRLR6225 product with 20 V rating identified in this chapter is suitable for all the SiC power MOSFET devices under development with ratings of 600–10,000 V because it operates in their gate drive loop. However, the DC gate voltage ($V_{DC,PROG}$) for the Si EMM must be adjusted based on the desired peak short-circuit current ($I_{P,SC}$) for each SiC MOSFET product to achieve the desirable short-circuit time of 10 μs.

References

1. A. Kanale, B.J. Baliga, Selection methodology for Si power MOSFETs used to enhance SiC power MOSFET short-circuit capability with the BaSIC(EMM) topology. IEEE Trans. Power Electron. **36**, 8243–8252 (2021)
2. https://www.infineon.com/cms/en/product/power/mosfet/
3. B.J. Baliga, *Fundamentals of Power Semiconductor Devices*, 2nd edn. (Springer-Science, New York, 2018)

Chapter 14
BaSIC(DMM) vs BaSIC(EMM) for Silicon Carbide Power MOSFETs

Excellent performance of the BaSIC topology has been demonstrated in previous chapters by employing silicon depletion mode MOSFETs (DMMs) and silicon enhancement mode MOSFETs (EMMs) as the non-linear resistances placed in series with the SiC power MOSFET. These two approaches are contrasted in this chapter [1]. Trade-off curves between the improvement in short-circuit time and the increase in the on-resistance or the switching losses are developed for the same SiC power MOSFET with the two approaches.

14.1 Device Selection

The silicon carbide power MOSFET chosen for the comparison of the two approaches was the 1.2 kV rated Wolfspeed Gen-2 device C2M0160120D with a typical on-state resistance of 160 mΩ at a gate bias of 20 V. The silicon DMM power device was the 100 V rated IXYS device IXTH16N10D2 with a typical on-resistance of 65 mΩ at zero gate bias. The silicon EMM power device was the 40 V rated Infineon OptiMOS device IP170N04S4 with a typical on-resistance of 6 mΩ at a gate bias of 10 V. The characteristics of these silicon devices used as the non-linear elements have been described in previous chapters.

14.2 BaSIC Topologies Used for Comparison

All the characterization was performed first for the stand-alone SiC power MOSFET, referred to as the M_1 and M_2 device in this chapter. These results served as the benchmark for assessment of the enhancements achieved using the BaSIC topology. This data is referred to as Case 1a and Case 2a in this chapter.

B. J. Baliga, A. Kanale, *The BaSIC Topology*,
https://doi.org/10.1007/978-3-031-86630-2_14

The BaSIC(DMM) topology was evaluated using three variations: Case 1b—Composite MOSFET with one DMM placed in series with the SiC power MOSFET; Case 1c—Composite MOSFET with two DMMs in parallel placed in series with the SiC power MOSFET; and Case 1d—Composite MOSFET with three DMMs in parallel placed in series with the SiC power MOSFET.

The BaSIC(EMM) topology was evaluated using three variations: Case 2b—Composite MOSFET with $V_{DC,PROG}$ of 5.4 V applied to the silicon EMM device placed in series with the SiC power MOSFET; Case 2c—Composite MOSFET with $V_{DC,PROG}$ of 5.7 V applied to the silicon EMM device placed in series with the SiC power MOSFET; Case 2d—Composite MOSFET with $V_{DC,PROG}$ of 6.0 V applied to the silicon EMM device placed in series with the SiC power MOSFET.

14.3 High Voltage Output Characteristics

As discussed in Chap. 2, knowledge of the high-voltage, high-current output characteristics is invaluable for understanding the short-circuit capability of the power transistors. A novel method for extracting the high-voltage, high-current output characteristics of power transistors was developed by using the short-circuit test configuration [2]. The short-circuit test configuration is used with a small pulse width of 1.5 μs to allow the current to increase to a peak value that is representative of the saturation current at the DC supply voltage. The DC supply voltage is varied in small steps to generate the high-voltage, high-current output characteristic for each of the cases mentioned in the previous section.

The experimentally extracted high-voltage, high-current output characteristic at a gate bias of 20 V for the stand-alone SiC power MOSFET, labeled as M_1 and M_2, is shown in Fig. 14.1 with the red line. The saturated drain current is 136 A for the stand-alone SiC power MOSFET at a collector bias of 800 V. The experimentally extracted high-voltage, high-current output characteristics for the composite MOSFET with the three DMM configurations are shown with the purple, blue, and green dashed lines. The measured saturated drain currents are 19, 34, and 51 A for the Case 1b, Case 1c, and Case 1d at a collector bias of 800 V. The experimentally extracted high-voltage, high-current output characteristics for the composite MOSFET with the three EMM configurations are shown with the solid purple, blue, and green lines. The measured saturated drain currents are 62, 73, and 100 A for the Case 2b, Case 2c, and Case 2d at a collector bias of 800 V. These operating points for the 7 cases are highlighted with the points on the curves. It can be observed that the DMM cases have lower drain current saturation levels when compared with the EMM cases.

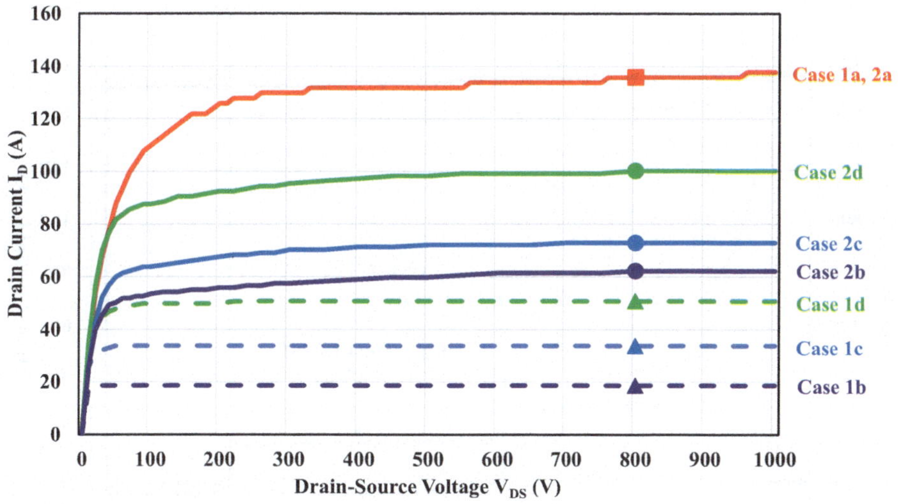

Fig. 14.1 High-voltage, high-current output characteristics using V_{GS} of 20 V: stand-alone SiC power MOSFET M_1, M_2; composite MOSFETs Case 1b, 1c, and 1d with DMMs; composite MOSFETs Case 2b, 2c, and 2d with EMMs

14.4 Short-Circuit Test Results

The short-circuit withstand time for all of the cases was obtained by gradually increasing the short-circuit pulse width while capturing the current waveform until the SiC power MOSFET failed. The DC supply voltage was maintained at the desired value of 800 V.

The measured short-circuit waveforms are shown in Fig. 14.2 for the cases with the silicon DMM devices. The red line was obtained for the stand-alone SiC power MOSFET M_1 which failed when the pulse width was increased to 3.5 μs with a peak short-circuit current of 136 A. The purple line represents Case 1b with a single DMM in series with the SiC power MOSFET. The peak short-circuit current is reduced to 19 A in Case 1b enabling an increase in the short-circuit withstand time t_{SC} to 27 μs. The blue line represents Case 1c with two DMM devices in parallel as the non-linear element placed in series with the SiC power MOSFET. The peak short-circuit current is reduced to 34 A in Case 1c enabling an increase in the short-circuit withstand time t_{SC} to 13.3 μs. The green line represents Case 1d with three DMM devices in parallel as the non-linear element placed in series with the SiC power MOSFET. The peak short-circuit current is reduced to 51 A in Case 1d enabling an increase in the short-circuit withstand time t_{SC} to 8.4 μs. The significant reduction of the peak short-circuit current with all the DMM cases produces a large increase in the short-circuit withstand time for the SiC power MOSFET.

The measured short-circuit waveforms are shown in Fig. 14.3 for the cases with the silicon EMM devices. The red line was obtained for the stand-alone SiC power MOSFET M_2 which failed when the pulse width was increased to 3.5 μs with a peak

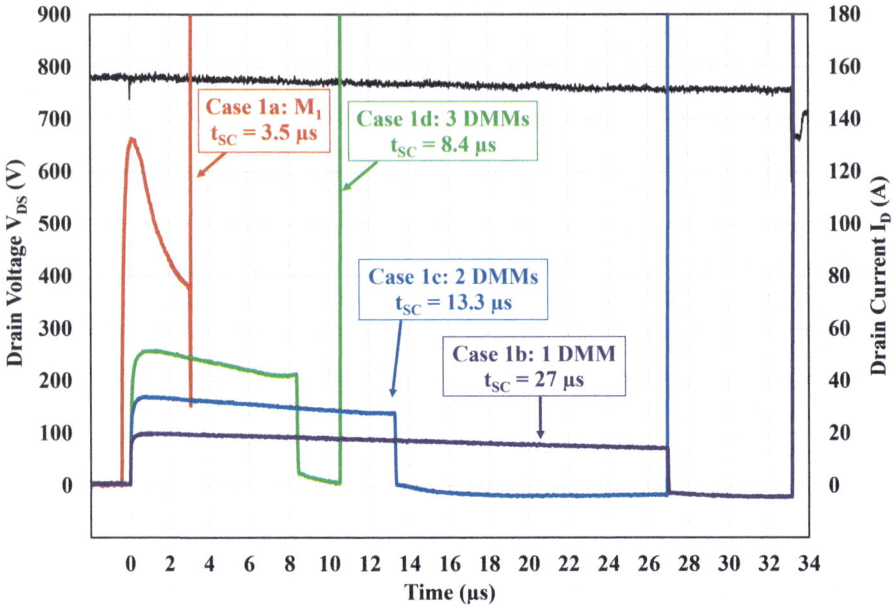

Fig. 14.2 Measured short-circuit waveforms: stand-alone SiC power MOSFET M_1, and composite transistors Case 1b, Case 1c, and Case 1d with silicon DMM devices

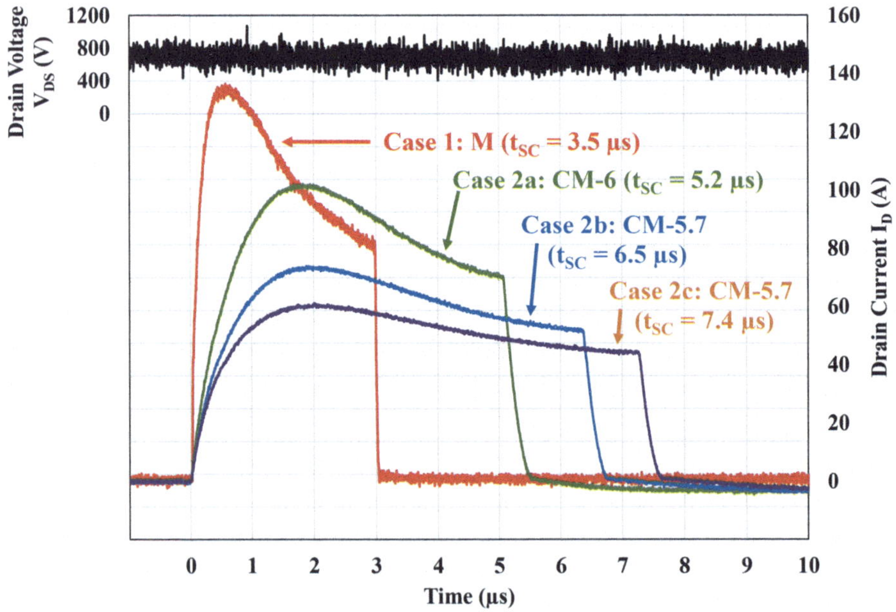

Fig. 14.3 Measured short-circuit waveforms: stand-alone SiC power MOSFET M_2, and composite transistors Case 2b, Case 2c, and Case 2d with silicon EMM devices operated at various DC gate biases $V_{DC,PROG}$

short-circuit current of 136 A. The purple line represents Case 2c with the EMM device, operated using a DC gate bias $V_{DC,PROG}$ of 5.4 V, in series with the SiC power MOSFET. The peak short-circuit current is reduced to 62 A in Case 2c enabling an increase in the short-circuit withstand time t_{SC} to 7.4 μs. The blue line represents Case 2b with the EMM device, operated using a DC gate bias $V_{DC,PROG}$ of 5.7 V, in series with the SiC power MOSFET. The peak short-circuit current is reduced to 73 A in Case 2b enabling an increase in the short-circuit withstand time t_{SC} to 6.5 μs. The green line represents Case 2a with the EMM device, operated using a DC gate bias $V_{DC,PROG}$ of 6.0 V, in series with the SiC power MOSFET. The peak short-circuit current is reduced to 100 A in Case 2a enabling an increase in the short-circuit withstand time t_{SC} to 5.2 μs. The reduction of the peak short-circuit current for all the EMM cases is not as large as for the silicon DMM cases. The increase in the short-circuit withstand time for the SiC power MOSFET for the silicon EMM device cases is consequently not as big as in the case of the silicon DMM devices.

All the measured short-circuit data can be combined to generate a basic trade-off graph, shown in Fig. 14.4, for the SiC power MOSFET used for this evaluation. As expected, the short-circuit withstand time increases when the peak short-circuit current becomes smaller irrespective of whether the DMM or EMM approach is utilized. A purple power-law trend line is shown in the figure with short-circuit time increasing in inverse proportion to the peak short-circuit current. All of the data points fall on this line. It can be concluded that the short-circuit withstand time increases in inverse proportion to the peak short-circuit current level consistent with

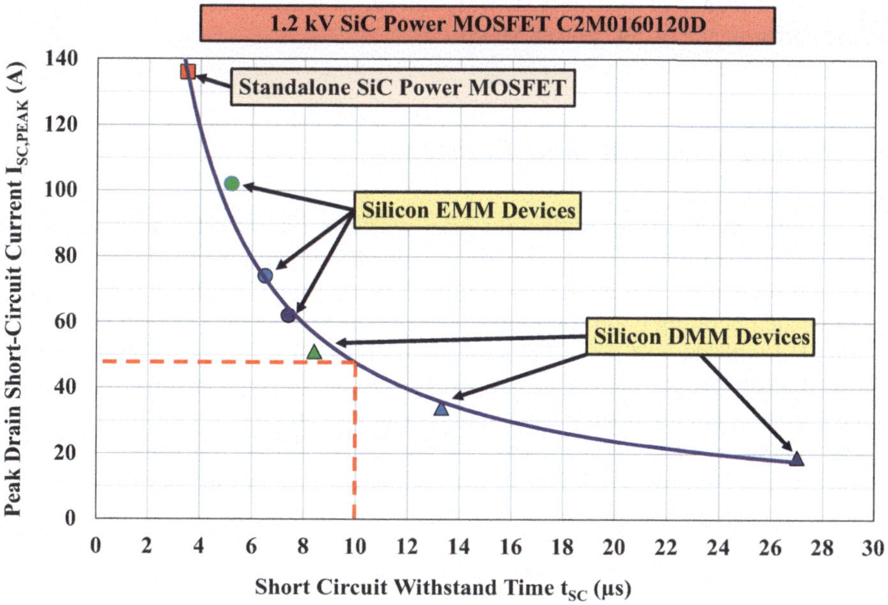

Fig. 14.4 Measured peak short-circuit current versus the observed short-circuit withstand time for the SiC power MOSFET Wolfspeed C2M0160120 product

the theory discussed in Chap. 2. From the red dashed lines in the figure, it can be concluded that a peak short-circuit current of 48 A is needed to achieve a short-circuit withstand time of 10 μs for the chosen SiC power MOSFET. The active area of the 1.2 kV Wolfspeed Gen-2 C2M0160120D device can be estimated to be 0.016 cm^2 based on a specific on-resistance of 2.5 mΩ cm^2. The peak short-circuit current density to achieve a short-circuit withstand time of 10 μs is then found to be 3000 A/cm^2 for the 1.2 kV Wolfspeed Gen-2 C2M0160120D device.

Based on the measured short-circuit data plotted in Fig. 14.4, it can be observed that the silicon DMM and silicon EMM approaches produce peak short-circuit currents and withstand times in different zones of the trade-off curve. The short-circuit test results demonstrate that a broad range of short-circuit withstand times can be obtained by using the BaSIC topology for SiC power MOSFETs by using the two approaches.

14.5 Impact on Normal Circuit Operation

The impact of the resistance of the Si DMM and EMM devices on the on-state resistance and switching behavior of the composite transistor within the BaSIC(DMM) and BaSIC(EMM) topologies with the SiC power MOSFET must be quantified. Trade-off curves between enhancing the short-circuit time and an increase on-state voltage drop and switching losses can then be developed to fully comprehend the relative performance of the BaSIC(DMM) and BaSIC(EMM) topologies to enhancing the performance of SiC power MOSFETs.

14.5.1 Impact on On-State Resistance

The on-state characteristics of the four cases with the silicon DMM devices, shown in Fig. 14.5, were obtained at lower drain voltages by using a curve tracer with a gate drive voltage of 20 V applied to the 1.2 kV Wolfspeed C2M0160120D product. The on-state operating point for all the devices was obtained at a current of 10 A as highlighted with the red lines. The on-state on-resistance for the stand-alone SiC power MOSFET is 177 mΩ. This value is larger than the typical value of 160 mΩ in the datasheet but less than the maximum value of 196 mΩ.

The measured on-state characteristics for the composite transistors (CTs) for various cases of the silicon DMM devices in series with the SiC power MOSFET are also shown in Fig. 14.5. The on-state operating point for each case can be determined at a current of 10 A. The on-state resistance at this operating point is 243 mΩ for Case 1b. This is consistent with the addition of the 65 mΩ on-resistance of a single DMM device to the on-resistance of the SiC power MOSFET. The on-state resistance at this operating point is 211 mΩ for Case 1c. This is consistent with the addition of the 34 mΩ on-resistance of two DMM devices operating in parallel to

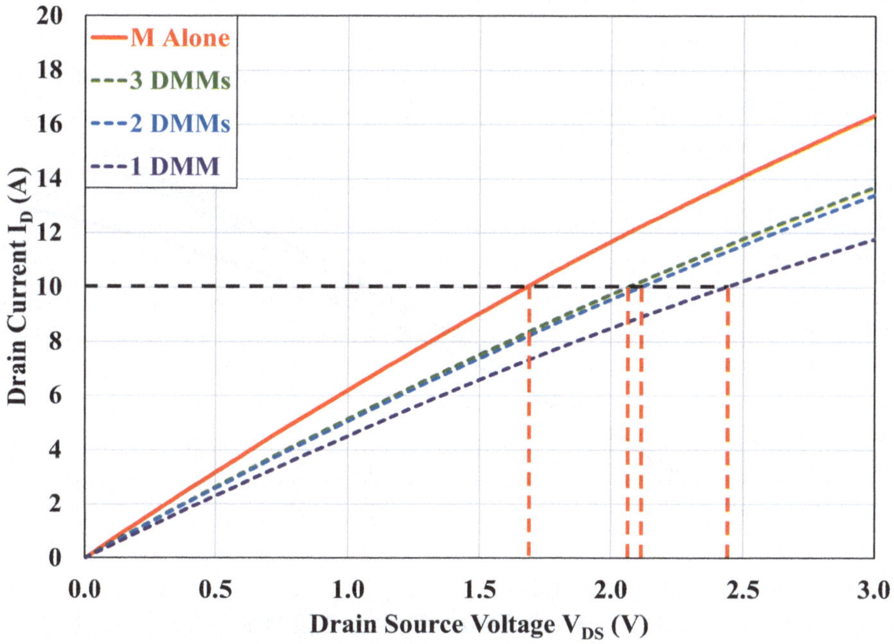

Fig. 14.5 Measured on-state characteristics: stand-alone SiC power MOSFET Wolfspeed C2M0160120 product, and composite transistor CT with various Si DMM cases

the on-resistance of the SiC power MOSFET. The on-state resistance at this operating point is 206 mΩ for Case 1d. This is slightly larger than the addition of the 22 mΩ on-resistance of three DMM devices operating in parallel to the on-resistance of the SiC power MOSFET. This disparity is due to the accuracy of the on-resistance measurement being limited to 3 mΩ due to the individually packaged devices.

The measured on-state characteristics for the composite transistors for various cases of the silicon EMM devices in series with the SiC power MOSFET are shown in Fig. 14.6. The on-state operating point for each case can be determined at a current of 10 A. The on-state on-resistance for the stand-alone SiC power MOSFET is 168 mΩ. This value is close to the typical value of 160 mΩ.

The on-state resistance at this operating point is 195 mΩ for Case 2b with a DC gate bias of 5.4 V applied to the Si EMM device. In order to understand the change in on-resistance, it is necessary to obtain the on-state characteristics of the Si EMM Infineon OptiMOS device IP170N04S4, which is shown in Fig. 14.7 for several DC gate bias values. It can be seen that this device enters its quasi-saturation region at a gate bias of 5.4 V and enters the drain current saturation mode at a gate bias of 5.0 V. The on-resistance of the Si EMM is about 25 mΩ for a DC gate bias of 5.4 V. The measured on-resistance of the composite MOSFET for Case 2b is therefore consistent with adding 25 mΩ to the on-resistance of the stand-alone SiC power MOSFET.

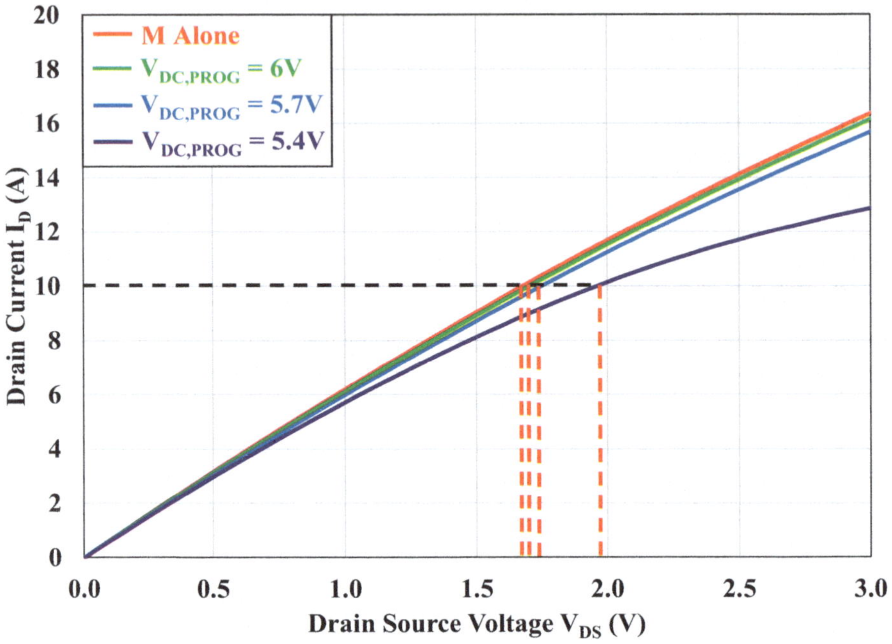

Fig. 14.6 Measured on-state characteristics: stand-alone SiC power MOSFET Wolfspeed C2M0160120 product, and composite transistor CT with various Si EMM cases

The measured on-state resistance at a drain current of 10 A is 175 mΩ for Case 2c with a DC gate bias of 5.7 V applied to the Si EMM device. This is consistent with a very small on-resistance of 7 mΩ of the Si EMM device at a DC gate bias of 5.7 V. Similarly, the measured on-state resistance at this operating point is 170 mΩ for Case 2d with a DC gate bias of 6.0 V applied to the Si EMM device. This is consistent with a very small on-resistance of 5 mΩ of the Si EMM device at a DC gate bias of 6.0 V. Note that the accuracy of the on-resistance measurement was limited to 3 mΩ due to the individually packaged devices.

14.5.2 Impact on Capacitances

The capacitances of the power MOSFET play a role in its switching performance [3]. The measured input, output, and reverse transfer capacitances for the BaSIC(DMM) cases were discussed in Chap. 5. It was found that composite MOSFETs in the BaSIC(DMM) topology have the same capacitance values as the SiC power MOSFET at all drain bias voltages. Similarly, the measured input, output, and reverse transfer capacitances for the BaSIC(EMM) cases were discussed in Chap. 10. It was found that composite MOSFETs in the BaSIC(EMM) topology

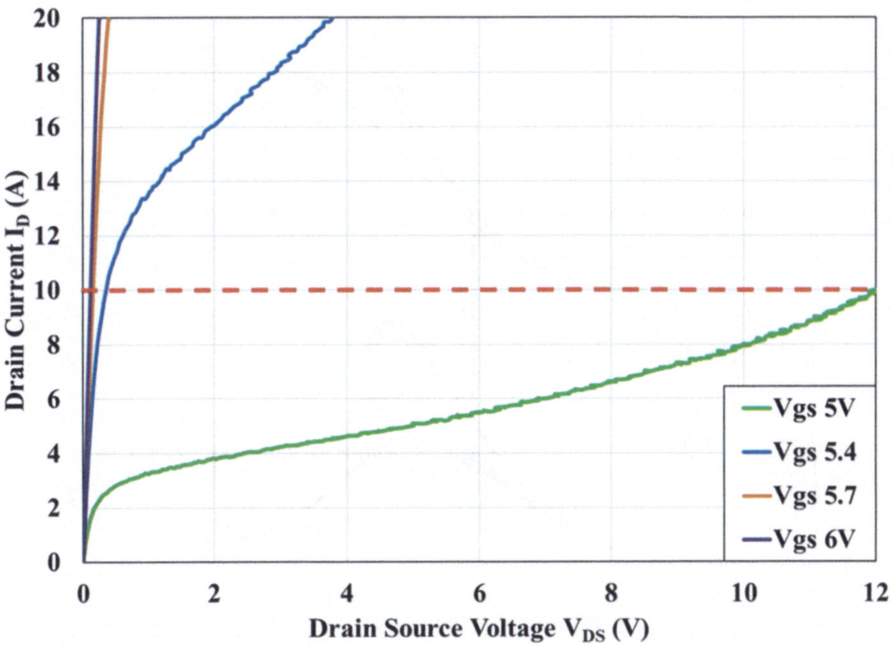

Fig. 14.7 Measured on-state characteristics of the Si EMM Infineon OptiMOS device IP170N04S4 at various DC gate bias voltages

have the same capacitance values as the SiC power MOSFET at all drain bias voltages. This is important for the switching performance of the BaSIC topology.

14.5.3 Impact on Transconductance

The transconductance g_M of the power MOSFET plays an important role in the switching transient for the collector voltage and current during operation with an inductive load [3]. Its value can be extracted from the transfer characteristics of the device. The transfer characteristics for all the composite MOSFETs with silicon DMM device cases were measured at a collector bias of 20 V and are shown in Fig. 14.8 together with the characteristics obtained for the stand-alone SiC power MOSFET C2M0160120 (M_1). The transconductance, extracted by taking a derivative of the collector current with respect to the gate voltage, is also shown in the figure. The value of the transconductance can be compared for the four cases at the same drain current level of 6 A by following the red dashed lines. It is given by the values at the points indicated on the transconductance curve.

The measured transconductance of the SiC power MOSFET M_1 product was 4.1 S. The transconductances of the composite transistors Case 1b, Case 1c, and

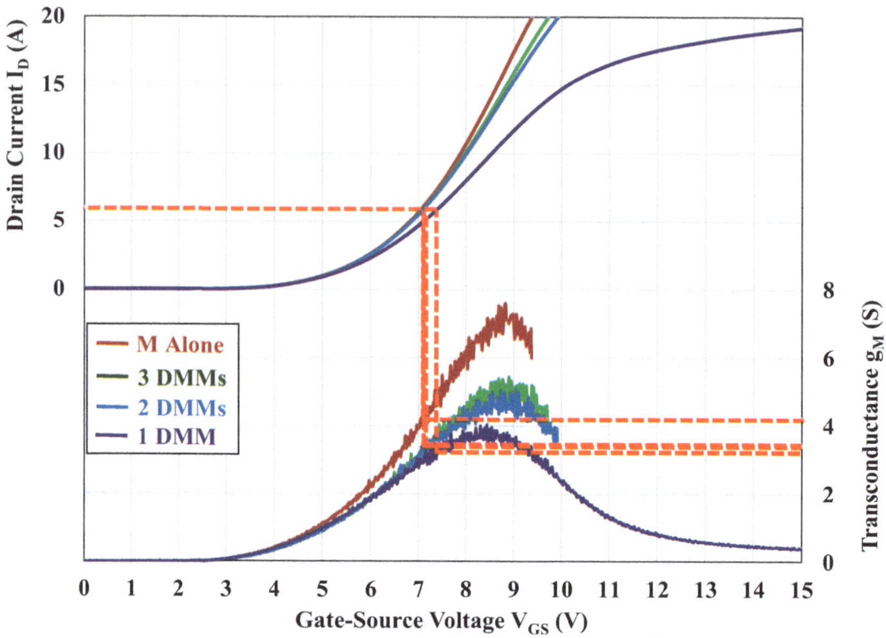

Fig. 14.8 Measured transconductances: stand-alone SiC power MOSFET Wolfspeed C2M0160120 product M_1, and composite MOSFETs CM with various Si DMM cases

Case 1d are 3.1, 3.6, and 3.8 S, respectively. The reduction of the transconductance is consistent with on-state resistances of the Si DMM cases and Eq. (5.3).

The transfer characteristics for all the composite MOSFETs with silicon EMM device cases were measured at a collector bias of 20 V and are shown in Fig. 14.9 together with the characteristics obtained for the stand-alone SiC power MOSFET C2M0160120 (M_2). The transconductance, extracted by taking a derivative of the collector current with respect to the gate voltage, is also shown in the figure. The value of the transconductance can be compared for the four cases at the same drain current level of 10 A by following the red dashed lines. It is given by the values at the points indicated on the transconductance curve.

The measured transconductance of the SiC power MOSFET M_2 product was 3.4 S. This value is much smaller than observed for product M_1. The transconductances of the composite transistors Case 2b, Case 2c, and Case 2d are 3.1, 3.2, and 3.3 S, respectively. The reduction of the transconductance is consistent with on-state resistances of the Si DMM cases and Eq. (5.3).

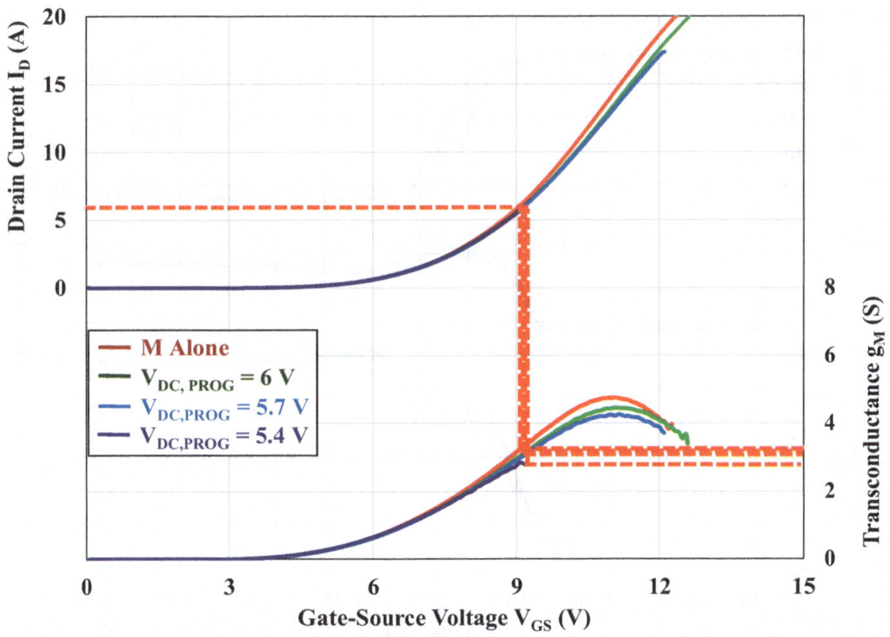

Fig. 14.9 Measured transconductances: stand-alone SiC power MOSFET Wolfspeed C2M0160120 product M_2, and composite transistor CT with various Si EMM cases

14.5.4 Impact on Switching Losses

The double-pulse test, whose circuit diagram was shown in Fig. 5.15, was used to evaluate the switching power losses of all the composite MOSFETs. The double-pulse testing of all the MOSFETs was performed using SiC JBS rectifiers as free-wheeling diodes. The test was performed at a collector current of 10 A and collector supply voltage of 800 V to match the conditions in the stand-alone SiC power MOSFET C2M0160120 datasheet. A gate drive resistance of 10 Ω was used for these tests.

The measured turn-off transients for composite MOSFETs with silicon DMM devices were found to be close to those observed for the stand-alone SiC power MOSFET M_1 and are therefore not shown here. The switching energy losses were determined by integrating the product of the voltage and current over the duration of the transient. The measured turn-off energy loss per cycle ranged from 74 to 78 μJ. Similarly, the measured turn-off transients for composite MOSFETs with silicon EMM devices were found to be close to those observed for the stand-alone SiC power MOSFET M_2. The measured turn-off energy loss per cycle ranged from 63 to 68 μJ.

The measured turn-on transients for the BaSIC(DMM) cases are shown in Fig. 14.10. The drain voltage transients take a shorter time when more silicon DMM

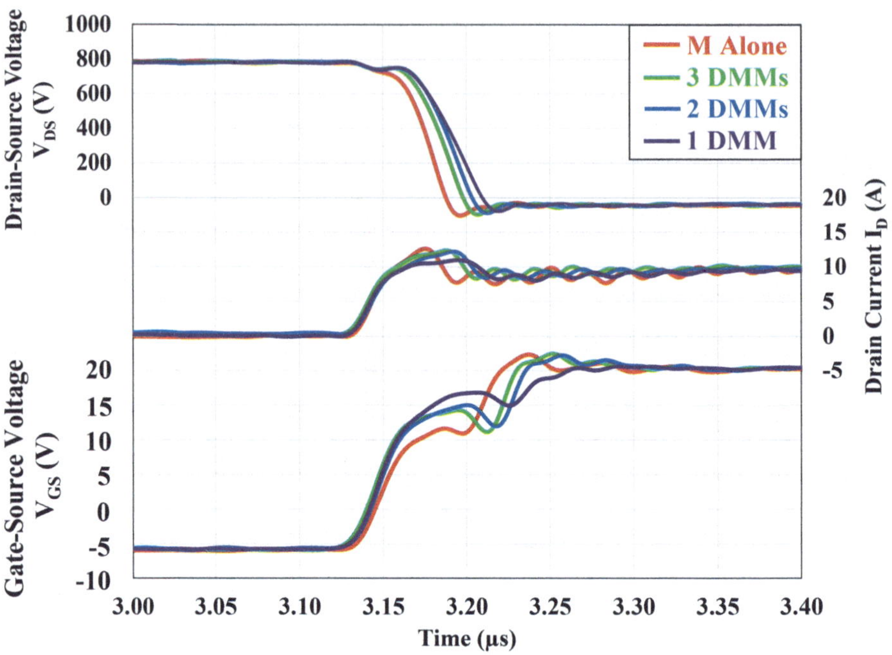

Fig. 14.10 Measured turn-on transients: stand-alone SiC power MOSFET Wolfspeed C2M0160120 product, and composite transistor CT with various Si DMM cases

devices are placed in parallel because of the reduced resistance. The switching energy losses were determined by integrating the product of the voltage and current over the duration of the transient. The measured turn-on switching loss per cycle for the stand-alone SiC power MOSFET M_1 was 247 μJ. The measured turn-on switching losses per cycle for Case 1b, Case 1c, and Case 1d were 358, 346, and 340 μJ, respectively. The turn-on switching losses decrease when the net resistance of silicon DMM is reduced by placing more of them in parallel. This correlates the increase in transconductance for these cases.

The measured turn-on transients for the BaSIC(EMM) cases are shown in Fig. 14.11. The drain voltage transients take a shorter time when a larger DC gate bias is applied to the silicon EMM device because this reduces its on-resistance. The switching energy losses were determined by integrating the product of the voltage and current over the duration of the transient. The measured turn-on switching loss per cycle for the stand-alone SiC power MOSFET M_2 was 314 μJ. This is larger than the value obtained for the stand-alone SiC power MOSFET M_1 due to the lower transconductance of device M_2. The measured turn-on switching losses per cycle for Case 2b, Case 2c, and Case 2d were 374, 368, and 322 μJ, respectively. The turn-on switching losses decrease when the resistance of silicon EMM is reduced by increasing the DC gate bias applied to the device. This correlates the increase in transconductance for these cases.

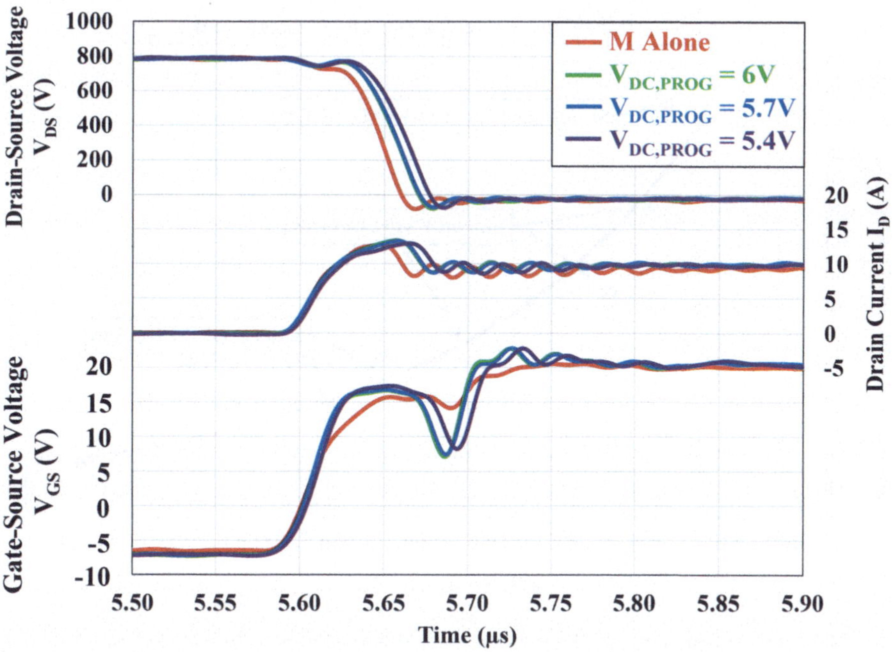

Fig. 14.11 Measured turn-on transients: stand-alone SiC power MOSFET Wolfspeed C2M0160120 product, and composite transistor CT with various Si EMM cases

14.6 Trade-Off Curves

Trade-off curves between short-circuit withstand time and the on-state voltage drop or the switching losses are essential to understand the relative performance of the BaSIC(DMM) and BaSIC(EMM) topologies for improving the short-circuit withstand capability of the SiC power MOSFET C2M0160120 product. It also beneficial to observe the fundamental trade-off curve between the drain saturation current and the on-resistance of the composite MOSFET created by the BaSIC topology.

14.6.1 Fundamental Trade-Off Curve

The information gathered in the measurements described in the previous sections can be combined to generate a fundamental trade-off curve between the drain saturation current and the on-resistance obtained by using the BaSIC topology for the SiC power MOSFET Wolfspeed C2M0160120 product. This plot is shown in Fig. 14.12 with a purple trend line. A single trend line represents both approaches with lower drain saturation currents produced using the Si DMM devices.

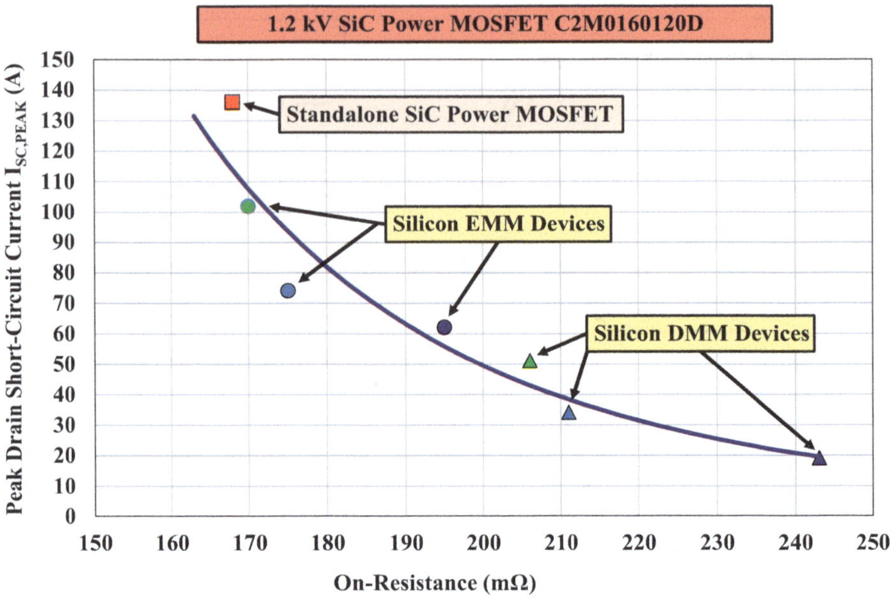

Fig. 14.12 Fundamental trade-off curve between drain saturation current and on-resistance obtained using the BaSIC topology for the SiC power MOSFET Wolfspeed C2M0160120 product

14.6.2 Trade-Off Curve Between Short-Circuit Withstand Time and On-Resistance

The relative performance of the BaSIC(DMM) and BaSIC(EMM) topologies can be best described by creating trade-off plots between the increase in on-resistance and enhancement of the short-circuit withstand time. The short-circuit withstand time data is combined with the measured increase in on-resistance from the previous sections in this chapter to produce the trade-off plot in Fig. 14.13 for the SiC power MOSFET Wolfspeed C2M0160120 product. It can be seen that the data for the BaSIC(DMM) and BaSIC(EMM) devices lie in different zones. The BaSIC(DMM) cases produce much larger short-circuit withstand times with relatively large increase in the on-resistance. In contrast, the BaSIC(EMM) cases produce a smaller enhancement in the short-circuit withstand time with relatively small increase in on-resistance. Both approaches produce the same short-circuit withstand time of 7.5 μs with 17% increase in on-resistance. If the same short-circuit withstand time of 6.5 μs is chosen, the BaSIC(EMM) case achieves the performance with only 4% increase in on-resistance compared with 12% for the BaSIC(DMM) case, as indicated by the dashed lines in the figure.

The trade-off curves shown in Fig. 14.13 were obtained using the particular silicon DMM and EMM devices used for the tests performed in the previous sections of this chapter. Based on the graph shown in this figure, it may be concluded that the BaSIC(EMM) approach cannot deliver a short-circuit withstand time of 10 μs due

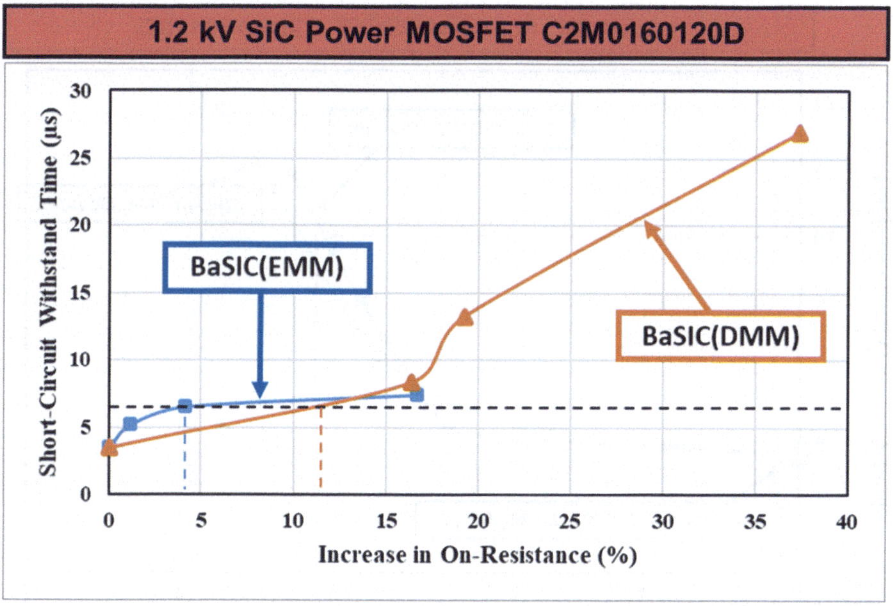

Fig. 14.13 Trade-off curve observed for the SiC power MOSFET Wolfspeed C2M0160120 product for the BaSIC(DMM) and BaSIC(EMM) topologies: on-state resistance

to very sharp increase in the on-resistance of the chosen Si EMM device and that the BaSIC(DMM) approach is the only way to achieve this short-circuit capability.

However, in Chap. 13, a selection methodology for silicon enhancement mode MOSFETs was described which greatly improves the performance achievable with this approach. The silicon IRF6225 device was found to be capable of delivering lower drain saturation current levels while maintaining a low on-resistance. In addition, in Chap. 9, it was shown that the silicon depletion mode MOSFET device can be significantly improved by using a 30 V rated trench-gate structure with thin gate oxide. It is insightful to compare the performance of the BaSIC(DMM) and BaSIC(EMM) topologies using these optimized solutions. The trade-off curves between short-circuit withstand time and increase in the on-resistance with the optimized silicon DMM and EMM devices are shown in Fig. 14.14. It is apparent from this figure that both approaches produce nearly identical trade-off curves after optimization of the silicon MOSFET devices. From the figure, it can be concluded that a short-circuit withstand time of 10 μs is possible for the SiC power MOSFET Wolfspeed C2M0160120 product with either BaSIC topology with only 3% increase in the on-resistance. The BaSIC(DMM) approach has the simplicity of implementation due to shorting the gate and source of the Si DMM device. The BaSIC(EMM) approach has the advantage of allowing programming of the DC gate bias applied to silicon EMM device by the power electronics designer to optimize performance for each application.

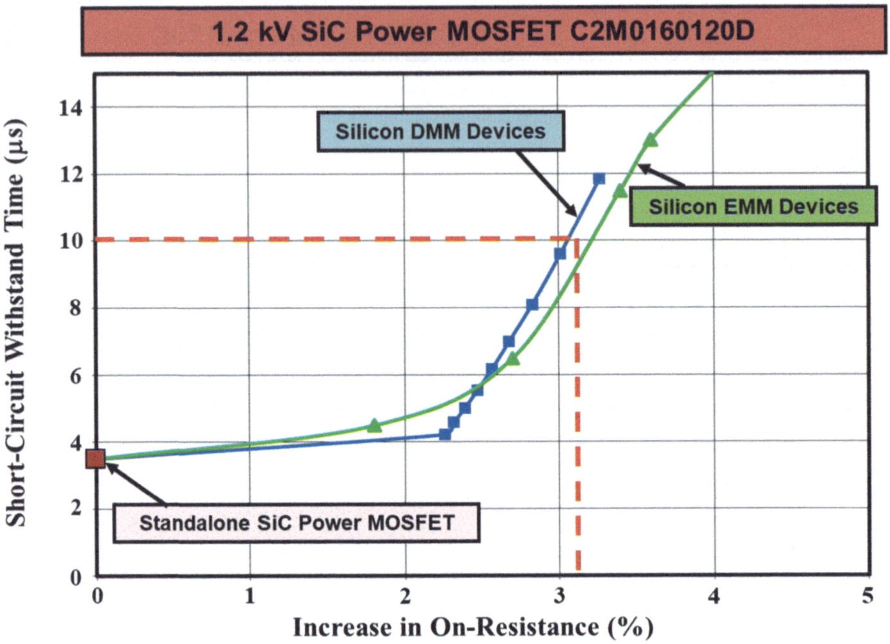

Fig. 14.14 Trade-off curve observed for the SiC power MOSFET Wolfspeed C2M0160120 product for optimized BaSIC(DMM) and BaSIC(EMM) topologies: on-state resistance

14.6.3 Trade-Off Curve Between Short-Circuit Withstand Time and Switching Power Loss

The trade-off curve between the short-circuit withstand time and the normalized total switching energy loss per cycle is shown in Fig. 14.15 for the SiC power MOSFET Wolfspeed C2M0160120 product when using the BaSIC(DMM) and BaSIC(EMM) topologies. A black dotted trend line is shown in the figure to achieve a short-circuit withstand time of 6.5 μs. The total switching energy loss per cycle increases by 13% for the BaSIC(EMM) case while it increases by 20% for the BaSIC(DMM) case.

It is estimated that the switching energy loss per cycle will increase by less than 5% for both approaches if the optimized silicon DMM and EMM devices are used. This is due to the very low on-resistance of the optimized silicon DMM and EMM devices which prevent degradation of the SiC power MOSFET transconductance.

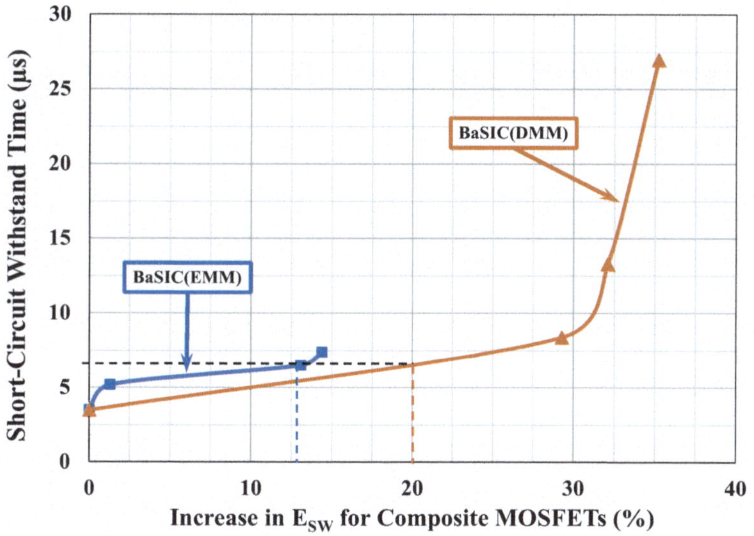

Fig. 14.15 Trade-off curve observed for the SiC power MOSFET Wolfspeed C2M0160120 product for the BaSIC(DMM) and BaSIC(EMM) topologies: switching energy loss

14.7 Summary

The relative benefits of using the BaSIC(DMM) and BaSIC(EMM) topologies for enhancing the short-circuit withstand capability of the SiC power MOSFET Wolfspeed C2M0160120 product have been described in this chapter. In the case of the silicon DMM and EMM devices used for the tests, it was found that the BaSIC(DMM) topology achieved larger increases in short-circuit withstand times than possible with the BaSIC(EMM) topology. However, in the case of optimized silicon DMM and EMM devices, it is found that the performance achieved by both methods is identical. The BaSIC(DMM) topology has the advantage of simplicity of implementation due to the shorting of the gate and source of the silicon DMM device. The BaSIC(EMM) topology has the advantage of allowing the power electronics designer to program the DC gate bias applied to the silicon EMM device to choose the best performance for each application.

References

1. A. Kanale, B.J. Baliga, Comparison of BaSIC(DMM) and BaSIC(EMM) topologies to enhance short-circuit capability in SiC power MOSFETs, in *IEEE Applied Power Electronics Conference* (2021), pp. 1275–1281

2. A. Kanale, K. Han, B.J. Baliga, S. Bhattacharya, Superior short circuit performance of 1.2kV SiC JBSFETs compared to 1.2kV SiC MOSFETs, in *12th European Conference on Silicon Carbide and Related Materials*, Paper WE.P.MO9 (2018)
3. B.J. Baliga, *Fundamentals of Power Semiconductor Devices*, 2nd edn. (Springer Science, New York, 2019)

Chapter 15
Unclamped Inductive Load Avalanche Capability

In the previous chapters, it has been demonstrated that the BaSIC topology enables achieving excellent short-circuit withstand time for commercially available SiC power MOSFETs with minimal increase in on-resistance and switching losses. Power MOSFETs in industrial and automotive applications can undergo avalanche breakdown under unclamped inductive load switching (UIS) conditions [1]. The avalanche capability under UIS has consequently become an important parameter for power MOSFETs [2]. The avalanche ruggedness of commercially available SiC power MOSFETs has therefore been evaluated under switching with an unclamped inductive load in many papers [3–5]. These evaluations have demonstrated that planar-gate SiC power MOSFETs have excellent avalanche withstand capability under UIS conditions. This parameter is now included in their datasheets [6].

It is not obvious whether the UIS capability for a SiC power MOSFET is degraded when the BaSIC topology is implemented. The acceptance of the BaSIC topology in industrial and automotive applications requires demonstration of good UIS capability. The UIS performance of the BaSIC(DMM) topology is analyzed in this chapter and compared to that of a stand-alone SiC power MOSFET product.

15.1 UIS Avalanche Testing Procedure

A schematic diagram of the test setup for evaluating the avalanche capability of a SiC power MOSFET under unclamped inductive load switching conditions is shown in Fig. 15.1. The inductor is first charged by turning-on the SiC power MOSFET. The current rises linearly with time as illustrated in Fig. 15.2 with a dI/dt determined by (V_{DC}/L_{AV}) until it reaches a maximum value of I_{LM} at time t_1. The SiC power MOSFET is turned-off at this time forcing the device into avalanche breakdown with a high current I_{LM} flowing through it.

B. J. Baliga, A. Kanale, *The BaSIC Topology*,
https://doi.org/10.1007/978-3-031-86630-2_15

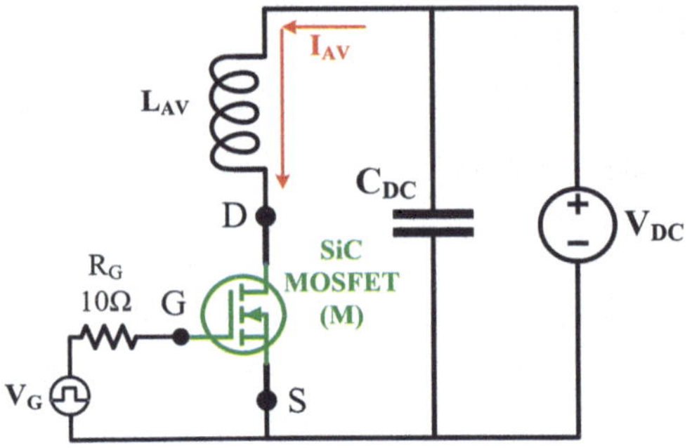

Fig. 15.1 UIS avalanche capability test setup for a SiC power MOSFET (M)

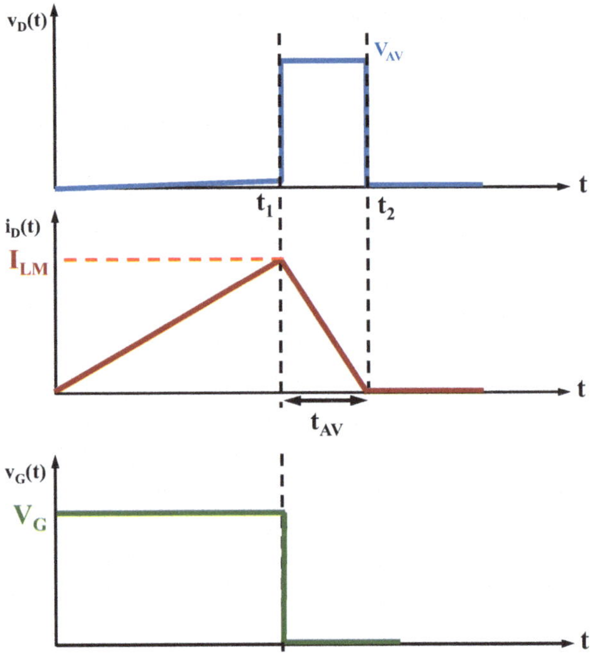

Fig. 15.2 Waveforms for SiC power MOSFET drain voltage and current during the UIS test. The gate voltage waveform is also shown

The avalanche breakdown voltage (V_{AV} in Fig. 15.2) under the UIS conditions is much larger than the rated blocking voltage of the transistor which is typically measured at 100 µA. For example, a 1.2 kV rated SiC power MOSFET has a breakdown voltage of about 1600 V. The UIS breakdown voltage has a much larger value of nearly 2000 V. The current in the inductor is therefore forced to zero at time t_2 due to the potential difference ($V_{AV} - V_{DC}$) across the inductor as shown in Fig. 15.2. The SiC MOSFET undergoes significant heating during the period ($t_{AV} = t_2 - t_1$) and can fail due to triggering of the internal N-P-N bipolar transistor at high temperatures [5]. The maximum energy that the SiC power MOSFET can tolerate before failure is provided in its datasheet as E_{AS} [6].

15.2 BaSIC(DMM) UIS Avalanche Test

The three-terminal composite MOSFET (CM) formed by the combination of the SiC power MOSFET and the Si DMM device in the BaSIC(DMM) topology replaces the stand-alone SiC power MOSFET in the power electronics applications. Its avalanche capability in the unclamped inductive load switching (UIS) case can be tested using the circuit schematic shown in Fig. 15.3.

The inductor is first charged by turning-on the SiC power MOSFET. The current rises linearly with time as illustrated in Fig. 15.2 with a dI/dt determined by (V_{DC}/L_{AV}) until it reaches a maximum value of I_{LM} at time t_1. The SiC power MOSFET is turned-off at this time forcing the device into avalanche breakdown with a high current I_{LM} flowing through it. During the avalanche breakdown time interval ($t_2 - t_1$) shown in Fig. 15.2, it is possible that both the SiC power MOSFET and the Si DMM device within the CM may undergo avalanche breakdown. The implications of this are analyzed in this chapter [7].

The output characteristics of the Si DMM device are illustrated in Fig. 15.4. Since the BaSIC(DMM) topology was created to improve the SC withstand time for the SiC power MOSFET, the Si DMM is chosen with a low on-resistance during the normal on-state current (I_{ON}) flow through the SiC MOSFET and a saturation current (I_{SAT}(DMM)) much smaller than that of the SiC power MOSFET. The operating point of the Si DMM during normal on-state operation of the SiC power MOSFET is shown by the green point in Fig. 15.4. The on-state voltage drop (V_{ON}) of the Si DMM is small at the on-state current level (I_{ON}) due to its low resistance.

During a short-circuit event, the Si DMM operates at the purple point with a voltage drop V_{SC}(DMM) across it, which is in the range of 5–10 V. This voltage reduces the effective gate bias (voltage between the gate and source of the SiC power MOSFET) to ($V_G - V_{SC}$(DMM)) for the SiC power MOSFET in the CM. Here, V_G is the gate drive voltage for the SiC power MOSFET (nominally 20 V). This brings down the saturation current of the SiC power MOSFET until it matches the saturation current of the Si DMM device automatically. The reduced saturation current for

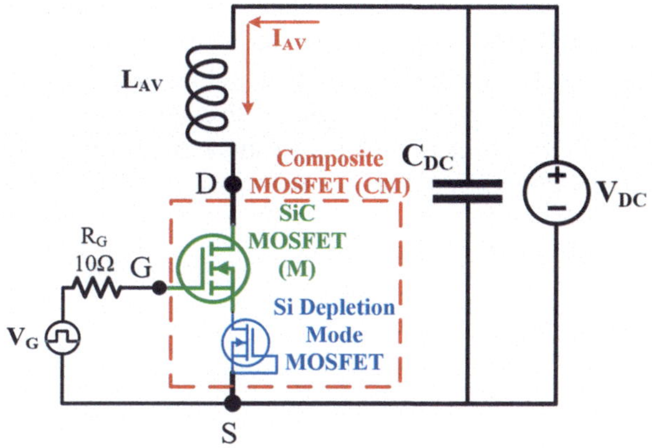

Fig. 15.3 UIS avalanche capability test setup for the composite MOSFET (CM) in the BaSIC(DMM) topology with a SiC power MOSFET (M)

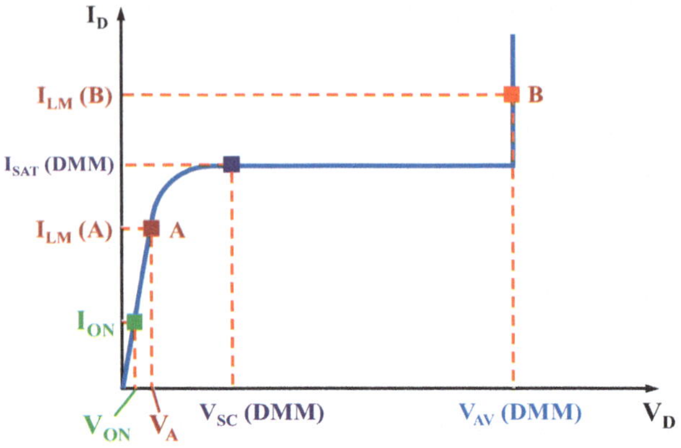

Fig. 15.4 Characteristics of Si depletion mode MOSFET (DMM)

the SiC power MOSFETs extends its short-circuit withstand time as discussed in Chap. 4.

The operating point for Si DMM device during the UIS avalanche condition is quite different from the short-circuit case. For the UIS case, the operating point can lie either at point A or point B on the Si DMM characteristics as shown in Fig. 15.4. On the one hand, if the Si DMM operates at point A in its linear regime with a low maximum current (I_{LM}(A)) during the UIS avalanche conditions, the voltage drop (V_A) across it will be small. It will then not undergo avalanche breakdown and only the SiC power MOSFET will be subjected to the avalanche breakdown stress.

On the other hand, if the Si DMM operates at point B due to a larger maximum current (I_{LM}(B)) during the UIS avalanche conditions that exceeds its saturation current (I_{SAT}(DMM)), the voltage drop across it will be its avalanche breakdown voltage (V_{AV}(DMM)). This will create a stress on the Si DMM that could lead to its failure. In addition, the avalanche breakdown voltage (V_{AV}(DMM)) typically exceeds 30 V, which will produce a negative voltage exceeding 30 V across the gate and source of the SiC power MOSFET because its gate bias is zero during the avalanche period as shown in Fig. 15.2. This could lead to rupture of the gate oxide and catastrophic failure of the SiC power MOSFET.

It is therefore important to determine the operating point for the Si DMM in the BaSIC(DMM) topology during the avalanche event in the UIS case. This will determine whether the BaSIC(DMM) topology allows the composite MOSFET to exhibit the desirable good avalanche withstand capability associated with SiC power MOSFETs. Experimental investigation of the avalanche capability for the BaSIC(DMM) topology is described in the next section with a typical SiC power MOSFET.

15.3 UIS Test Experimental Results

Experimental investigation of the BaSIC(DMM) topology during an avalanche breakdown event for the UIS case was performed using a commercially available 1.2 kV 4H-SiC planar-gate power MOSFET C2M0280120D with on-resistance of 280 mΩ [6]. The results obtained by extensive measurements of the BaSIC(DMM) topology with this device and a commercially available Si DMM IXTH16N10D2 [8] have already been published in several papers [9–11]. It has been reported in these papers that using this Si DMM device with the 280 mΩ SiC power MOSFET produces an increase in R_{on} by 17% and E_{SW} by 20%. These papers have also demonstrated that the voltage at the drain terminal of the Si DMM can be used to monitor or sense the current flowing through the SiC power MOSFET. These results were discussed in the previous book chapters.

The output characteristics of the IXTH16N10D2 Si DMM device are reproduced in Fig. 15.5 for reference. The characteristics marked with $V_{GS} = 0$ V is of interest here because the gate and source of the device are shorted in the BaSIC(DMM) topology. At current levels below 15 A, the device operates in its linear region with a maximum resistance of 64 mΩ according to its datasheet. At larger drain bias voltages, the current saturates at about 29 A (I_{SAT}(DMM) in Fig. 15.5) at a drain bias of 50 V as indicated by the purple line. However, the Si(DMM) operates at a drain bias of only 6 V during the short-circuit event [9] because this is sufficient to bring down

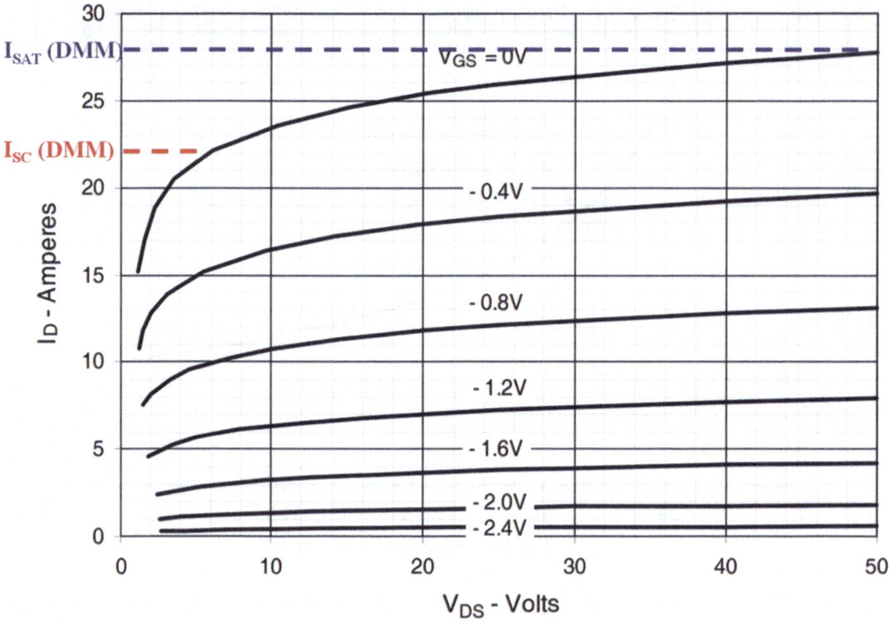

Fig. 15.5 Output characteristics of the IXTH16N10D2 Si depletion mode MOSFET (DMM)

the saturation current of the SiC power MOSFET to about 23 A while supporting 800 V. This point of operation is shown by the red line in Fig. 15.5.

The Si(DMM) device has a rated voltage of 100 V on the datasheet. Experimental measurements showed an actual breakdown voltage V_{AV}(DMM) exceeding 150 V for the device. Based on these Si DMM characteristics, it can be concluded that the BaSIC(DMM) topology will operate in a satisfactory manner if the current remains below 20 A during the UIS conditions but will undergo catastrophic failure of the SiC power MOSFET if the current exceeds 30 A.

The datasheet for the 1.2 kV SiC power MOSFET C2M0280120D includes a single pulse avalanche energy (E_{AS}) of 280 mJ obtained using a drain current of 6 A. It also provides a single pulse avalanche safe operating area (SOA) curve for the product that is reproduced in Fig. 15.6 for discussion here. It is worth pointing out that the SOA curve shown in the datasheet corresponds to an avalanche energy (E_{AV}) of 250 mJ which is smaller than the value of 280 mJ provided in the datasheet. It can be seen from the datasheet graph that the recommended avalanche currents for the SiC power MOSFET to avoid damaging the device are less than 10 A. The avalanche breakdown tests under UIS conditions were therefore performed using a maximum drain current (I_{LM} in Fig. 15.2) of 6 A to be consistent with the manufacturers datasheet.

The output characteristics of the stand-alone SiC power MOSFET and of the composite MOSFET in the BaSIC(DMM) topology were measured as a function of

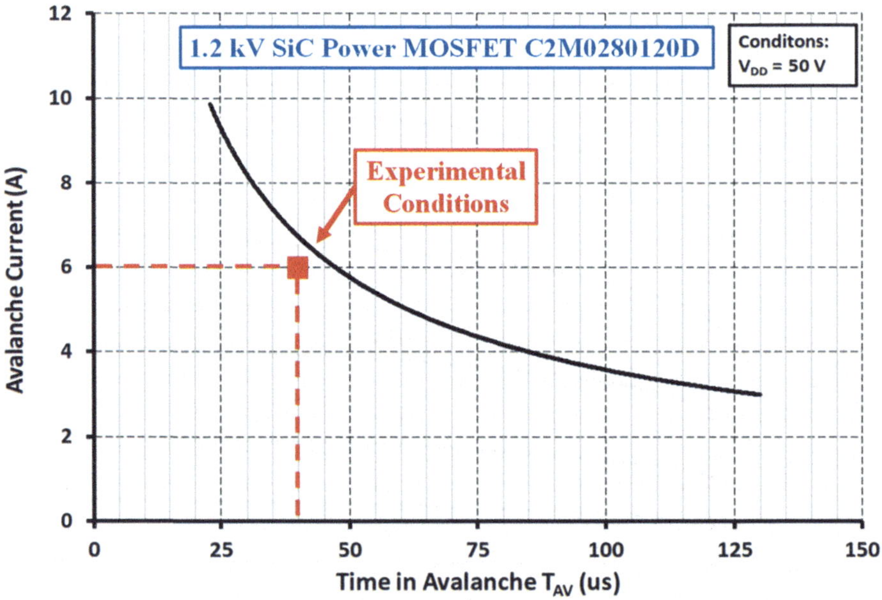

Fig. 15.6 Single avalanche safe operating area curve for 1.2 kV SiC power MOSFET C2M0280120D given in its datasheet. The chosen UIS test condition is shown by the red point on the graph

gate bias. These output characteristics can be compared in Fig. 15.7. It can be concluded from these measured characteristics that the composite MOSFET in the BaSIC(DMM) topology can be treated like three-terminal MOSFET device. The measured R_{on} for the composite MOSFET at a gate bias of 20 V was 413 mΩ compared with 360 mΩ for the stand-alone SiC power MOSFET. The observed increase in value of 53 mΩ is due to the on-resistance contributed by the Si DMM device to

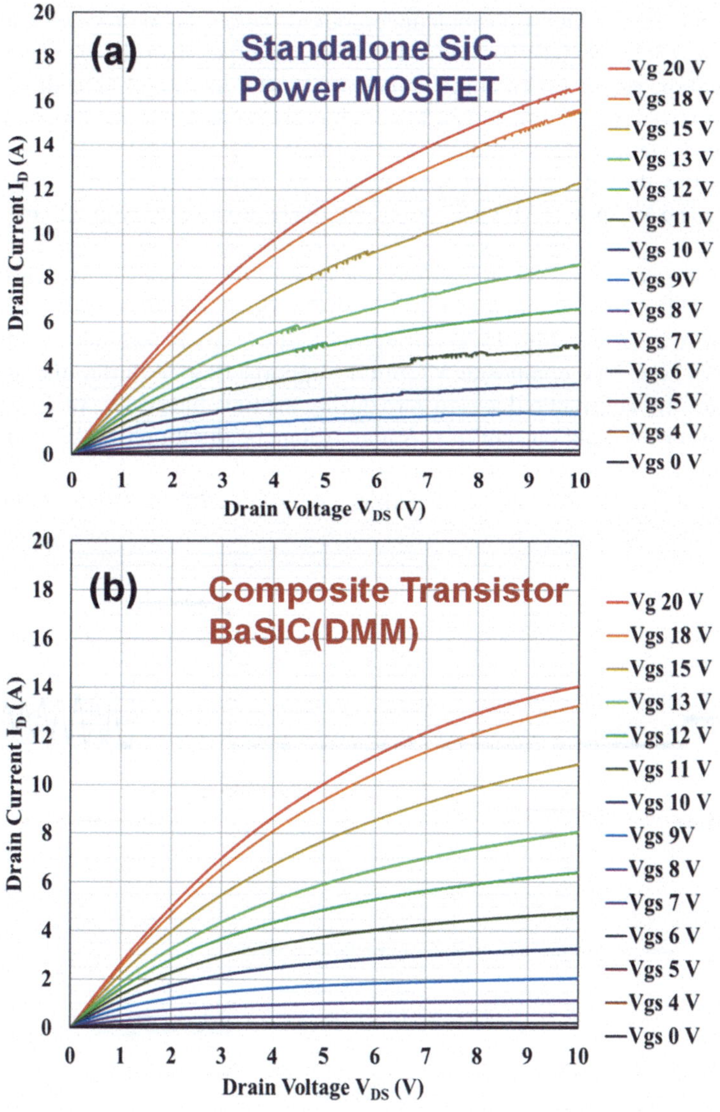

Fig. 15.7 Measured output characteristics of (**a**) the stand-alone SiC power MOSFET and (**b**) the composite transistor in the BaSIC(DMM) topology

the composite transistor. The drain saturation current for the composite MOSFET is smaller than that observed for the SiC power MOSFET due to impact of the voltage drop across the Si DMM device at larger current levels.

The avalanche breakdown tests under UIS conditions were performed using an inductor (L_{AV} in Fig. 15.1) of 10.6 mH and a DC supply voltage (V_{DC} in Fig. 15.1) of 350 V. The experimental waveforms recorded for the case of the stand-alone SiC power MOSFET are shown in Fig. 15.8. The current was allowed to increase to 6 A (I_{LM} in Fig. 15.2) during the first 180 μs at which point the gate bias was switched to zero. The SiC power MOSFET then undergoes avalanche breakdown with a voltage (V_{AV} in Fig. 15.2) of 1830 V. This value is much larger than the rated voltage of 1200 V and the measured blocking voltage of 1650 V at 100 μA. The drain current reduces to zero during a time interval of 39.4 μs. This combination of a drain current of 6 A and avalanche duration of 39.4 μs is shown in Fig. 15.6 by the red square. It lies just below the safe operating area curve as recommended in the manufacturer's datasheet. The SiC power MOSFET was able to absorb an energy of 220 mJ without failure.

The experimental results obtained for the case of the composite MOSFET in the BaSIC(DMM) topology are shown in Fig. 15.9. The current was again allowed to increase to 6 A (I_{LM} in Fig. 15.2) during the first 180 μs before the gate bias was switched to zero. The composite MOSFET supports 1830 V during the avalanche period. The drain current reduces to zero during the time interval t_{AV} of 39.2 μs. This combination of a drain current of 6 A and avalanche duration of 39.2 μs is essentially identical to that observed for the stand-alone SiC power MOSFET. The composite MOSFET was also able to absorb an energy of 220 mJ without failure. This

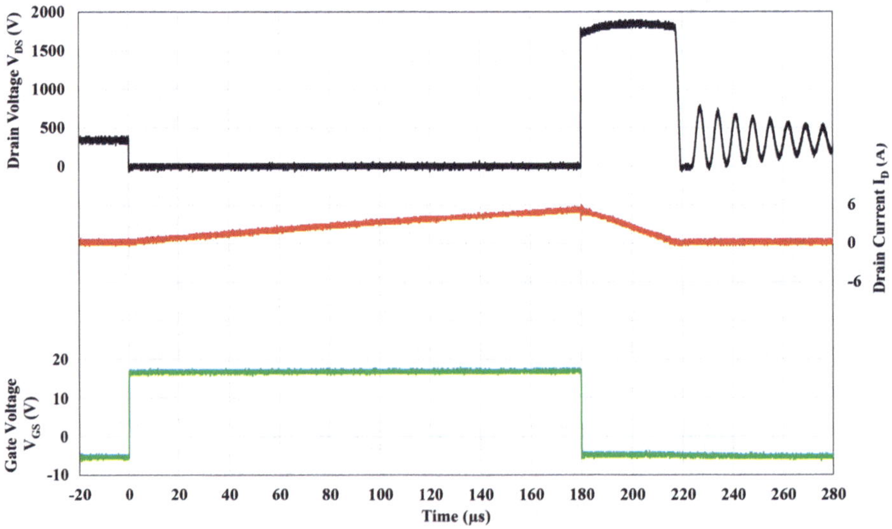

Fig. 15.8 Measured avalanche characteristics of the stand-alone SiC power MOSFET under UIS conditions with an inductor of 10.6 mH

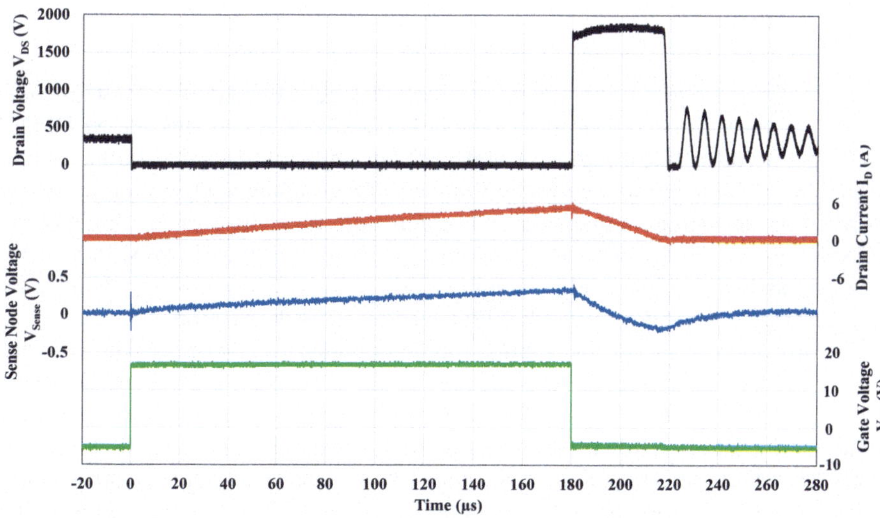

Fig. 15.9 Measured avalanche characteristics of the composite transistor in the BaSIC(DMM) topology under UIS conditions with an inductor of 10.6 mH

test demonstrates that the BaSIC(DMM) topology does not compromise the avalanche capability of the SiC power MOSFET under UIS conditions.

In the previous publications on the BaSIC(DMM) topology [9–11], it has been shown that that the drain terminal of the Si DMM device can be used as a sense node whose voltage (V_{Sense}) is proportional to the current flowing through the SiC power MOSFET. It can also be used to sense the occurrence of the short-circuit event with a voltage rise to about 5 V.

The sense node voltage (V_{Sense}) was monitored during the avalanche test and is shown in blue in Fig. 15.9. It can be seen that the voltage V_{Sense} increases proportional to the drain current magnitude. It reaches a value of 0.32 V at the maximum drain current (I_{LM}) of 6 A. This corresponds to a resistance of 53 mΩ, which is identical to the resistance of the Si DMM extracted using the output characteristics in Fig. 15.7. These values show that the Si DMM remains in the linear region of operation (operating point A in Fig. 15.4) during the avalanche tests and that it does not undergo avalanche breakdown. The experimental results conclusively demonstrate that the small voltage of 0.32 V developed across the Si DMM will not stress the gate oxide of the SiC power MOSFET avoiding its failure. Consequently, the BaSIC(DMM) topology does not compromise the ruggedness of the SiC power MOSFET.

15.4 Summary

Analysis of the BaSIC(DMM) topology under avalanche breakdown with UIS conditions has been discussed in this chapter. The analysis shows that the Si DMM could operate in its linear region or under avalanche breakdown depending on the magnitude of the maximum current during the UIS conditions. Operation of the Si DMM in its linear region does not compromise the ruggedness of the SiC power MOSFET inside the BaSIC(DMM) topology. Operation of the Si DMM in its avalanche breakdown regime could result in its failure as well as producing damage to the gate oxide of the SiC power MOSFET leading to its destruction.

The operation of the BaSIC(DMM) topology under avalanche breakdown with UIS conditions for a typical practical implementation using a 280 mΩ 1.2 kV SiC power MOSFET product with a 100 V Si DMM product was performed. It was found that the BaSIC(DMM) topology functions identically to the stand-alone SiC power MOSFET and does not compromise its avalanche ruggedness. This is an important outcome that allows the use of the BaSIC(DMM) topology to achieve the desired enhanced short-circuit withstand capability in industrial and electric vehicle motor drives with SiC power MOSFETs.

References

1. D. Schleisser, D. Ahlers, M. Eicher, M. Purschel, Repetitive avalanche of automotive MOSFETs, in *Proc. 15th Eur. Conf. Power Electron. Appl.* (2013), pp. 1–7
2. B.J. Baliga, *Fundamentals of Power Semiconductor Devices*, 2nd edn. (Springer-Science, 2019)
3. L. Yang, A. Fayyaz, A. Castellazzi, Characterization of high-voltage SiC MOSFETs under UIS avalanche stress, in *IET Int. Conf. Power Electronics, Machines, and Drives* (2014), pp. P1–P5
4. M.D. Kelley, B.N. Pushpakaran, S.B. Bayne, Single-pulse avalanche mode robustness of commercial 1200 V/80 mΩ SiC MOSFETs. IEEE Trans. Power Electron. **32**, 6405–6415 (2017)
5. N. Ren, K.L. Wang, J. Wu, H. Xu, K. Sheng, Failure mechanism analysis of SiC MOSFETs in unclamped inductive switching conditions, in *IEEE Int. Symp. Power Semiconductor Devices and ICs* (2019), pp. 183–186
6. 1.2 kV SiC Power MOSFET Datasheet: Wolfspeed product page. https://www.wolfspeed.com/c2m0280120d
7. A. Kanale, B.J. Baliga, S. Bhattacharya, Unclamped inductive load (UIS) avalanche capability of the BaSIC(DMM) topology, in *IEEE Power Electronics and Drive Systems* (2023)
8. IXYS Power Semiconductor product page. https://www.littelfuse.com/products/power-semiconductors/ixys-power-semiconductors-and-ics.aspx
9. A. Kanale, B.J. Baliga, Enhancing short circuit capability of 1.2 kV SiC power MOSFETs using a gate-source-shorted Si depletion-mode MOSFET in series with the source, in *IEEE 13th Int. Conf. Power Electronics and Drive Systems* (2019), pp. 1–6
10. A. Kanale, B.J. Baliga, Comparison of current suppression methods to enhance short circuit capability of 1.2 kV SiC power MOSFETs: A new approach using a series-connected, gate-source-shorted Si depletion-mode MOSFET vs use of a series resistance, in *IEEE Workshop on Wide Bandgap Power Devices and Applications* (2019), pp. 53–58

11. A. Kanale, B.J. Baliga, Comparison of current suppression methods to enhance short circuit capability of 1.2 kV SiC power MOSFETs: A new approach using a series-connected, gate-source-shorted Si depletion-mode MOSFET vs reduced gate bias operation, in *IEEE Industrial Electronics Conference* (2019), pp. 5114–5119

Chapter 16
Repetitive Short-Circuit Protection

The protection of SiC power MOSFETs during a short-circuit event requires turning off the gate drive voltage before the device heats up sufficiently to fail. The short-circuit withstand time for a typical 1.2 kV SiC power MOSFET has been found to be about 3.5–4.0 μs. This time interval is too short for employment of widely available IGBT gate drives designed to turn-off devices within 10 μs. It has been proposed that the response time of gate drive circuits can be enhanced to solve this problem. This may be feasible for single short-circuit events. However, it is demonstrated in this chapter that the SiC power MOSFET can still fail after repetitive short-circuit events and that this problem can be solved by using the BaSIC topology [1]. It is the only approach that can address this failure mode for SiC power MOSFET at this time.

16.1 Repetitive Short-Circuit Events

SiC power MOSFETs used in industrial and automotive applications may experience multiple short-circuit events during their lifetime if the gate drive circuit is designed to protect them from failure during a single short-circuit event. However, SiC MOSFETs have been found to degrade when subjected to repetitive, non-destructive short-circuit events with short-circuit duration of less than the failure time [2–4]. It has been observed that the threshold voltage increases after multiple short-circuit events. This creates a reliability issue for SiC power MOSFETs during use in industrial motor drives and in electric vehicles.

The junction temperature of SiC power MOSFETs increases rapidly to high values during the short-circuit event due to the large power dissipation caused by the high saturation current and DC supply voltage [5]. The temperature at the upper surface of the SiC power MOSFET can then exceed the melting point of the aluminum metallization used for the top contacts [6]. When the aluminum contact melts,

B. J. Baliga, A. Kanale, *The BaSIC Topology*,
https://doi.org/10.1007/978-3-031-86630-2_16

it can penetrate the upper junction structure of the SiC MOSFET leading to loss of ability to support the DC bias voltage. However, it has been observed that destructive failure may not occur even after melting of the aluminum if the device is turned-off soon after the aluminum reaches its melting point. In addition, it has been observed that cracks can form in the silicon dioxide inter-metal dielectric layer used between the source metal and the gate electrode due to mechanical stress [7]. Penetration of the aluminum through the crack occurs producing an increase in gate leakage current and eventually a gate-source short circuit under repetitive short-circuit stress. The penetration of the aluminum progresses after each successive short-circuit event eventually leading to failure of the SiC power MOSFET when subjected to repetitive short-circuit events.

It has been shown in the previous chapters that the BaSIC topology enables achieving excellent short-circuit withstand time for commercially available SiC power MOSFETs with minimal increase in on-resistance and switching losses. A feature of the BaSIC topology is the significantly reduced current level during the short-circuit event. This leads to a much smaller power dissipation during the short-circuit event. One benefit of the smaller power dissipation is the increase in short-circuit withstand time as described in the previous chapters. Another benefit of the smaller power dissipation during the short-circuit event is a reduction of the junction temperature at the point when the gate drive voltage is turned-off to protect the SiC power MOSFET. This can prevent the melting of the aluminum metallization and consequently suppress the failure of the SiC power MOSFET after multiple short-circuit events.

16.2 Repetitive Short-Circuit Testing Procedure

The test setup, shown in Fig. 16.1, for evaluating the repetitive short-circuit failure of a SiC power MOSFET is the same as that used for evaluating single event failures. In repetitive short-circuit evaluation, the gate pulse on-duration is kept smaller than the single event short-circuit failure time for the SiC power MOSFET. The short-circuit events are repeatedly applied until failure of the device is detected. All the basic electrical characteristics of the SiC power MOSFET are monitored after each short-circuit event to observe any degradation of the device.

The Wolfspeed 1.2 kV SiC power MOSFETs C2M0280120D [8] were chosen as the SiC device labeled M in Fig. 16.1a. The composite MOSFET in Fig. 16.1b of the BaSIC topology was formed using this SiC power MOSFET together with a IXYS 100 V Si depletion mode MOSFETs IXTH16N10D2 in series to limit the short-circuit current level [9].

The short-circuit waveform for a single event for the stand-alone SiC power MOSFET (M) can be compared with that for the composite MOSFET (CM-1) in Fig. 16.2 for the case of a DC supply voltage of 800 V. The same gate bias of 20 V was used in both cases as shown in Fig. 16.1. The short-circuit time duration was chosen as 3.5 μs since improved gate drive circuits have been proposed to prevent

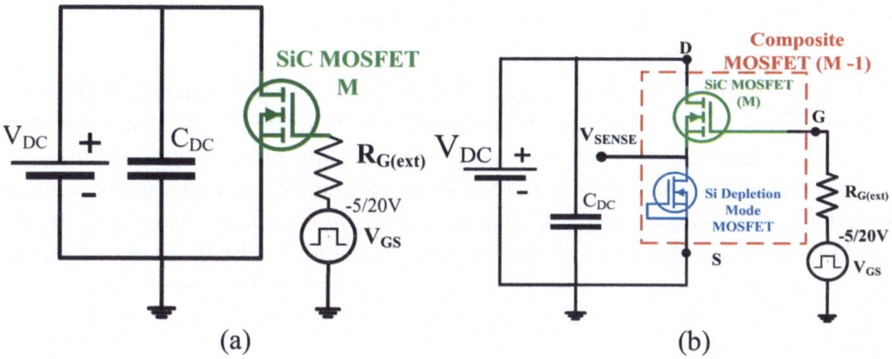

Fig. 16.1 Repetitive short-circuit test setup for: (**a**) the stand-alone SiC power MOSFET (M); (**b**) the BaSIC topology **c**omposite MOSFET containing the SiC power MOSFET

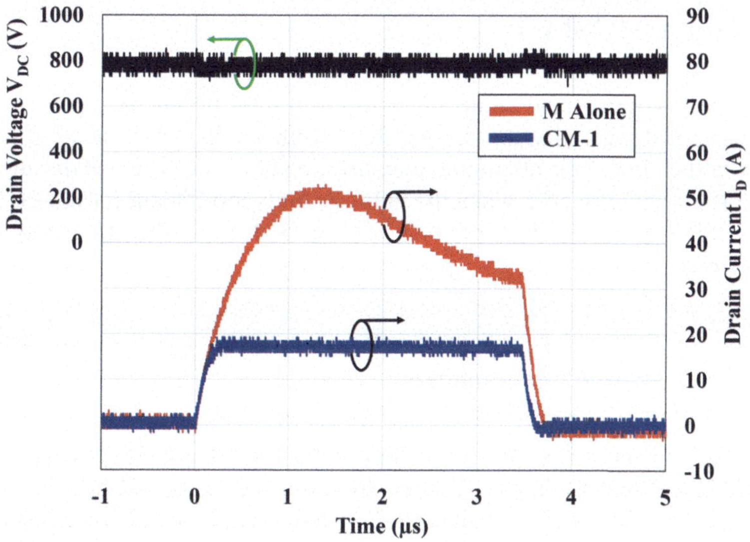

Fig. 16.2 Short-circuit waveforms observed for the stand-alone SiC power MOSFET (red) and the composite MOSFET (CM) in the BaSIC(DMM) topology with the SiC power MOSFET (blue)

failure of SiC power MOSFETs within this time. The current for the stand-alone SiC power MOSFET (M) increases to a peak value of 53 A and then reduces to 32 A at the end of the gate pulse. In contrast, the current for the composite MOSFET (CM-1) remains steady at a value of only 19 A. This clearly demonstrates the improved conditions achieved using the BaSIC topology.

16.3 Repetitive Short-Circuit Experiments

Multiple short-circuit pulses were applied to the devices with an interval of approximately 1 s between the pulses to allow the SiC power MOSFETs to cool down. The stress test sequences were interrupted after completing different sets of pulses to perform static characterization of the SiC power MOSFET and look for any signs of degradation. The device characterization tests were conducted using a Keysight B1505A curve tracer to measure the on-resistance, threshold voltage, gate leakage current, and body diode forward characteristics. The repetitive short-circuit pulse sequences were continued until the SiC power MOSFET showed a drastic change in the electrical characteristics.

16.4 Experimental Results: Stand-Alone SiC Power MOSFET

The degradation of the stand-alone SiC power MOSFET during the repetitive short-circuit pulses is discussed in this section. It is worth noting that evidence of melting and re-solidification of the aluminum metallization has been observed during single event short-circuit tests even when the pulse width is insufficient for device failure [6]. The aluminum metal texture changes from a shiny surface to a gray-brown rough texture.

16.4.1 On-State Characteristics

The on-state *i–v* curves measured for the stand-alone SiC MOSFET (M) at a gate bias of 20 V are shown in Fig. 16.3 after stresses of 100, 200, and 210 short-circuit pulses together with initial characteristics. No change in the on-resistance is observed with repetitive short-circuit stress of up to 200 pulses. The stand-alone SiC MOSFET exhibited a very high on-resistance (seen as a flat purple line with 0 A) after 210 short-circuit pulses, indicating device failure.

16.4.2 Transfer Characteristics

The measured transfer characteristics for the stand-alone SiC power MOSFET (M), at a drain bias of 10 V are shown in Fig. 16.4 at various stages of the repetitive SC pulse sequence. A small change in the transfer characteristics is observed after repetitive short-circuit stress of up to 200 pulses. There is a shift in the threshold voltage after the short-circuit stress as also observed in previous papers [2]. However,

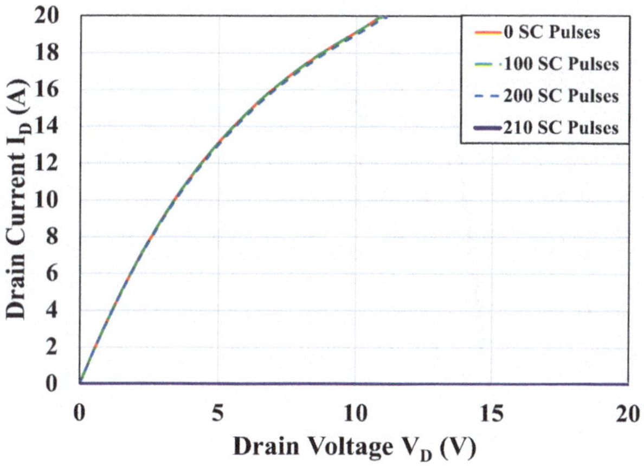

Fig. 16.3 Measured on-state characteristics of the stand-alone SiC power MOSFET during the repetitive short-circuit pulse tests

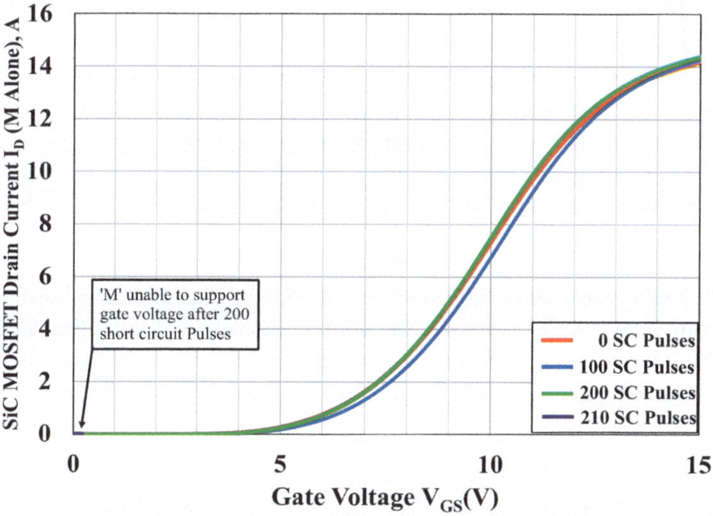

Fig. 16.4 Measured transfer characteristics of the stand-alone SiC power MOSFET during the repetitive short-circuit pulse tests

the device showed no current flow after 200 short-circuit pulses indicating failure of the gate structure.

16.4.3 Gate Characteristics

The measured gate leakage current I_{GSS} for the stand-alone SiC MOSFET (M), with zero drain bias, is shown in Fig. 16.5. Very low leakage currents of less than 0.2 nA are observed until 200 short-circuit pulses. The leakage current is observed to

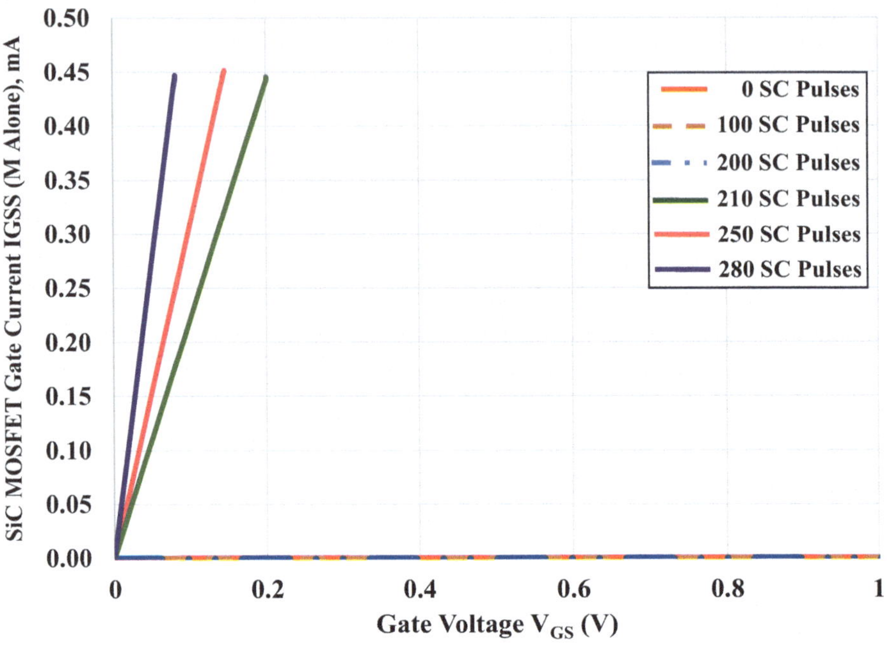

Fig. 16.5 Measured gate leakage characteristics of the stand-alone SiC power MOSFET during the repetitive short-circuit pulse tests

increase to the mA level after 200 short-circuit pulses. This indicates formation of a current path between the gate and the source electrodes due to the short-circuit stress, which is consistent with the discussion in Sect. 16.1 with the aluminum gate metal melting and penetrating through the inter-metal dielectric layer.

16.4.4 Third-Quadrant Body-Diode Characteristics

The measured third-quadrant body diode forward characteristics at different stages of the repetitive short-circuit pulse sequence are shown in Fig. 16.6. The body diode forward voltage drop is observed to increase when the stand-alone MOSFET (M) is subjected to only 50 repetitive short-circuit pulses. It becomes worse when the number of repetitive short-circuit pulses is increased to 200. This is due to a degradation of the contact between the source metal and the P^+ plug used to contact the P^+ shielding region. This contact is made using a nickel silicide layer which is dissolved by the melted aluminum source contact metal during the repetitive short-circuit events.

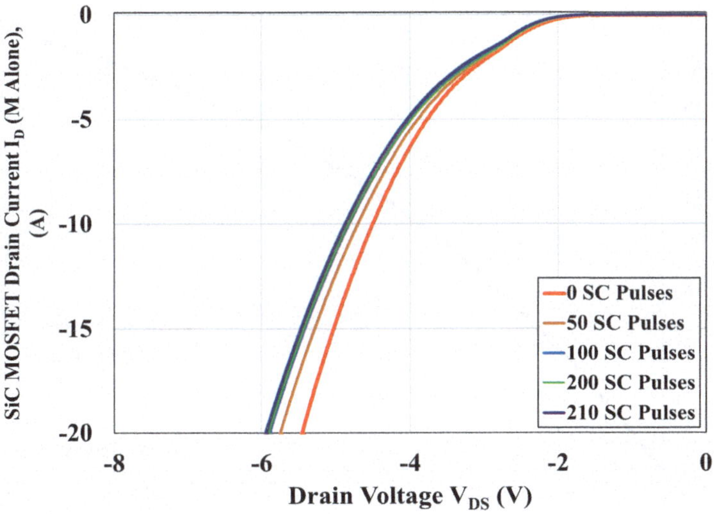

Fig. 16.6 Measured body-diode characteristics of the stand-alone SiC power MOSFET during the repetitive short-circuit pulse tests

16.5 Experimental Results: Composite MOSFET

The repetitive short-circuit pulse test results obtained with the composite MOSFET formed by the BaSIC(DMM) topology are discussed in this section. Here, the internal SiC power MOSFET characteristics were monitored during the repetitive short-circuit tests. No degradation of the internal Si depletion mode MOSFETs was observed as verified in a subsequent section.

16.5.1 On-State Characteristics

The measured on-state characteristics for the SiC power MOSFET within the composite MOSFET (CM-1) with a gate bias of 20 V are shown in Fig. 16.7 at different stages of the repetitive short-circuit pulses. These tests were performed up to 3000 pulses. In this case, the on-state characteristics of the SiC power MOSFET do not change even after being subjected to 3000 short-circuit pulses of 3.5 μs duration. This indicates that the lower short-circuit current level in the BaSIC topology is sufficiently reducing the power dissipation to prevent melting of the aluminum metallization up to a pulse width of 3.5 μs.

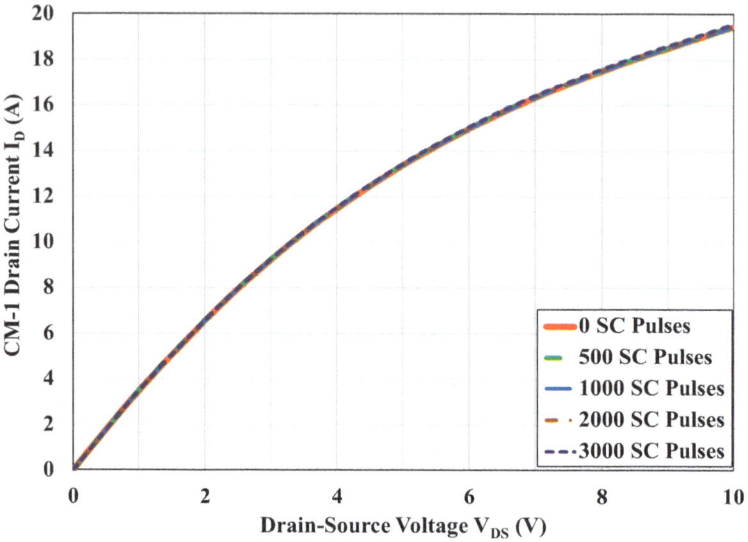

Fig. 16.7 Measured on-state characteristics of the SiC power MOSFET within the composite MOSFET during the repetitive short-circuit pulse tests

16.5.2 Transfer Characteristics

The measured transfer characteristics for SiC power MOSFET within the composite MOSFET (CM-1) are shown in Fig. 16.8 during the repetitive short-circuit stress tests. The overlapping plots demonstrate that no degradation of the SiC power MOSFET is occurring even after 3000 short-circuit pulses. The transfer characteristics could be measured up to a gate bias of 20 V indicating integrity of the gate structure. There is no evidence of a shift in the threshold voltage or degradation of the channel mobility due to reduced short-circuit stress in the case of the BaSIC topology.

16.5.3 Gate Characteristics

The measured leakage currents for the SiC power MOSFET in the composite MOSFET CM-1 at different stages of the repetitive short-circuit pulse sequence are shown in Fig. 16.9. The gate leakage current remains below 0.2 nA, which is the measurement noise limit, even after 3000 short-circuit pulses. No current path forms between the gate and source region of the SiC power MOSFET in this case even after 3000 short-circuit pulses. This provides compelling evidence that the BaSIC(DMM) topology prevents damage to the gate structure under repetitive short-circuit stress because of the reduced power dissipation.

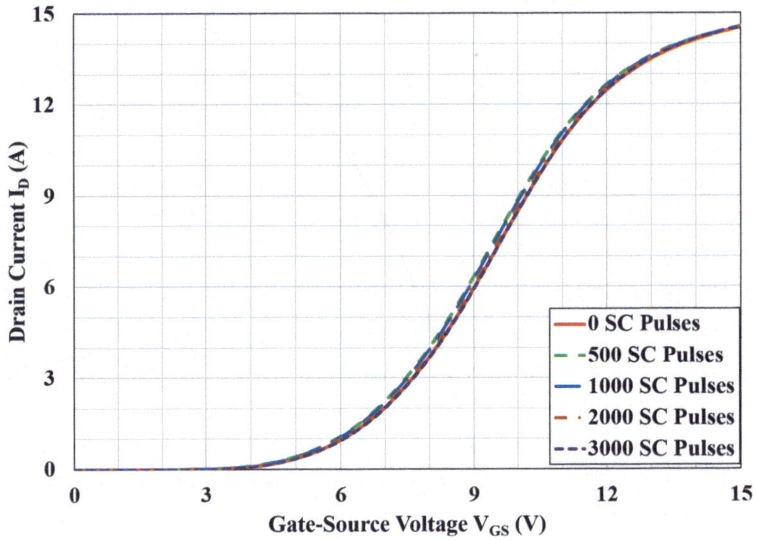

Fig. 16.8 Measured transfer characteristics of the SiC power MOSFET within the composite MOSFET during the repetitive short-circuit pulse tests

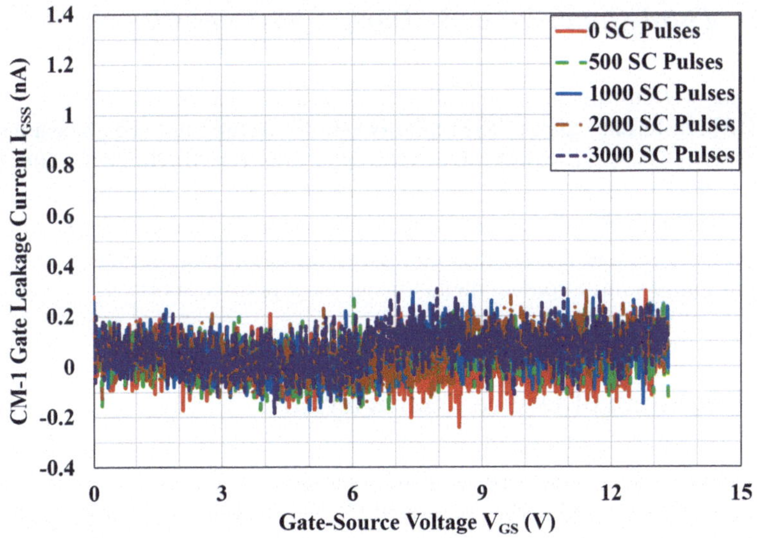

Fig. 16.9 Measured gate leakage characteristics of the SiC power MOSFET within the composite MOSFET during the repetitive short-circuit pulse tests

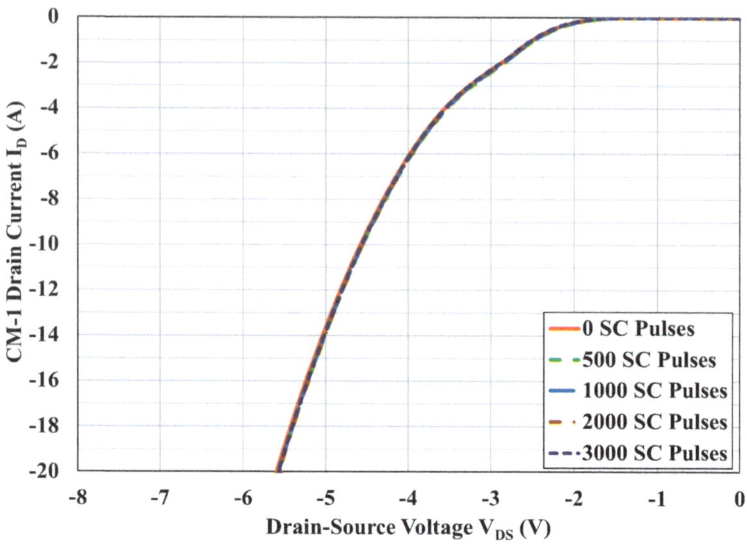

Fig. 16.10 Measured body-diode characteristics of the SiC power MOSFET within the composite MOSFET during the repetitive short-circuit pulse tests

16.5.4 Third-Quadrant Body-Diode Characteristics

The measured third-quadrant body diode forward characteristics for the composite MOSFET at different stages of the repetitive short-circuit pulse sequence are shown in Fig. 16.10. The overlapping plots show that the body diode of the SiC power MOSFET does not degrade even after 3000 short-circuit pulses in the BaSIC topology. This indicates that the contact between the source metal and the P^+ plug used to contact the P^+ shielding region is not being degraded because there is no melting of the aluminum source contact metal during the repetitive short-circuit events.

16.6 Evolution of Device Characteristics

The evolution of the characteristics of the stand-alone SiC power MOSFET and the SiC power MOSFET within the composite MOSFET of the BaSIC topology is contrasted in this section. The measured values for the on-resistance, threshold voltage, and gate leakage current are mapped as function of the number of applied short-circuit pulses.

The evolution of the measured on-state resistance of the stand-alone SiC power MOSFET and the SiC power MOSFET within the composite MOSFET is plotted in Fig. 16.11 as a function of the number of applied short-circuit pulses. There is small difference between the initial on-resistance of the two SiC power MOSFETs within the parametric distribution in the vendors datasheet. The stand-alone SiC MOSFET (M) exhibits severe increase in on-resistance after 200 short-circuit pulses. In contrast, the on-state resistance of the SiC MOSFET within the composite MOSFET

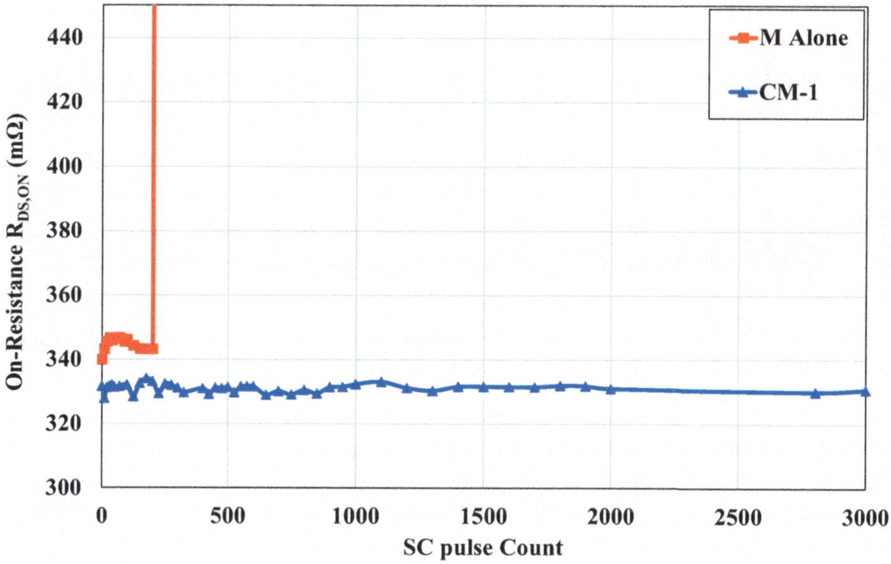

Fig. 16.11 Evolution of the on-state resistance of the stand-alone SiC power MOSFET and the SiC power MOSFET within the composite MOSFET during the repetitive short-circuit pulse tests

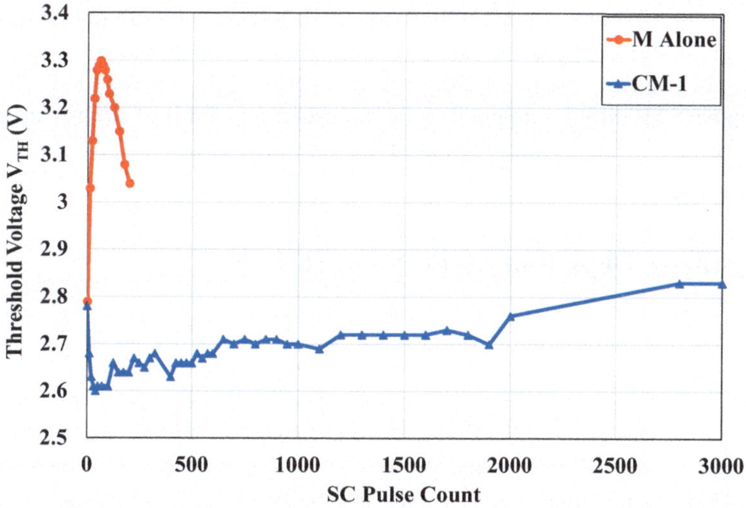

Fig. 16.12 Evolution of the threshold voltage of the stand-alone SiC power MOSFET and the SiC power MOSFET within the composite MOSFET during the repetitive short-circuit pulse tests

(CM-1) shows no change even after 3000 short-circuit pulses. It is obvious from this data that the BaSIC topology is able to prevent degradation of the SiC power MOSFET during repetitive short-circuit stress events.

The measured threshold voltages through the repetitive short-circuit pulses are shown in Fig. 16.12. The stand-alone SiC power MOSFET exhibits a change in

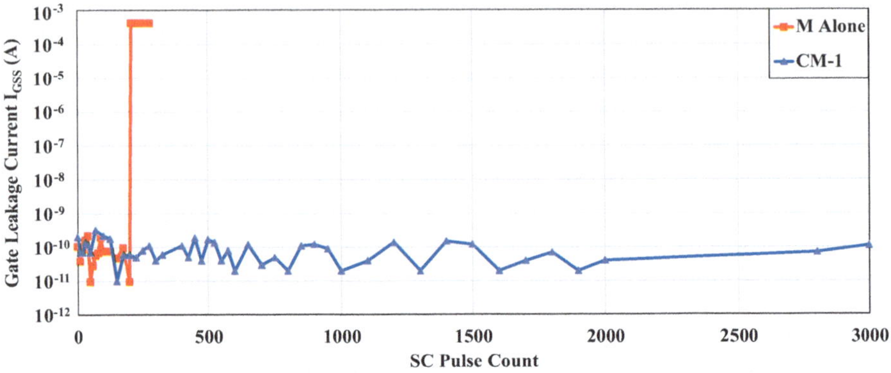

Fig. 16.13 Evolution of the gate leakage current of the stand-alone SiC power MOSFET and the SiC power MOSFET within the composite MOSFET during the repetitive short-circuit pulse tests

threshold voltage (V_{TH}) of about 0.3–0.5 V during the 200 short-circuit pulses before failure of the device. In contrast, the threshold voltage of the SiC power MOSFET within the composite MOSFET was stable within the measurement error of about 0.1 V up to 3000 short-circuit pulses.

The gate leakage current at the on-state gate bias of 20 V was also monitored during the repetitive short-circuit tests. A large increase in gate leakage current was observed for the stand-alone MOSFET (M) after 200 short-circuit pulses as shown in Fig. 16.13 indicating penetration of aluminum through the inter-metal gate dielectric. In contrast, the gate leakage current for the SiC power MOSFET within the composite MOSFET remained at the measurement limit of 0.2 nA even after 3000 pulses indicating no degradation of the gate structure.

16.7 Silicon Depletion Mode MOSFET

In order to demonstrate stability of the composite MOSFET of the BaSIC topology, it is also important to show that the internal Si depletion mode MOSFET (DMM) does not undergo any degradation during the repetitive short-circuit events. In previous sections, it was pointed out that the power dissipation in the Si DMM is much smaller than for the SiC power MOSFET even though they operate at the same current level during the short-circuit event. The Si DMM device operates with only 6–8 V across it when compared with 800 V for the SiC power MOSFET. The 100-fold smaller power dissipation keeps its junction temperature in the 150–175 °C range preventing any degradation.

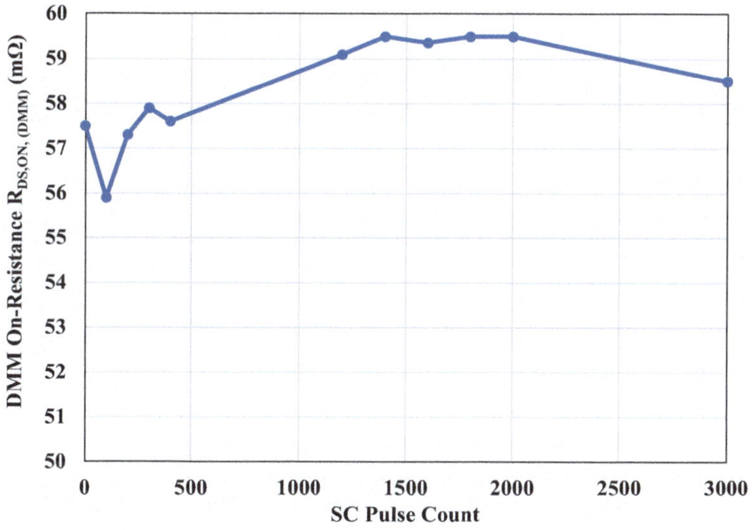

Fig. 16.14 Evolution of the on-resistance of the Si depletion mode MOSFET within the composite MOSFET during the repetitive short-circuit pulse tests

There are two important parameters of the Si DMM device that determines the performance of the composite MOSFET in the BaSIC topology. The first is the on-state resistance, and the second is its saturation current. Any degradation (or increase) in the on-state resistance of the Si DMM device would increase the resistance of the composite MOSFET making its performance worse after repetitive short-circuit events. The on-state resistance of the Si DMM was therefore also measured after the repetitive short-circuit events. It can be seen from plot in Fig. 16.14 that the on-resistance of the Si DMM device remains close to its initial value of 58 mΩ within ±2 mΩ.

The saturation current of the Si DMM determines the current flowing through the SiC power MOSFET during the short-circuit event. It must remain steady after repetitive short-circuit stress pulses to ensure protection of the SiC power MOSFET. The plot in Fig. 16.15 shows that the short-circuit current of the Si DMM device remains very steady at a value of 19–20 A even after 3000 short-circuit pulses. This result ensures excellent protection of the SiC power MOSFET within the BaSIC topology.

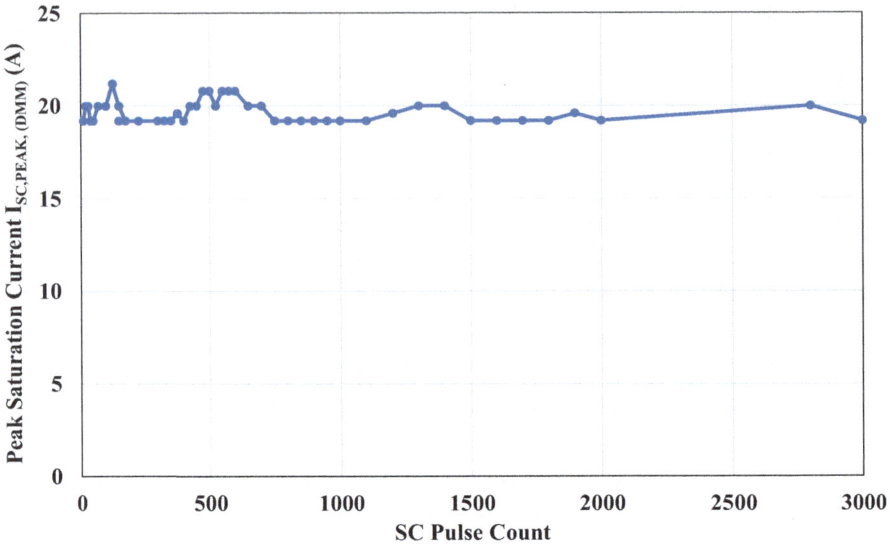

Fig. 16.15 Evolution of the saturation current of the Si depletion mode MOSFET (DMM) within the composite MOSFET during the repetitive short-circuit pulse tests

16.8 Thermal Analysis

The temperature rise in the SiC power MOSFET during the short-circuit event can be obtained using the adiabatic heating model [5]:

$$\Delta T = \frac{V_S \int_0^t I_D(t)\,dt}{A_A \cdot W_{SiC} \cdot C_V(T)} \tag{16.1}$$

where V_S is the drain supply voltage, $I_D(t)$ is the instantaneous drain current during the short-circuit event, A_A is the device active area, W_{SiC} is the thickness of the SiC MOSFET drift region where the heat is generated, and $C_V(T)$ is the volumetric specific heat for SiC. The increase in junction temperature can be computed using this equation with the short-circuit waveform for each device case. These waveforms were shown in Fig. 16.2 for the stand-alone SiC power MOSFET and the SiC power MOSFET within the BaSIC topology.

The temperature rise calculated for the two cases is shown in Fig. 16.16 (dashed lines) using the above equation including the dependence of the specific heat for SiC on temperature. The short-circuit current waveforms for the two cases are also shown in this figure for completeness. The red line in the figure demarks the melting point of aluminum (660 °C). It can be observed that the temperature for the stand-alone SiC power MOSFET (M) increases to 760 °C at 3.5 μs which is sufficient to melt the aluminum. This explains the observed degradation of the stand-alone SiC power MOSFET during the repetitive short-circuit tests. In contrast, the temperature

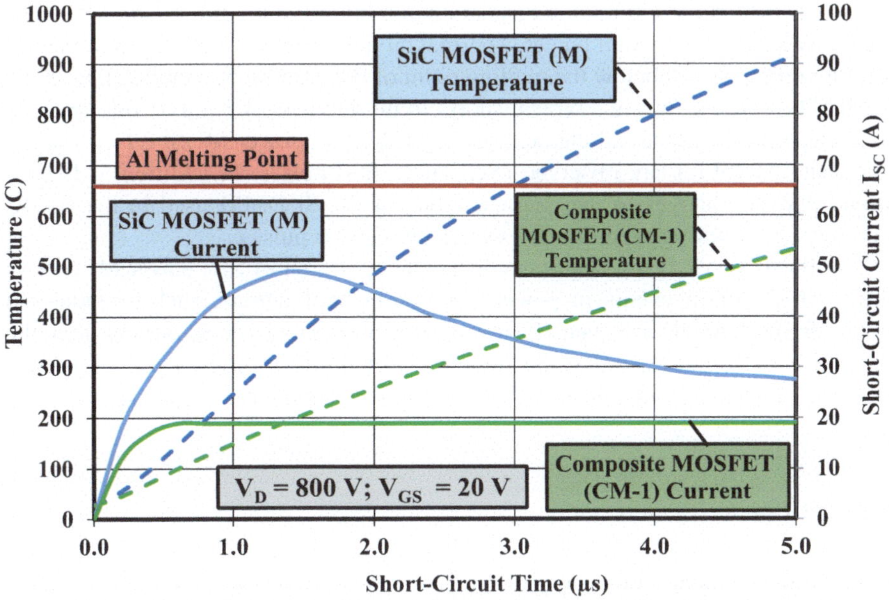

Fig. 16.16 Evolution of the saturation current of the Si depletion mode MOSFET (DMM) within the composite MOSFET during the repetitive short-circuit pulse tests

for the SiC power MOSFET in the composite MOSFET (CM-1) reaches only 400 °C, which is well below the melting point for aluminum. This eliminates the degradation and failure of the SiC power MOSFET in the BaSIC topology under repetitive short-circuit events.

16.9 Summary

It has been demonstrated in this chapter that the stand-alone 1.2 kV SiC power MOSFET is prone to degradation and failure under repetitive short-circuit events with pulse time of 3.5 μs, when operated at application level drain bias of 800 V with a gate bias of 20 V. A large peak short-circuit current is observed for these devices during the short-circuit duration. Thermal modeling indicates that the failure of the device is related to the melting of the source contact aluminum metallization, which is commonly used in these products, during a typical short-circuit duration of 3.5 μs. The test results show a dramatic increase in gate leakage current after 200 short-circuit pulses. This is most likely due to penetration of the molten aluminum through the inter-metal dielectric (SiO_2) used between the source metal and the silicided poly-silicon gate.

In contrast, the BaSIC(DMM) approach, implemented with a gate-source-shorted silicon depletion mode MOSFET in series with the source of the SiC power

MOSFET, is found to eliminate degradation and failure of the SiC power MOSFET even after 3000 short-circuit pulses. Thermal modeling shows that the junction temperature remains well below the melting point of the aluminum metallization in the BaSIC(DMM) case during a typical short-circuit duration of 3.5 μs. Consequently, no changes in the electrical characteristics of the SiC power MOSFET within the composite MOSFET are observed even after 3000 pulses. In addition, it is confirmed that no changes in the characteristics of the Si DMM used in the BaSIC topology occur during repetitive SC events up to 3000 pulses.

The BaSIC(DMM) topology ensures stable operation of SiC power MOSFETs during short-circuit events in power conversion applications, such as industrial motor drives and electric vehicle inverters. This chapter demonstrates that the BaSiC(DMM) topology solves an important reliability limitation for SiC power MOSFETs, which could constrain their adoption in commercial applications.

References

1. A. Kanale, B.J. Baliga, Eliminating repetitive short-circuit degradation and failure of 1.2 kV SiC power MOSFETs. IEEE J. Emerg. Sel. Top. Power Electron. **9**, 6773–6779 (2021)
2. J. Sun, J. Wei, Z. Zheng, Y. Wang, K.J. Chen, Short circuit capability and short circuit induced V_{TH} instability of a 1.2 kV SiC power MOSFET. IEEE J. Emerg. Sel. Top. Power Electron. **7**, 1539–1546 (2019)
3. X. Zhou, H. Su, Y. Wang, R. Yue, G. Dai, J. Li, Investigations on the degradation of 1.2 kV 4H-SiC MOSFETs under repetitive short-circuit tests. IEEE Trans. Electron Devices **63**, 4346–4352 (2016)
4. J. Wei, S. Liu, L. Yang, J. Fang, T. Li, S. Li, W. Sun, Comprehensive analysis of electrical parameters degradation for SiC power MOSFETs under repetitive short-circuit stress. IEEE Trans. Electron Devices **64**, 5440–5447 (2018)
5. B.J. Baliga, *Fundamentals of Power Semiconductor Devices*, 2nd edn. (Springer-Science, 2019)
6. K. Han, A. Kanale, B.J. Baliga, B. Ballard, A. Morgan, D. Hopkins, New short circuit failure mechanism for 1.2 kV 4H-SiC MOSFETs and JBSFETs, in *IEEE Workshop on Wide Bandgap Power Devices and Applications* (2018), pp. 108–113
7. J. Liu, G. Zhang, B. Wang, W. Li, J. Wang, Gate failure physics of SiC MOSFETs under short-circuit stress. IEEE Electron Device Lett. **41**, 103–106 (2020)
8. Wolfspeed product information. https://www.wolfspeed.com/c2m0280120d
9. IXYS power semiconductor product. https://www.littelfuse.com/products/power-semiconductors/ixys-power-semiconductors-and-ics.aspx

Chapter 17
Overcurrent Protection

During the 1990s, IGBTs were engineered with short-circuit withstand times of more than 10 μs for use in motor drive applications. This provided sufficient time to detect the occurrence of short circuits in the motor windings, send a signal back to the gate drive circuit, and then initiate a shut off of the gate drive voltage. Prevention of destructive failure of the inverter allowed repair of the motors without expensive replacement of the power electronics. This has now become an industry standard practice. Consequently, the replacement of Si IGBTs with SiC power MOSFETs to reduce switching losses will be impeded if these devices are unable to be safely shut off during short-circuit events.

The commercialized SiC power MOSFETs have a poor short-circuit withstand time (<5 μs) compared to Si IGBTs [1]. This vulnerability has motivated detailed research into developing improved overcurrent protection circuits for wide bandgap power devices [2, 3]. Overcurrent protection methods can be derived from the mechanism used to measure the short-circuit current through the power device.

17.1 DESAT Detection Method

The DESAT protection circuit approach demonstrated in the 1990s [4] has become the most popular method for identification of the short-circuit event for motor drives using Si IGBTs in the inverter. Si IGBTs were provided with overcurrent protection circuits based on detecting the short-circuit current exceeding a predetermined threshold and turning-off the gate drive. The most common adaptation of this mechanism, known as the desaturation, or DESAT detection method, has typically been implemented with a turnaround time of 10 μs from onset of the fault until the IGBT gate shut off. The 10 μs duration includes the current buildup in the power device, detection and signal propagation delays in the gate driver IC, and a blanking time to

B. J. Baliga, A. Kanale, *The BaSIC Topology*,
https://doi.org/10.1007/978-3-031-86630-2_17

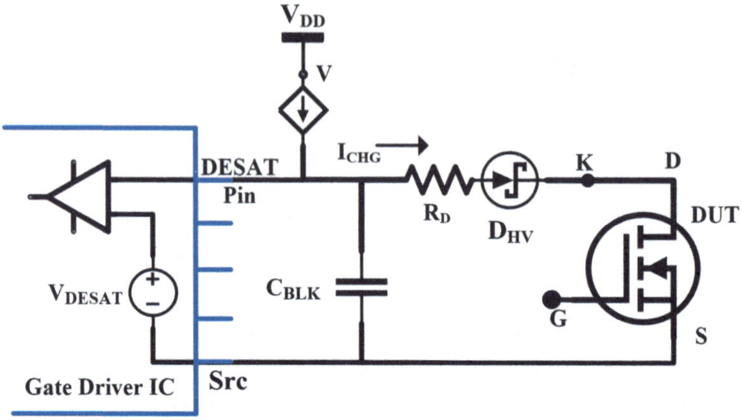

Fig. 17.1 Schematic diagram of DESAT-based protection circuit for power transistors

allow transient noise to settle down before the fault signal is flagged. Consequently, the 10 μs short-circuit withstand time (SCWT) became accepted for industrial applications.

The DESAT detection method, shown in Fig. 17.1, relies on the voltage drop across the path K-V exceeding the fixed DESAT detection voltage threshold of the IC. This method requires sensing the voltage drop of the power device at a node undergoing high-voltage fluctuations, i.e., the drain of a power MOSFET as shown in the figure. The DESAT diode D_{HV} must have the same voltage rating as the power device. It must have fast switching speed which is more difficult as the voltage rating increases. Consequently, the DESAT approach increases system cost and imposes clearance requirements for the board layout of gate drive components.

17.2 Series Resistance Method

An alternative to the DESAT approach is with the use of a resistor in series with the source of the power MOSFET [3]. The series resistance method for overcurrent protection is shown in Fig. 17.2. The series resistance R_S must be large enough to product a noise-free DESAT signal distinguishable by the drive circuit. Depending on the power device on-resistance or voltage drop, this can result in additional power losses and have adverse impacts on the trade-off between short-circuit withstand time, on-state performance, and switching behavior, as described in Chap. 7.

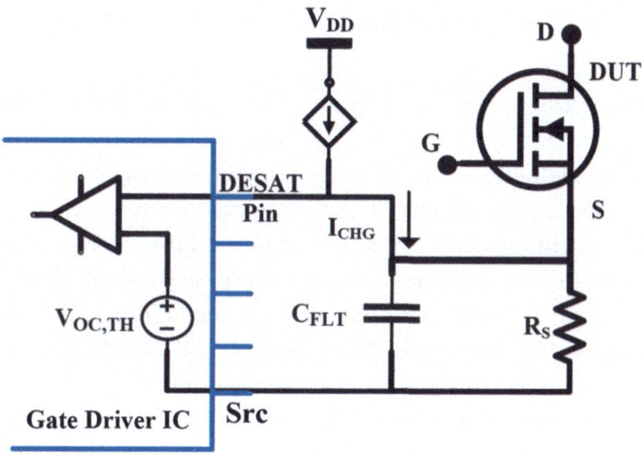

Fig. 17.2 Schematic diagram of series resistance-based protection circuit for power transistors

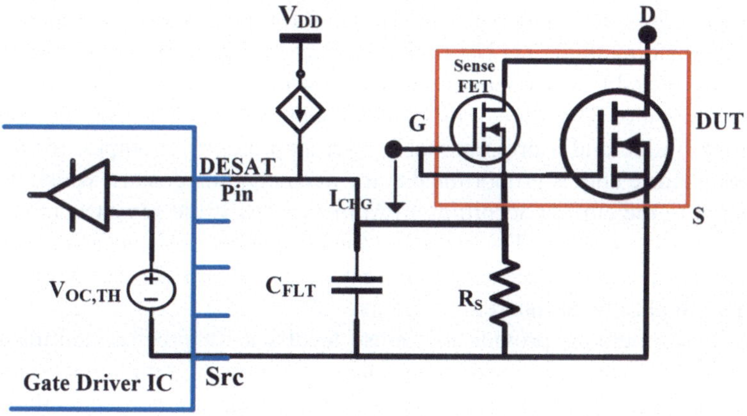

Fig. 17.3 Schematic diagram of current sensing-based protection circuit for power transistors

17.3 Current Sensing Method

The SenseFET current sensing scheme is shown in Fig. 17.3. This method requires the integration of small MOSFET, called the SenseFET, in parallel with main power device [5]. A scaled-down fraction of the full short-circuit current (typically <1%) flows through the SenseFET during the event. The current through the SenseFET device is passed through a sense resistor R_S to derive a voltage which can be monitored to detect short-circuit faults. Consequently, the current sensing does not add to the resistance of the main SiC power MOSFET.

The presence of a parallel shunt without additional parasitic inductance in the power loop eliminates noise. The inclusion of a monolithic SenseFET with the main power device consumes semiconductor area which increases chip cost. The device with a SenseFET requires a four-terminal package adding to the system cost.

17.4 Other Methods

Alternative overcurrent protection methods involve non-contact monitoring of the current through Rogowski coil-based sensors [6]. The switch current sensor consists of an inductor with an integrator which converts di/dt information to the current flowing in the transistor. An integrated current sensor has been built using a multi-layer printed circuit board. The printed circuit board winding and shielding during its layout can minimize the noise due to the high dv/dt observed during the switching events. An OpAmp-based integrated circuit allows obtaining high accuracy and bandwidth. This method can protect a SiC MOSFET from short-circuit faults within 1.17 μs. One of the drawbacks that limits the switching frequency is the non-ideal performance of the integrator. This method has been found to have a shorter response time than the DESAT method. However, the Rogowski coils occupy space on the gate drive boards.

Magnetic sensor ICs, which provide isolation for the gate drive components, are another alternate method [7]. The tunnel magnetoresistance-based sensor exhibits high sensitivity and low temperature drift with small footprint allowing integration inside the module. The tunnel magnetoresistance current sensor consists of a Wheatstone bridge and a programmable gain amplifier. The output of the tunnel magnetoresistance chip is proportional to the magnetic flux density, which in turn is proportional to the current according to Ampere's law if the sensor is mounted on the transistor current path. The output of tunnel magnetoresistance sensor is then proportional to the current flowing in the SiC power MOSFET. The response time for this method has been found to be 1.5 μs.

While these methods provide low noise feedback for precise identification of fault events, they consume area on the printed circuit board. Additional circuitry for soft-turn off of the power device also consumes board space, consequently limiting the power density achievable in a power converter system.

17.5 BaSIC Overcurrent Protection

The short-circuit detection is performed for the BaSIC topology using the same approach as shown in Fig. 17.2 with the sensing resistor replaced by either the Si gate-source-shorted depletion mode MOSFET or the Si enhancement mode MOSFET. The voltage across these devices (at the voltage sense node) abruptly increases during the short-circuit event as shown in previous chapters, allowing detection of the fault. The voltage at the sense node in the BaSIC topology always remains below 10 V eliminating the need for the high-voltage diode D_{HV} in the typical DESAT implementation. The response time for the BaSIC topology is shorted due to absence of a blanking time required for the typical DESAT method. This allows turning-off the SiC MOSFET in a shorter duration reducing its temperature rise and stress. Experimental demonstration of overcurrent protection with the BaSIC topology is discussed in the next sections.

17.5.1 Device Module Implementation

The BaSIC topology was implemented using commercially available Wolfspeed 1200 V, 160 mΩ, bare die SiC MOSFETs CPM2-1200-0160B with the USCi 35 V, 1.1 mΩ Si enhancement mode MOSFETs (EMM) AW1060. Two stand-alone SiC MOSFET modules and two BaSIC(EMM) composite MOSFET modules were fabricated. All modules had the same substrate and footprint. A Kelvin-sense termination was added to improve extraction of the voltages. Consequently, the SiC MOSFET modules had four terminals as shown in Fig. 17.4a. The BaSIC(EMM) modules had six terminals as shown in Fig. 17.4b. The die-attach, wire-bonded assembly, along with an image of the final encapsulated modules is shown in these figures.

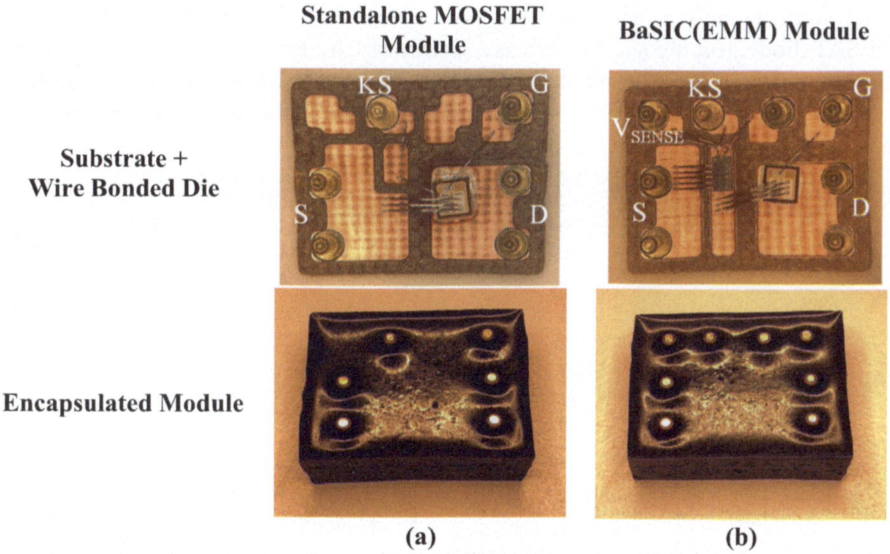

Fig. 17.4 Modules prepared for overcurrent protection experiments: (**a**) Stand-alone SiC MOSFET. (**b**) BaSIC topology with SiC and Si MOSFETs

17.5.2 Stand-Alone SiC MOSFET DESAT Implementation

The SiC MOSFET used in this study has a short-circuit withstand time of 3.4 μs as shown in previous chapters. Consequently, the standard DESAT circuit was adopted for the stand-alone SiC MOSFET module, with a designed turnaround time of 2 μs from fault occurrence to the turning off of the gate drive. The schematic of the DESAT circuit for the stand-alone SiC MOSFET module is shown in Fig. 17.5. It utilizes a high-voltage ultrafast diode STTH112A, also called the DESAT diode D_{HV}, whose cathode is connected to the drain of the stand-alone SiC MOSFET. A 1 kΩ resistor (R_{11} in Fig. 17.5) is connected in series with the DESAT diode. A blanking capacitor is connected between the DESAT pin and the source of the SiC power MOSFET. The blanking capacitor C_{BLK} (C_{27} in Fig. 17.5) has a Zener diode (D_7 in Fig. 17.5) in parallel to clamp the voltage across it. The gate driver IC UCC21750DW was used to drive the SiC power MOSFET. Its datasheet mentions a DESAT voltage threshold of 9 V and a DESAT current injection capability of about 450 μA.

Under normal conditions, the DESAT current I_{CHG} (shown as I_{CHG} (Normal)) flows into the SiC power MOSFET through the DESAT diode. Under short-circuit conditions, the voltage across the SiC power MOSFET rises and reverse biases the DESAT diode, forcing I_{CHG} (shown as I_{CHG} (Short-Circuit)) to flow into the blanking capacitor C_{BLK}. Once the blanking capacitor is fully charged, the DESAT current I_{CHG} stops flowing and the IC signals a fault, simultaneously turning-off the gate drive voltage to protect the SiC power MOSFET.

The time taken to charge the blanking capacitor can be calculated by [3]

$$t_{BLK} = \frac{C_{BLK} V_{DESAT}}{I_{CHG}} \tag{17.1}$$

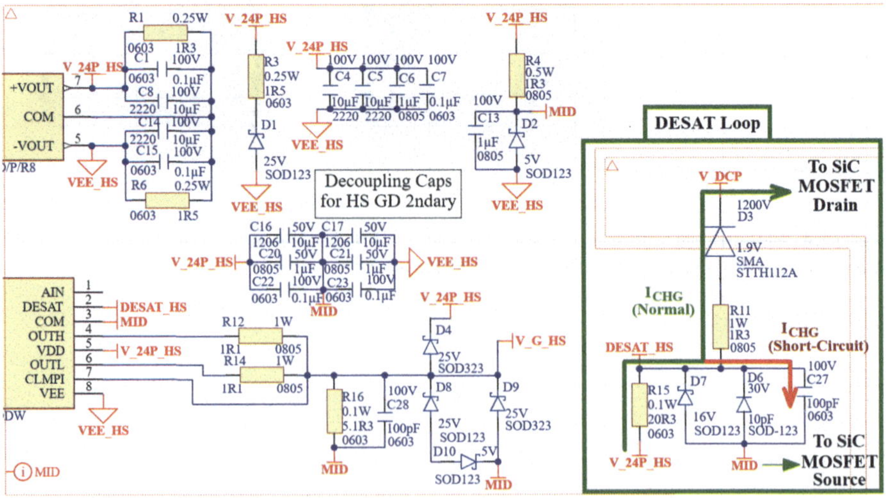

Fig. 17.5 Schematic of DESAT circuit implemented for the stand-alone SiC power MOSFET

A blanking time of 2 μs is obtained using a blanking capacitor value of 100 pF with the V_{DESAT} value of 9 V and a DESAT current I_{CHG} of 450 μA. The DESAT protection turnaround time can be changed by adjusting the blanking capacitor value. The DESAT resistor R_D (R_{11} in Fig. 17.5) can be changed to alter the voltage of the SiC MOSFET at which the DESAT detection is triggered. The shortest blanking time is limited by the large voltage swing at the drain of the SiC power MOSFET using normal operation leading to false triggering of the over-protection circuit.

17.5.3 BaSIC Topology DESAT Implementation

The BaSIC(EMM) topology produces a voltage V_{SENSE} at the sense node highlighted in Fig. 17.4b. In previous chapters, it has been demonstrated that V_{SENSE} has a low value proportional to the current flowing through the SiC power MOSFET under normal operating conditions. It has also been shown in previous chapters that V_{SENSE} increases to 5–8 V under short-circuit conditions.

Consequently, an overcurrent protection circuit similar to the DESAT circuit was implemented for the BaSIC(EMM) topology. There are two main contrasting features of the BaSIC overcurrent protection circuit compared with the standard DESAT circuit: (1) the DESAT loop K-V (see Fig. 17.1) is connected to the sense node in the BaSIC topology case instead of the SiC power MOSFET drain terminal in the conventional DESAT case; (2) the DESAT loop is equipped with a series of low voltage Schottky diodes in the BaSIC topology case instead of a single high-voltage diode (STTH112A) in the DESAT case.

The schematic of the BaSIC overcurrent protection circuit is shown in Fig. 17.6. The aforementioned features allow the adoption of the standard DESAT circuit with the same gate driver IC, without the high-voltage isolation for the gate drive

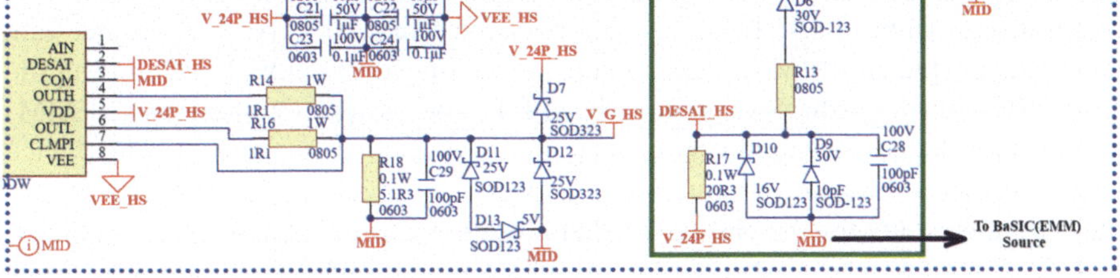

Fig. 17.6 Schematic of BaSIC overcurrent protection circuit for internal SiC power MOSFET

components, to the BaSIC topology. This enables the realization of a more compact gate drive layout, which is beneficial for increasing the system power density for power converters.

17.6 Experimental Validation of Overcurrent Protection

Two half-bridge circuit boards were assembled with the footprints for the stand-alone SiC MOSFET and BaSIC(EMM) modules. These boards were used for both short-circuit and dynamic characterization experiments. Board-1 for the stand-alone SiC MOSFET was equipped with the standard DESAT overcurrent protection circuit shown in Fig. 17.5, while Board-2 for the BaSIC(EMM) module was equipped with the BaSIC overcurrent protection circuit shown in Fig. 17.6.

Static characterization for the modules was conducted on the Keysight B1505A Curve Tracer. Measured parameters included on-resistance, threshold voltages, and transconductance. Dynamic characterization was conducted using the double-pulse method with an external 150 μH solenoid inductor connected across the high-side device as load together with a SiC JBS antiparallel diode.

Short-circuit tests were conducted by turning on the SiC power MOSFETs using a −5/+20 V pulse with a DC bus voltage of 800 V. Different DC gate bias voltages were applied to the Si EMM device in the BaSIC(EMM) topology. The gate pulse widths were increased until the DESAT or BaSIC overcurrent protection signal was triggered.

17.6.1 High-Voltage, High-Current Output Characteristics

High-voltage, high-current output characteristics for the stand-alone SiC MOSFET module and the BaSIC(EMM) module were obtained by conducting short-circuit tests with 1.5 μs-wide gate pulses of gate voltage −5/+20 V at various DC bus voltages applied to the drain, as discussed in previous chapters. The peak short-circuit current observed during the pulse was recorded as the saturation current for the corresponding DC bus voltage. The output characteristic shown in red in Fig. 17.7 for the stand-alone SiC power MOSFET exhibits a saturation current of 136 A at a DC bus voltage of 800 V. For the same DC bus voltage, the BaSIC(EMM) module has a saturation current of 132 A for $V_{DC,PROG}$ between 5 and 10 V. At these higher gate bias voltages, the Si EMM device operates in its linear region without limiting the current flow through the SiC power MOSFET. The current for the BaSIC(EMM) module is reduced for smaller gate bias voltages applied to the Si EMM device because it enters the drain saturated current regime of operation. The observed saturated drain currents for the BaSIC(EMM) module were 115 A for $V_{G,EMM}$ of 4.5 V, 97 A for $V_{G,EMM}$ of 4.0 V, 72 A for $V_{G,EMM}$ of 3.7 V and 50 A $V_{G,EMM}$ of 3.3 V. These

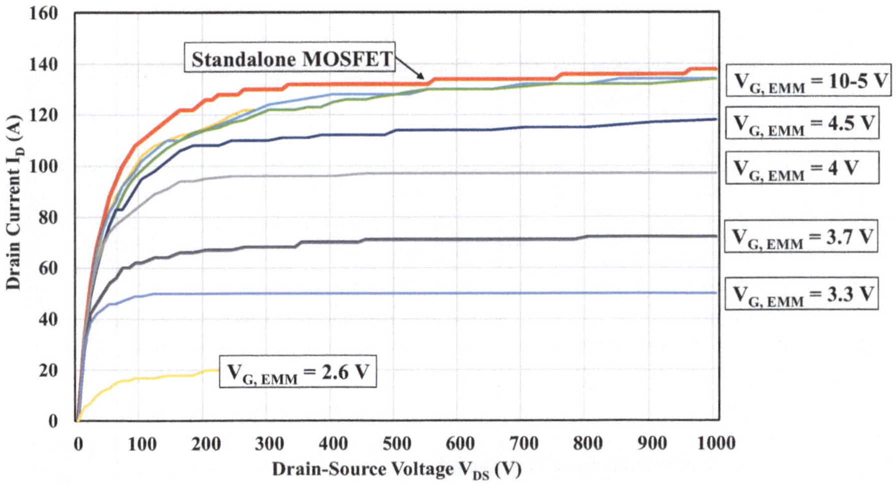

Fig. 17.7 High-voltage, high-current output characteristics measured for the stand-alone SiC power MOSFET (red line) and the BaSIC(EMM) topology. The DC gate bias applied to Si EMM device is indicated as $V_{G,EMM}$

results indicate that the BaSIC(EMM) topology with a $V_{G,EMM}$ of 3.7 V can increase the short-circuit withstand time of the SiC MOSFET by a factor of 1.9×.

17.6.2 *Short-Circuit Overcurrent Protection: Stand-Alone SiC Power MOSFET*

The overcurrent protection for the stand-alone SiC power MOSFET was performed using a DC bus voltage of 800 V and a gate pulse width of 2 μs in the short-circuit test setup. The measured short-circuit test waveforms for the stand-alone SiC MOSFET are shown in Fig. 17.8. The gate drive signal is shown by the green trace. The drain current shown by the red trace increases rapidly when the SiC power MOSFET is turned on. The peak short-circuit current is 136 A. Due to blanking time, the DESAT circuit detects the short-circuit event in 1.36 μs and then safely shuts off the SiC power MOSFET.

The same test was repeated after the blanking capacitor C_{BLK} was reduced from 100 to 33 pF. From the waveforms in Fig. 17.9, it is observed that the DESAT circuit activation reduces from 1.36 to 0.92 μs. The peak short-circuit current is still high at 130 A. These results demonstrate that the DESAT circuit implemented for the stand-alone SiC power MOSFET works as expected and can be modified to detect SC events within a shorter duration.

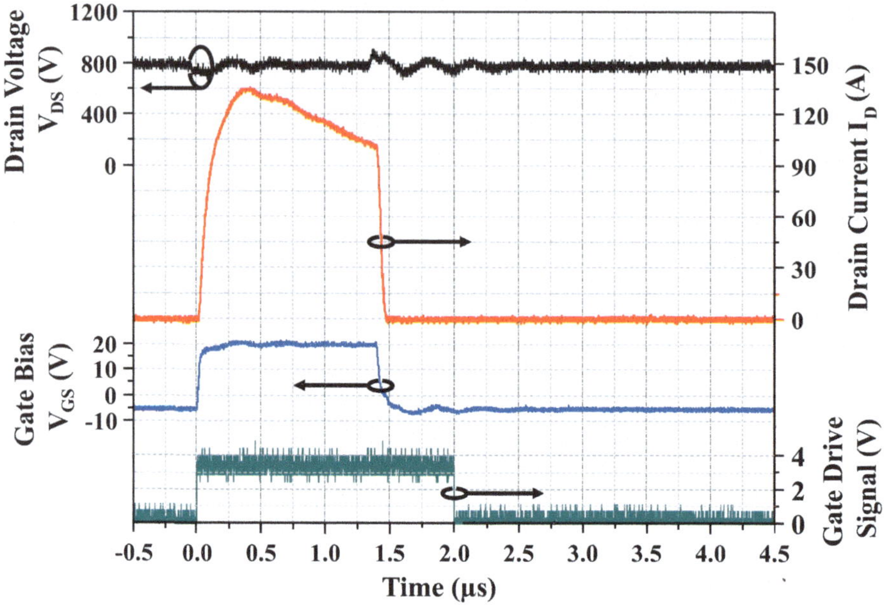

Fig. 17.8 Measured overcurrent protection waveforms: Stand-alone SiC power MOSFET using conventional DESAT protection with C_{BLK} of 100 pF

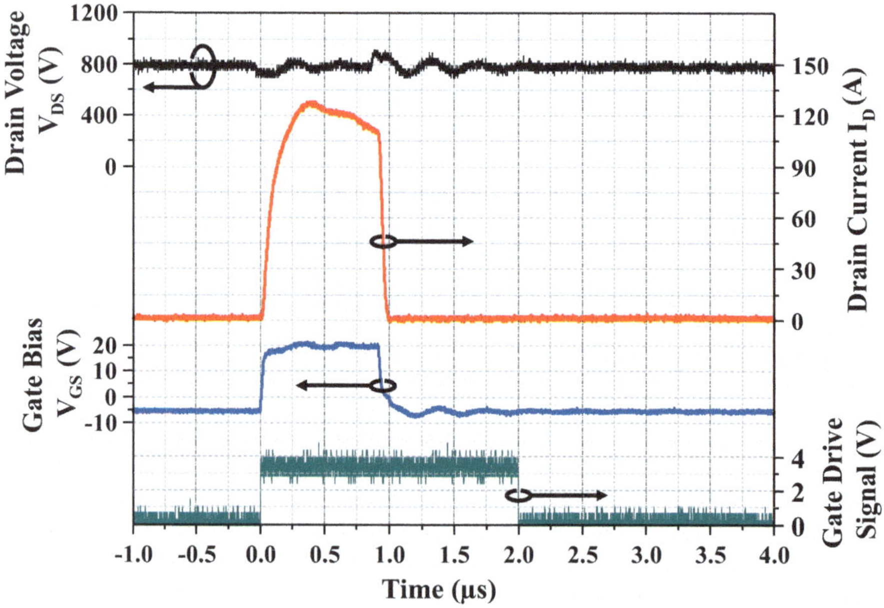

Fig. 17.9 Measured overcurrent protection waveforms: Stand-alone SiC power MOSFET using conventional DESAT protection with C_{BLK} of 33 pF

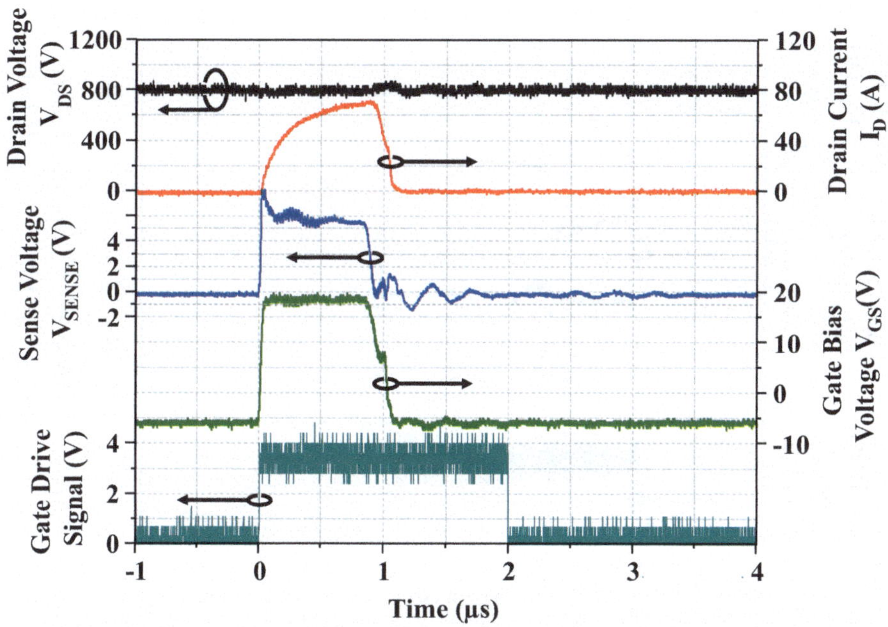

Fig. 17.10 Measured overcurrent protection waveforms: BaSIC(EMM) topology using overcurrent protection with C_{BLK} of 100 pF

17.6.3 Short-Circuit Overcurrent Protection: BaSIC(EMM) Topology

The measured short-circuit test waveforms for the BaSIC(EMM) module at a DC bus voltage of 800 V and a gate pulse width of 2 μs are shown in Fig. 17.10 for the case of a DC gate bias $V_{G,EMM}$ of 3.7 V applied to the Si EMM device. A blanking capacitor of 100 pF was used in this test. The short-circuit event is detected by the BaSIC overcurrent protection circuit at 0.85 μs, and the composite MOSFET is fully turned-off in 1.1 μs. The maximum current is greatly reduced to 67 A in this case.

The BaSIC overcurrent protection circuit can be modified to react faster by reducing the blanking capacitor C_{BLK} from 100 to 33 pF. The measured short-circuit waveforms with the BaSIC overcurrent protection circuit are shown in Fig. 17.11 with a C_{BLK} of 33 pF. The maximum current is greatly reduced to 67 A in this case as well. The BaSIC overcurrent protection action shuts off the gate drive in 0.68 μs, and the composite MOSFET completely turns off by 0.89 μs.

An alternate option for reducing the BaSIC overcurrent protection detection time is by increasing the resistance R_{13} in Fig. 17.6. The larger voltage drop across R_{13} triggers the BaSIC overcurrent protection action at a lower value of the sense voltage, which shortens the response time. The measured short-circuit waveforms with the resistor R_{13} changed from 1 to 5 kΩ are shown in Fig. 17.12 with C_{BLK} of

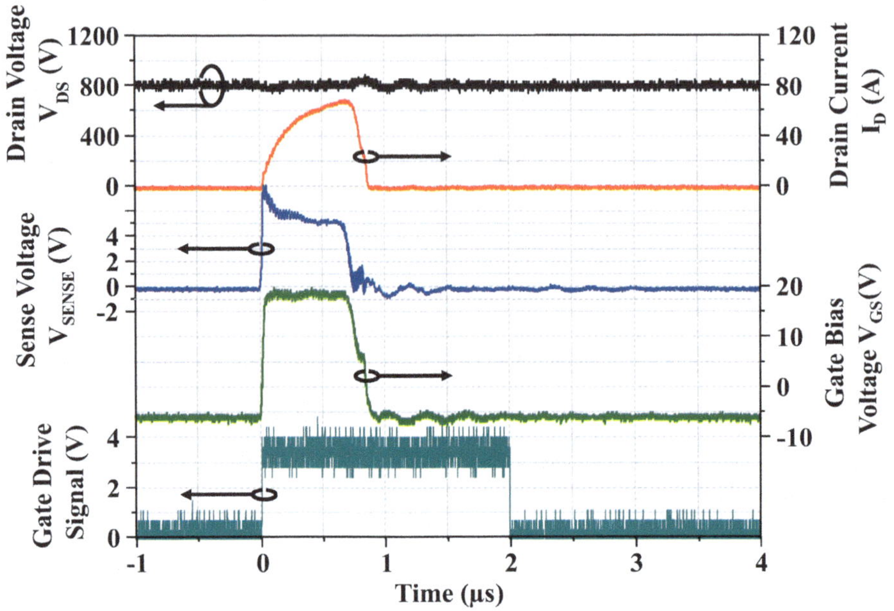

Fig. 17.11 Measured overcurrent protection waveforms: BaSIC(EMM) topology using overcurrent protection with C_{BLK} of 33 pF

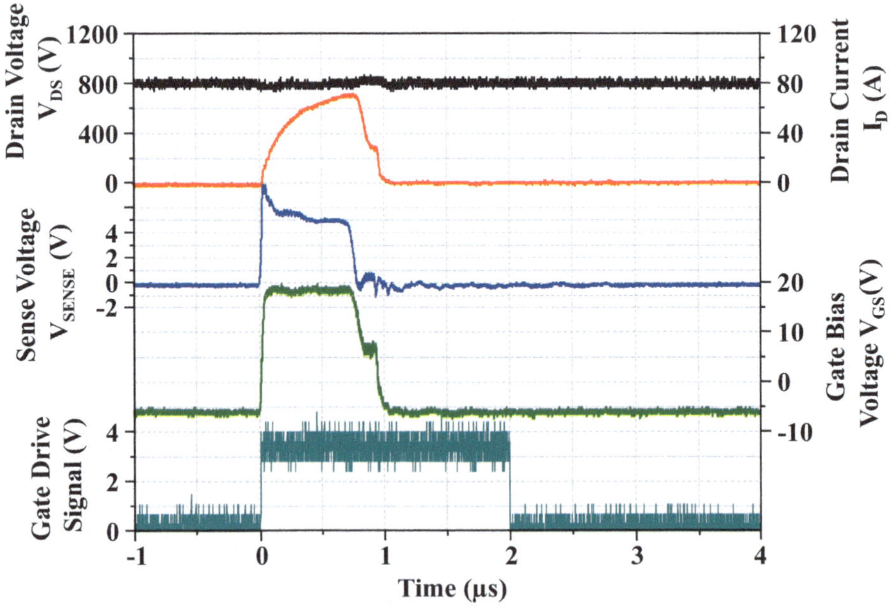

Fig. 17.12 Measured overcurrent protection waveforms: BaSIC(EMM) topology using overcurrent protection with R_{11} increased to 5 kΩ

100 pF. The maximum current is greatly reduced to 67 A in this case as well. The BaSIC overcurrent protection shuts off the gate drive at 0.7 μs and the composite MOSFET completely turns off by 1.0 μs.

The BaSIC(EMM) overcurrent protection action does not change when the EMM DC gate bias $V_{G,EMM}$ is changed between 3.3 and 4.5 V. but the peak short circuit increases when the DC gate bias for the Si EMM device is increased. However, it is worth pointing out that the BaSIC(EMM) overcurrent protection action does not occur when $V_{G,EMM}$ is increased beyond 4.5 V since the EMM operates in its linear region at the 136 A saturated drain current flowing through the SiC power MOSFET.

These results demonstrate that the BaSIC(EMM) overcurrent protection circuit can be used to detect short-circuit events and successfully turn-off the composite MOSFET between 0.6 and 1.0 μs. This is a shorter response time that 0.9–1.36 μs obtained using the conventional DESAT method.

17.7 Impact on Normal Circuit Operation

For completeness, it is necessary to document the impact of the BaSIC(EMM) topology on the static and dynamic characteristics of the SiC power MOSFET. Measurements of the on-resistance, threshold voltage, and transconductance were performed on the stand-alone SiC power MOSFET module and the BaSIC(EMM) module. Their switching loss characteristics were evaluated using double-pulse testing.

17.7.1 On-State Characteristics

The measured on-state characteristic for the stand-alone SiC MOSFET is shown in Fig. 17.13 for a gate bias of 20 V (topmost trace). For a gate bias of 20 V, the stand-alone SiC MOSFET module had an $R_{DS,ON}$ of 186 mΩ. The measured on-state characteristics for the BaSIC(EMM) module for a gate bias of 20 V are also shown in Fig. 17.13 with the various DC gate biases given in the figure applied to the gate of the Si EMM device. The on-state characteristics of the BaSIC(EMM) module become close to that of the stand-alone SiC power MOSFET when the DC gate bias for the Si EMM device is increased above 5 V. The on-state characteristics begin to deviate from that of the SiC power MOSFET when the DC gate bias applied to the Si EMM device is reduced to the 3.4–4.5 V range. The composite MOSFET in the BaSIC(EMM) module operates in its linear regime at up to 10 A for these DC gate bias voltages applied to the Si EMM device. This no longer holds true when the DC gate bias voltage applied to the Si EMM device is reduced below 3.4 V.

The extracted on-resistance ($R_{DS,ON}$) at a current of 10 A is plotted in Fig. 17.14. The stand-alone SiC MOSFET module has an $R_{DS,ON}$ of 186 mΩ at a gate bias of

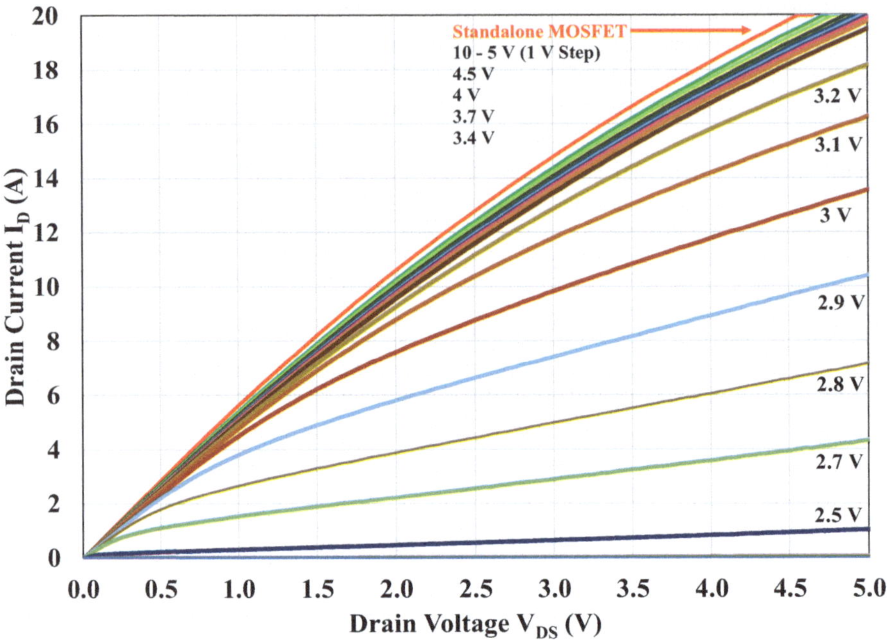

Fig. 17.13 Measured on-state characteristics of the stand-alone SiC power MOSFET and the composite MOSFET in the BaSIC(EMM) topology

20 V. The on-resistance of the BaSIC(EMM) module increases gradually from 191 to 209 mΩ with reduction of the Si EMM DC gate bias ($V_{G,EMM}$) from 10 to 3.7 V. The on-resistance then increases very rapidly because the Si EMM no longer operates in the linear region at 10 A for DC gate bias voltages ($V_{G,EMM}$) below 3.5 V. The measured values for the on-resistance are documented in Table 17.1.

17.7.2 *Threshold Voltage*

Threshold voltage was measured at a drain voltage of 0.1 V and a drain current of 1 mA. The measured values are documented in Table 17.1 It increases slightly when the Si EMM on-resistance goes up at reduced DC gate bias voltages $V_{G,EMM}$.

17.7.3 *Transconductance*

The transfer characteristics of the transistors were obtained using a drain voltage of 20 V by varying the gate bias applied to the SiC power MOSFET. The transconductance was extracted at a drain current of 10 A. The measured values are documented

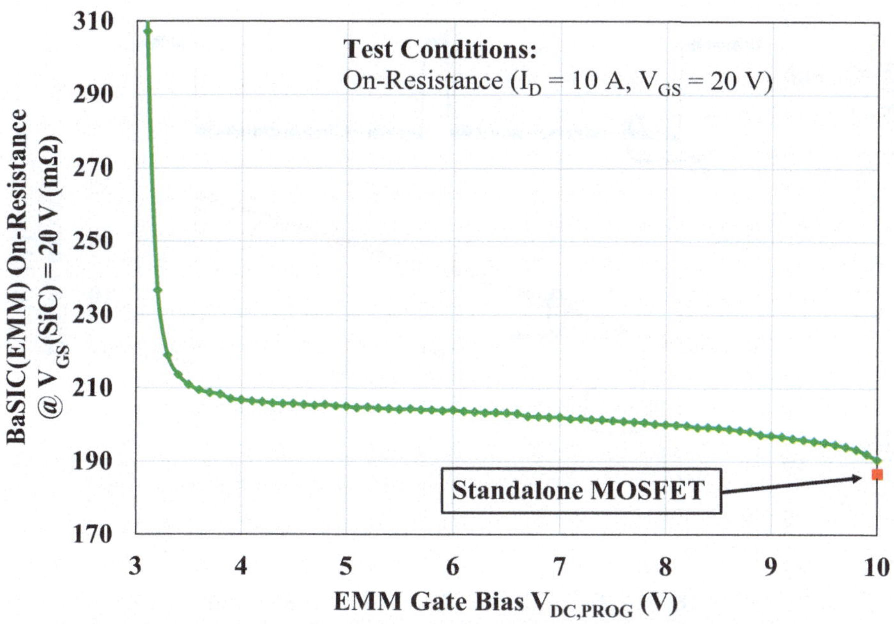

Fig. 17.14 Measured on-state resistance for the stand-alone SiC power MOSFET and the composite MOSFET in the BaSIC(EMM) topology

Table 17.1 Transistor characteristics

Parameter	SiC MOSFET	BaSIC(EMM), $V_{G,EMM}$					
		10	7.0	5.0	4.0	3.7	3.4
$R_{DS,ON}$ (mΩ)	186	191	202	205	207	209	214
V_{TH} (V)	3.14	3.24	3.38	3.40	3.40	3.40	3.40
g_M (S)	6.76	6.12	6.13	6.08	6.08	5.98	4.53

in Table 17.1. The transconductance is reduced for the composite MOSFET in the BaSIC(EMM) topology due to resistance of the Si EMM device as discussed in previous chapters. This produces an increase in the switching losses as discussed in the next section.

17.7.4 Switching Power Losses

The measured double-pulse test switching waveforms for the stand-alone SiC MOSFET are shown in Fig. 17.15 for current rising to 15 A at the end of the first gate pulse. They were obtained using a DC bus voltage of 800 V and SiC MOSFET gate bias of −5/+20 V.

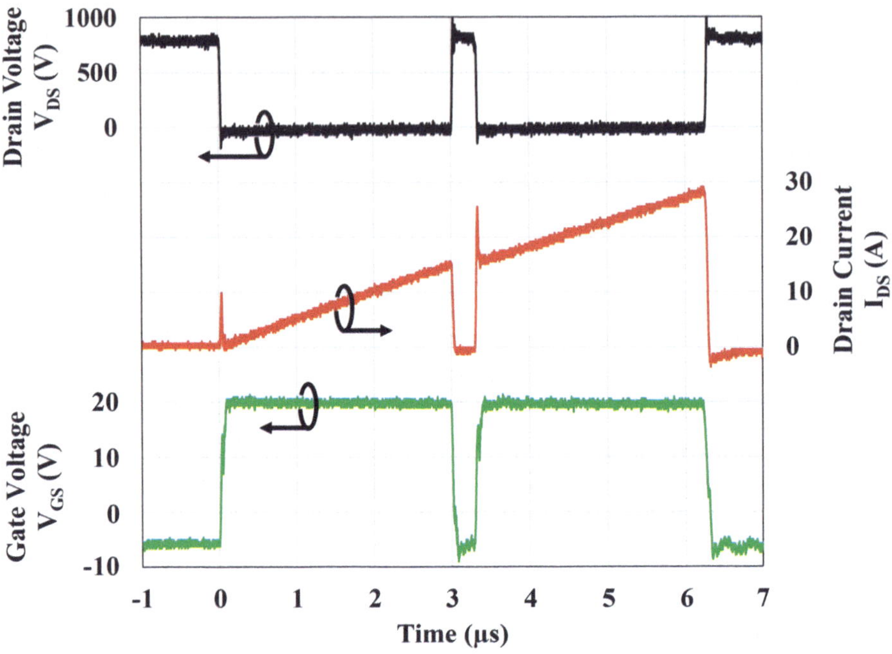

Fig. 17.15 Measured double-pulse waveforms for the stand-alone SiC power MOSFET

The measured double-pulse test switching waveforms for the composite transistor in the BaSIC(EMM) topology are shown in Fig. 17.16 for current rising to 15 A at the end of the first gate pulse. They were obtained using a DC bus voltage of 800 V and SiC MOSFET gate bias of −5/+20 V. The waveforms are similar to the stand-alone SiC MOSFET but the switching losses were found to be larger as discussed in previous chapters. The voltage at the sense node is shown for completeness.

The turn-on, turn-off, and total switching losses are plotted in Figs. 17.17, 17.18, and 17.19, respectively. They are given as a function of the DC gate bias $V_{G,EMM}$ applied to the Si EMM device. The measured turn-off switching loss for the BaSIC(EMM) topology is the same as that measured for the stand-alone SiC power MOSFET (see Fig. 17.18). The turn-off switching loss for the BaSIC(EMM) topology increases when the DC gate bias for the Si EMM device is reduced. This is due to the reduction of the transconductance of the composite MOSFET in the BaSIC(EMM) topology. As a consequence, the total switching power loss for the BaSIC(EMM) topology also increases when the DC gate bias for the Si EMM device is reduced. The BaSIC(EMM) implementation exhibits a 40% increase in total switching loss at a $V_{G,EMM}$ of 3.7 V.

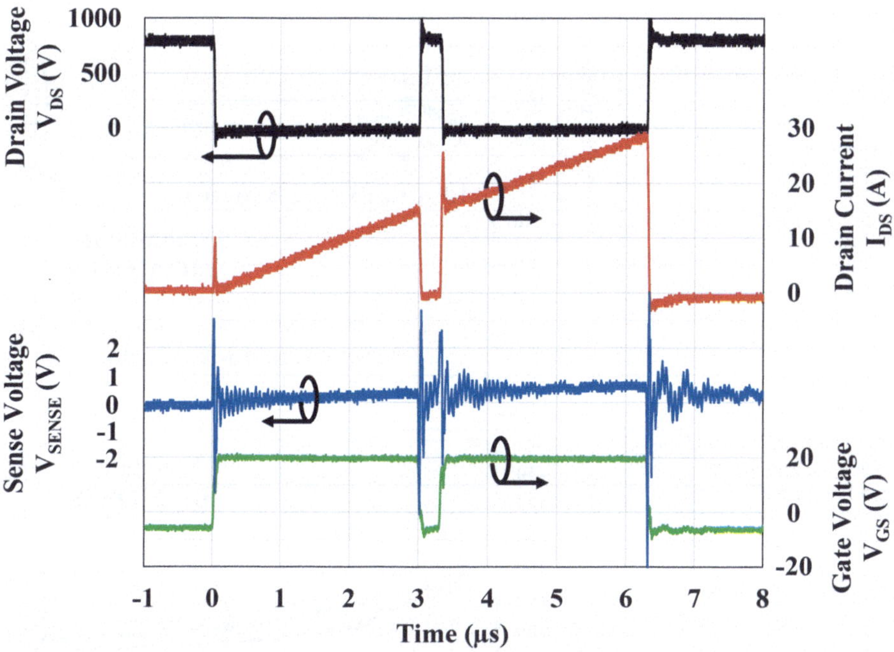

Fig. 17.16 Measured double-pulse waveforms for the composite MOSFET in the BaSIC(EMM) topology

Fig. 17.17 Measured turn-on switching loss for the stand-alone SiC power MOSFET and BaSIC(EMM) module with various DC gate bias voltages

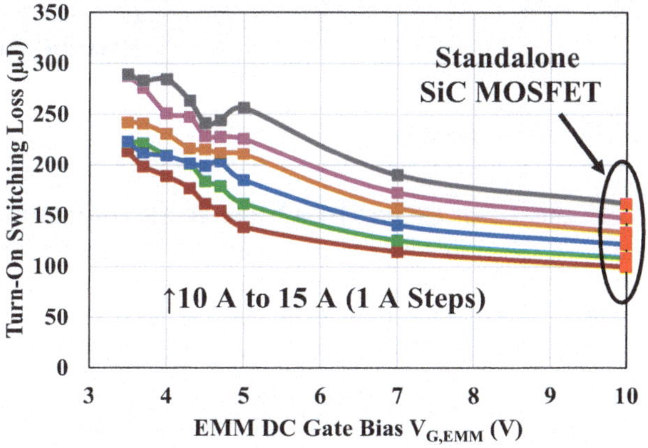

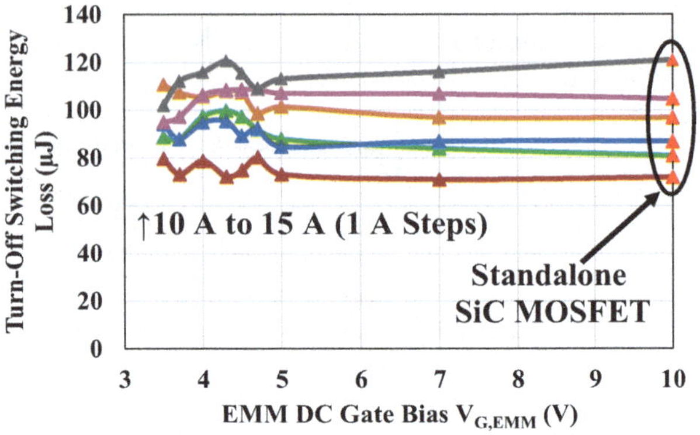

Fig. 17.18 Measured turn-off switching loss for the stand-alone SiC power MOSFET and BaSIC(EMM) module with various DC gate bias voltages

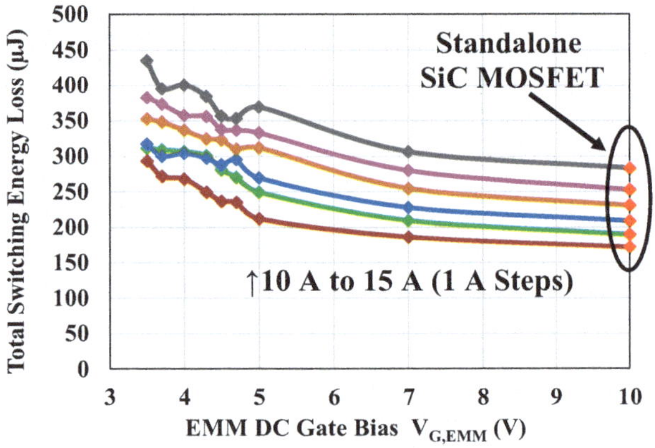

Fig. 17.19 Measured total switching loss for the stand-alone SiC power MOSFET and BaSIC(EMM) module with various DC gate bias voltages

17.8 Discussion

As discussed in Chaps. 10–13, the BaSIC(EMM) topology requires a trade-off between the peak short-circuit current suppression and an increase in composite MOSFET on-resistance and an increase in total switching loss compared to the stand-alone SiC MOSFET. In the case of the devices used to build the BaSIC(EMM) module discussed in this chapter, the BaSIC(EMM) topology suppresses the peak short-circuit current by a factor of 1.9× in the case of $V_{G,EMM}$ of 3.7 V at the expense of an increase in composite MOSFET on-resistance by 12% and an increase in total switching losses by 40% when compared to the stand-alone SiC MOSFET.

When building the BaSIC(EMM) modules for the overcurrent protection studies, a UnitedSiC Si MOSFET (AW1060) was utilized due to availability of bare die. It is beneficial to examine the trade-off curve between the saturated drain current and the on-resistance for the Si EMM device. The trade-off curve between on-state resistance at 25 °C and saturation current under short-circuit conditions for the UnitedSiC Si MOSFET AW1060 is plotted as the red line in Fig. 17.20. The on-resistance values for the trade-off curve were obtained from the datasheet graph at a

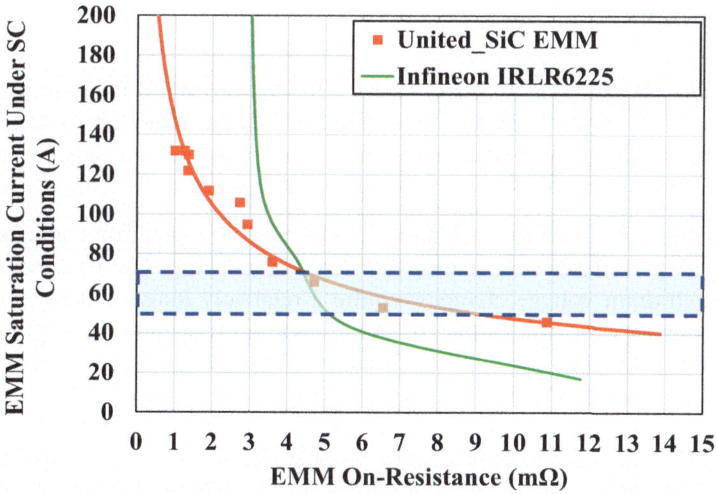

Fig. 17.20 Trade-off curve between saturated drain current and on-resistance

drain bias of 0.1 V for the UnitedSiC Si MOSFET AW1060 device. The saturation drain current values for the UnitedSiC Si MOSFET AW1060 device were obtained using the peak SC current values observed during the high-voltage, high-current output curve characterization as discussed in Sect. 17.5.1. The SiC power MOSFET CPM2-1200-0160B has a short-circuit withstand time of 3.5 μs as shown in Chap. 10 due to a saturated drain current of 136 A. Based on these values, it is necessary to reduce the saturated drain current by a factor of 2.86× in order to increase the short-circuit withstand time of the SiC power MOSFET to 10 μs. The saturated drain current for the Si EMM must therefore be 47.6 A. A short-circuit withstand time of 6.8 μs can be achieved if the Si EMM saturated drain current is 70 A which may also be satisfactory. The 47–70 A window of interest is highlighted in Fig. 17.20 by the transparent blue box. Using the trade-off graph for the UnitedSiC Si MOSFET AW1060 device (red line), the on-resistance is then increased by 4.5–9.5 mΩ, which is 2.4–5.1% of the SiC power MOSFET on-resistance. This is a relatively small increase in the on-resistance.

The selection methodology described in Chap. 13 identified the Si EMM device with the best trade-off curve to be the IRF6225 (now Infineon EMM IRLR6225 due to acquisition of International Rectifier by Infineon). The trade-off curve between on-state resistance at 25 °C and saturation current at 175 °C for the Infineon EMM IRLR6225 was previously shown in Fig. 13.9. This data is reproduced in Fig. 17.20 as the green line for comparison with the UnitedSiC Si MOSFET AW1060 device. For the 47–70 A window of interest, the on-resistance of the Infineon EMM IRLR6225 device ranges from 4.5 to 5 mΩ, which is 2.4–2.7% of the SiC power MOSFET on-resistance. This is superior to the performance of the UnitedSiC Si MOSFET AW1060 device.

17.9 Summary

The standard DESAT protection circuit for a SiC power MOSFET requires the use of a high-voltage diode to isolate the control circuit from the large excursions in drain voltage during normal circuit operation. This method has three major shortcomings:

(a) High-voltage rated, fast responding diodes with low junction capacitance are not readily available commercially. When available, they can be expensive.
(b) The high drain voltage changes require a relatively long blanking time. This limits the response time of the DESAT circuit for protection of medium-voltage and high-voltage power MOSFETs and IGBTs.
(c) The drain of the SiC power MOSFET (or the Collector of Si IGBTs) undergoes large voltage fluctuations during its operations. Since the DESAT diode taps this node, adequate clearance must be provided for the board layout of gate drive components. In case of power converters with multiple switches, the additional circuit board area limits their power density resulting in bulky system assemblies.
(d) Implementation of the DESAT cannot be readily scaled up for power devices with voltage ratings about 6.5 kV due to a lack of DESAT diodes with similar voltage ratings, as compared to the abundance of low voltage EMMs and Schottky/Zener diodes.

The overcurrent protection circuit for the BaSIC(EMM) topology circumvents the above challenges due to utilization of the voltage sense node:

(a) The voltage sense node in the BaSIC(EMM) topology does not undergo high voltage fluctuations. Consequently, the BaSIC overcurrent protection circuit, which is connected to the sense node, can be operated with a shorted blanking time.
(b) Since the BaSIC overcurrent protection circuit is connected to the sense node operating at less than 10 V. it does not require board layout clearances or specialized gate drive ICs. It can be implemented with compact form factors.
(c) The BaSIC overcurrent protection circuit utilizes low voltage Schottky diodes which are available in abundance. These fast responding, low capacitance diodes can be readily connected to the voltage sense node in the BaSIC overcurrent protection circuit to realize rapid fault detection.
(d) The BaSIC(EMM) topology uses low voltage enhancement mode MOSFETs that are commercially available in abundance. The EMM voltage rating is only 20–30 V, determined by the maximum gate drive applied to the SiC power MOSFET. Consequently, the BaSIC(EMM) topology and the BaSIC overcurrent protection circuit can be easily scaled to power transistors of a wide range of rated voltages. This is beneficial to deployment of SiC power MOSFET with voltage ratings of 6.5 kV and above, and SiC IGBT with even higher blocking voltage ratings of 20 kV and above.

In conclusion, the BaSIC overcurrent protection circuit provides superior fault detection capabilities to the DESAT method while simultaneously enhancing the short-circuit withstand capability of the SiC power MOSFET. Consequently, simple gate drive circuits along with the BaSIC overcurrent protection circuit are sufficient to operate the SiC power MOSFETs under normal operating conditions and protect them under short-circuit fault events.

References

1. A. Bolotnikov et al., Overview of 1.2 kV–2.2 kV SiC MOSFETs targeted for industrial power conversion applications, in *IEEE Applied Power Electronics Conference and Exposition* (2015), pp. 2445–2452
2. D. Sadik et al., Short circuit protection circuits for silicon carbide power transistors. IEEE Trans. Ind. Electron. **63**, 1995–2004 (2016)
3. Texas Instruments, Understanding the short-circuit protection for silicon carbide MOSFETs. TI TechNotes, Jan 2018, pp. 1–4
4. V. John, B.-S. Suh, T.A. Lipo, Fast-clamped short-circuit protection of IGBTs. IEEE Trans. Ind. Appl. **35**, 477–486 (1999)
5. K. Han, B.J. Baliga, 1.2 kV 4H-SiC SenseFET with monolithically integrated sensing resistor. IEEE Electron Device Lett. **41**, 437–440 (2020)
6. S. Mosevic, J. Wang, R. Burgos, D. Boroyevich, M. Jaksic, C. Stancu, B. Peaslee, Comparison between desaturation sensing and Rogowski coil current sensing for short-circuit protection of 1.2 kV SiC MOSFET module, in *IEEE Applied Power Electronics Conference and Exposition* (2018), pp. 2666–2672
7. Y. Feng, S. Shao, J. Du, Q. Chen, Z. Zhang, X. Wu, Short-circuit and over-current fault detection for SiC MOSFET modules based on tunnel magnetoresistance with predictive capabilities. IEEE Trans. Power Electron. **37**, 3719–3723 (2022)

Chapter 18
BaSIC(DMM) for Si IGBTs

Silicon IGBTs are ubiquitously used in power conversion applications due to their high on-state current density, gate voltage control with low driving energy, gate voltage dependent saturated-current output characteristics, high blocking voltages, large current handling capability, and ruggedness to fault conditions [1]. A trade-off between on-state voltage drop ($V_{CE,SAT}$) and turn-off switching loss (E_{OFF}) for IGBTs was recognized from the inception [2]. Subsequently, short-circuit withstand capability for IGBTs became a highly desired requirement for motor drive applications. A trade-off between short-circuit withstand time (t_{SC}), on-state voltage drop ($V_{CE,SAT}$), and switching losses (E_{OFF}) has been therefore commonly performed by device manufacturers [3].

The trench-gate structure for IGBTs was introduced in the 1990s to reduce the on-state voltage drop [1]. These devices were found to exhibit poor short-circuit withstand times t_{SC} due to a very high saturated collector current related to their large channel density [4]. Satisfactory short-circuit withstand times t_{SC} for trench-gate IGBTs had to be achieved by reducing the channel density at the expense of larger on-state voltage drop $V_{CE,SAT}$ [5].

Manufacturers now supply two types of IGBTs: (a) Type 1 devices with short-circuit withstand times t_{SC} exceeding 10 μs for motor drive applications and (b) Type 2 devices with small short-circuit withstand time t_{SC} but lower on-state voltage drop $V_{CE,SAT}$ for resonant converter applications such as induction cooktops. Application engineers need to choose between these options when designing their circuits.

As in the case of the SiC power MOSFETs discussed in Chap. 5, the short-circuit withstand capability, t_{SC}, of the IGBT is defined as the maximum time duration before the onset of device failure under short-circuit (SC) conditions. This stress condition is referred to as the SC Type 1 fault condition [6].

In the datasheet of typical commercially available trench-gate Si IGBTs, the short-circuit withstand time, $t_{SC,}$ is measured at a DC bus voltage of 80% of their rated voltage. The short-circuit withstand time can be well below the industry accepted standard of 10 μs for devices optimized for achieving low $V_{CE,SAT}$ as

B. J. Baliga, A. Kanale, *The BaSIC Topology*,
https://doi.org/10.1007/978-3-031-86630-2_18

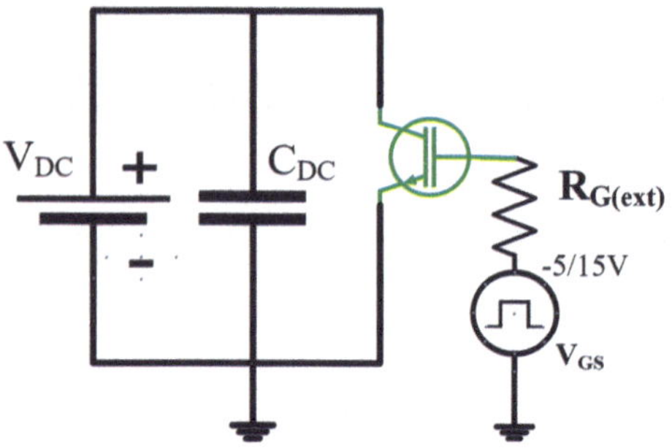

Fig. 18.1 Short-circuit test configuration for the stand-alone Si IGBT

demonstrated in this chapter when using standard gate drive voltage of 15 V for on-state operation. However, for motor drive applications, a t_{SC} of over 10 μs is required to allow detection of the fault and subsequent turn-off of the gate bias before onset of device failure with the commonly available gate drivers that utilize the DESAT short-circuit detection method.

In this chapter, the application of the BaSIC topology with a gate-source shorted (GSS), depletion mode Si low breakdown voltage power MOSFET (DMM) connected in series with the emitter of the IGBT device is investigated to improve the short-circuit withstand time for the Type 2 IGBTs [7]. The approach is first utilized for enhancing the performance of a commercially available Type 2 1.2 kV IGBT. It is then utilized to achieve enhanced performance for a Type 2 1.7 kV IGBT to demonstrate scalability.

18.1 Short-Circuit Configuration: Stand-Alone Si IGBT

The test circuit for obtaining the short-circuit withstand time for the stand-alone Si IGBT is shown in Fig. 18.1. It consists of a DC power supply with voltage V_{DC}, a capacitor C_{DC} across the DC power supply to deliver the short-circuit current into the Si IGBT, and the gate drive voltage V_{GS} with series resistor R_G. The gate drive voltage for the Si IGBT was set at 15 V in accordance with its datasheet.

18.2 Short-Circuit Configuration: BaSIC(DMM) Topology

The short-circuit tests for the Si IGBTs with the BaSIC(DMM) topology were performed with a silicon gate-source-shorted (GSS) Si depletion mode MOSFET (DMM) connected in series with the emitter electrode of the Si IGBT. This testing configuration is shown in Fig. 18.2 with the composite IGBT labeled as CT-1. It is worth pointing out that the gate drive voltage is applied to the gate of the Si IGBT as usually done in its power circuit applications. The silicon DMM device only serves as the non-linear resistance Q_{LV} with no gate bias applied to its gate terminal.

Fig. 18.2 Short-circuit test configuration for Si IGBT with BaSIC(DMM) topology: case CT-1 with single DMM device

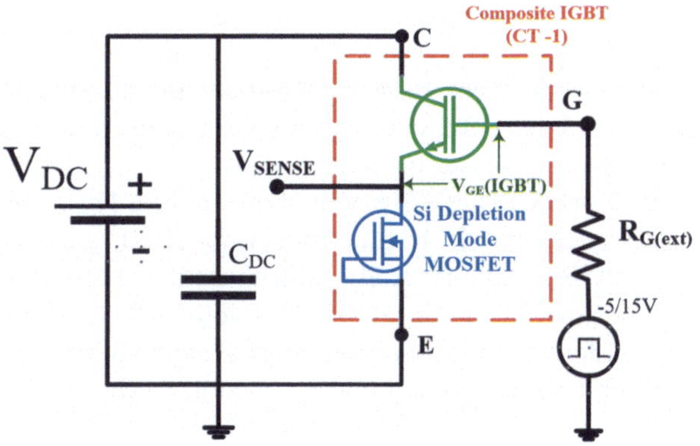

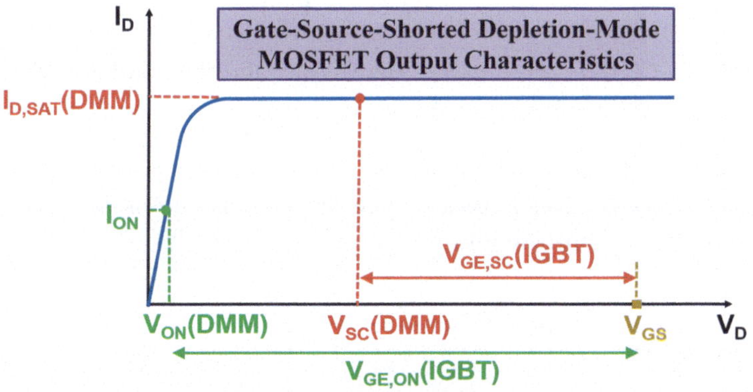

Fig. 18.3 Typical output characteristics of the Si DMM with nomenclature defined

18.3 Operating Principles of BaSIC(DMM) Topology for Si IGBTs

In the BaSIC(DMM) topology for Si IGBTs, the GSS Si DMM operates as a two-terminal device because no bias is applied to its gate terminal. Consequently, the Si IGBT with the series GSS Si DMM behaves as a three terminal device referred to as a "Composite Transistor (CT)" in this chapter. The Si power DMMs must be designed with very low on-resistance to minimize the increase in the forward voltage drop of the composite IGBT.

The representative output characteristics of the GSS Si DMM are shown in Fig. 18.3 with nomenclature used in the discussion. It behaves as a non-linear resistance with low resistance under normal on-state current (I_{ON}) operating conditions of the IGBT. Its resistance increases at higher current levels until the current becomes saturated at a value $I_{D,SAT}$ (DMM). This saturated current of the GSS Si DMM determines the short-circuit current for the composite transistor.

The BaSIC(DMM) topology employs a lower current saturation level of the GSS Si DMM than for the Si IGBT to increase the short-circuit withstand time t_{SC}. During a short-circuit event, the current through the IGBT and the DMM increases until the Si DMM enters the current saturation regime. The drain voltage of the GSS Si DMM then increases from the on-state value V_{ON}(DMM) of ~0.5 V to a much larger value V_{SC}(DMM) of ~6–8 V as shown in Fig. 18.3. This reduces the gate-to-emitter voltage for the Si IGBT from $V_{GE,ON}$(IGBT) to $V_{GE,SC}$(IGBT). The value for $V_{GE,SC}$(IGBT) automatically adjusts until the saturation current of the Si IGBT matches that of the Si DMM. This automatic regulation of the IGBT gate voltage during short-circuit conditions and balancing of the current within the IGBT and the MOSFET devices is an important feature of the BaSIC(DMM) topology.

The saturation current for the IGBT is given by:

$$I_{C,SAT} = K\left(V_{GE} - V_{TH}\right)^2 \tag{18.1}$$

where $K = Z\mu_{CH}C_{OX}/[2L_{CH}(1 - \alpha_{PNP})]$. Here, V_{TH} = threshold voltage, Z = channel width, μ_{CH} = channel mobility, C_{OX} = gate oxide capacitance, L_{CH} = channel length, and α_{PNP} = gain of the PNP transistor within the IGBT. The value for $I_{C,SAT}$ for the IGBT at the typical gate drive voltage (V_{GE} = 15 V) is very large for devices optimized with low $V_{CE,SAT}$. This leads to a large short-circuit current I_{SC}.

The short-circuit withstand time for a Si IGBT is given by [1]

$$t_{SC}\left(\text{IGBT}\right) = \frac{\left(T_{CR} - T_{HS}\right)W_{Si}A_A C_V}{I_{SC}V_{DC}} \tag{18.2}$$

where T_{CR} is the critical temperature for failure of the IGBT, T_{HS} is the heat sink temperature, W_{Si} is the wafer thickness, A_A is the IGBT chip active area, C_V is the volumetric specific heat for Si, and V_{DC} is the collector supply voltage. The short-circuit withstand time is inversely proportional to the magnitude of I_{SC}, the short-circuit current magnitude. This value to approximately the same as $I_{C,SAT}$ in Eq. (18.1). Large values for the $I_{C,SAT}$ are therefore detrimental for achieving the desired short-circuit withstand time of 10 μs.

In the presence of the GSS Si DMM, the IGBT current becomes limited by the saturated drain current of the GSS Si DMM. Consequently:

$$I_{D,SAT}\left(\text{IGBT}\right) = K\left(V_{GE,SC}\left(\text{IGBT}\right) - V_{TH}\right)^2 = I_{D,SAT}\left(\text{DMM}\right) \tag{18.3}$$

Rearranging Eq. (18.3):

$$V_{GE,SC(IGBT)} = \sqrt{I_{D,SAT}\left(\text{DMM}\right)/K} + V_{TH} \tag{18.4}$$

The voltage at the drain of the GSS Si DMM is then

$$V_{SC}\left(\text{DMM}\right) = V_{GS} - V_{GE,SC}\left(\text{IGBT}\right) \tag{18.5}$$

This voltage defines the operating point of the GSS Si DMM during the SC event. It is in the range of 6–8 V.

The on-state voltage drop $V_{CE,SAT}$(CT) of the composite IGBT consists of the voltage drop across the IGBT plus the voltage drop across the GSS Si DMM device. The on-state voltage drop of the IGBT in the BaSIC(DMM) topology is increased because the voltage drop across the GSS Si DMM decreases the IGBT gate drive voltage during on-state operation:

$$V_{GE,ON}(IGBT) = V_{GS} - V_{ON}(DMM) \tag{18.6}$$

as indicated in Fig. 18.3. The on-state voltage drop of the composite IGBT CT-1 is therefore:

$$V_{ON}(CT-1) = V_{CE,SAT}\left[IGBT@V_{GE} = V_{GE,ON}(IGBT)\right] + V_{ON}(DMM) \tag{18.7}$$

with the voltage drop across the GSS Si DMM obtained using:

$$V_{ON}(DMM) = I_{ON}R_{ON}(DMM) \tag{18.8}$$

The resistance of the GSS Si DMM also has a detrimental impact on the transconductance of the composite IGBT which impacts its switching performance. Due to the series source resistance R_{ON}(DMM), the transconductance of the composite transistor CT-1 is given by

$$g_M(CT-1) = g_M / \left[1 + g_M R_{ON}(DMM)\right] \tag{18.9}$$

A reduction in g_M slows down the turn-on transient for the collector voltage of the composite transistor CT-1 [1]:

$$\frac{dV_{CE}}{dt}(CT-1) = \frac{dV_{CE}}{dt}(IGBT) - \frac{I_{ON}R}{R_G C_{GC}(V_{CE})} \tag{18.10}$$

The reduced value of (dV_{CE}/dt)(CT-1) increases the collector voltage transient time resulting in larger switching losses.

18.4 BaSIC(DMM) Topology for Si IGBTs with Multiple DMM Devices

It is beneficial to reduce the resistance of the non-linear resistance Q_{LV} in any BaSIC topology implementation. This can be achieved by paralleling the silicon DMM devices. The case with two DMM devices in parallel is shown in Fig. 18.4 with the composite IGBT labeled as CT-2. The case with three DMM devices in parallel is shown in Fig. 18.5 with the composite IGBT labeled as CT-3. The paralleling of the

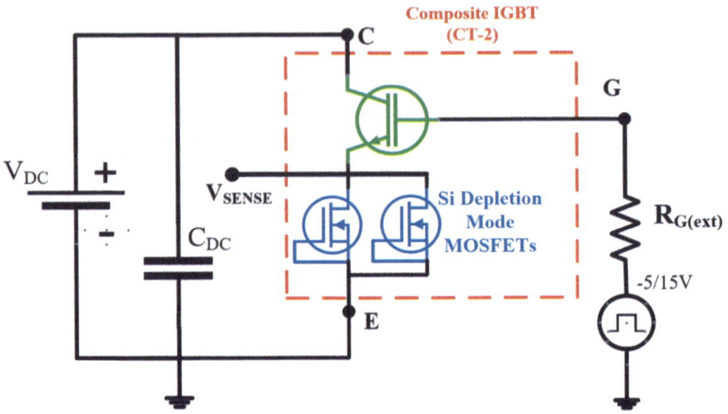

Fig. 18.4 The Si IGBT with BaSIC(DMM) topology: case CT-2 with two DMMs in parallel

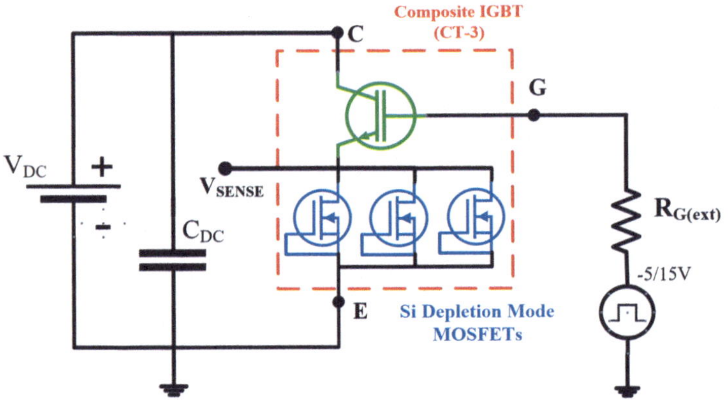

Fig. 18.5 The Si IGBT with BaSIC(DMM) topology: case CM-3 with three DMMs in parallel

Si DMM devices reduces the on-state voltage drop of the composite IGBT. However, this also increases the saturation current for the composite IGBT, which will shorten the short-circuit withstand time. A trade-off curve for the Si IGBT in a BaSIC(DMM) topology can therefore be created by changing the number of DMM devices connected in parallel.

18.5 Experimental Results: A&O 1.2 kV IGBT

The Alpha and Omega AOK20B120E1 IGBT product is rated as a 1.2 kV, 20 A device with a gate drive voltage of 15 V. It comes with an anti-parallel diode for use in H-bridge applications. According to its datasheet, the device is optimized for induction cooking applications where low $V_{CE,SAT}$ and turn-off losses are desirable.

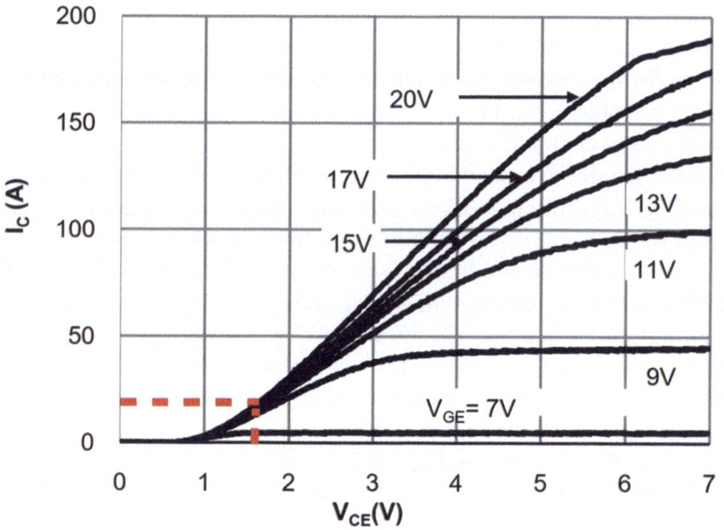

Fig. 18.6 The Alpha and Omega AOK20B120E1 IGBT product on-state characteristics from its datasheet

No information on its short-circuit withstand capability is provided in the datasheet. This device was therefore chosen as a good candidate for the BaSIC(DMM) topology to examine if its short-circuit withstand capability can be enhanced with minimum impact on the on-state voltage drop and switching losses.

The on-state characteristics of the A&O IGBT at 25 °C are reproduced in Fig. 18.6 from the datasheet. The on-state operating point is indicated using the red dashed lines. The on-state voltage drop $V_{CE,SAT}$ at 20 A is 1.6 V as stated in the datasheet at a gate bias of 15 V. It can be seen that the on-state voltage drop is not a very sensitive function of the gate drive voltage near 15 V. This is beneficial for operation of this device in the BaSIC(DMM) topology when the effective gate drive voltage is reduced by about 0.5 V due to the resistance of the DMM device.

18.5.1 High Voltage Output Characteristics

As explained in Chap. 2, the power transistor operates in its current saturation mode during the short-circuit event while supporting a high voltage equal to the DC supply voltage for the load. This operating point is part of the power transistor's output characteristics. Knowledge of the high-voltage, high-current output characteristic is therefore invaluable for understanding the short-circuit capability of power transistors. However, these characteristics are not provided in the A&O IGBT datasheet.

A novel method for extracting the high-voltage, high-current output characteristics of power transistors was developed by using the short-circuit test configuration

[8]. The short-circuit test configuration shown in Fig. 18.1 was used with a small pulse width of 1.5 μs. During this time, the current increases to a peak value that is representative of the saturation current of the transistor at the DC supply voltage. The DC supply voltage can be varied in small steps to generate the high-voltage, high-current output characteristic for the chosen gate bias.

The experimentally extracted high-voltage, high-current output characteristic at a gate bias of 15 V for the stand-alone IGBT, and the composite transistors CT-1 and CT-2 are shown in Fig. 18.7. The composite transistor CT-3 with three Si DMM devices in parallel was not evaluated. The measured saturation current values at a collector bias of 960 V, for the stand-alone IGBT, composite transistor CT-2, and composite transistor CT-1 are 72, 56, and 31 A, respectively. The high-voltage, high-current output characteristic for the stand-alone IGBT was also extracted with gate bias values chosen to obtain the same saturation currents as the composite transistors CT-1 and CT-2. These gate bias values were found to be 12 and 8.5 V for the stand-alone Si IGBT as shown in Fig. 18.7.

18.5.2 Short-Circuit Test Results

The short-circuit withstand time for the power transistors was obtained by using the test configuration shown in Figs. 18.1, 18.2, and 18.4 using a gate drive voltage of 15 V. The short-circuit pulse width was gradually increased while capturing the current waveform until the IGBT device failed. The DC supply voltage is monitored as

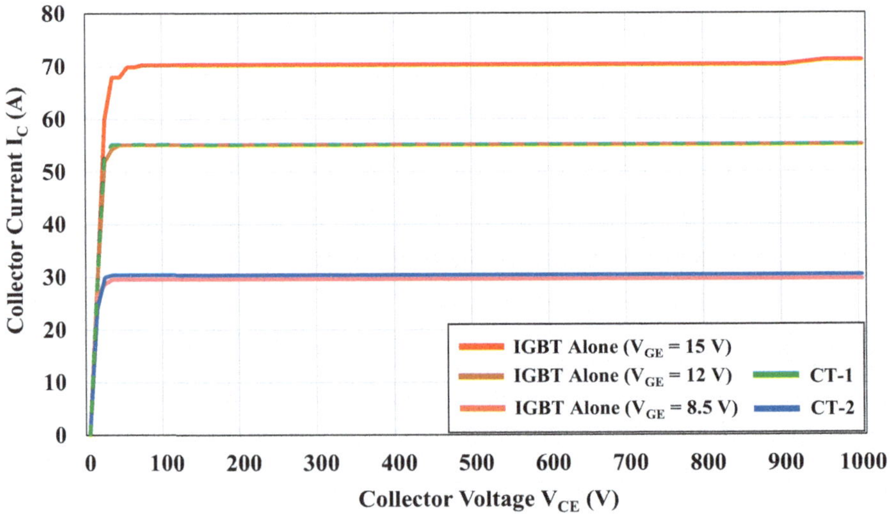

Fig. 18.7 High-voltage, high-current output characteristics: stand-alone Si A&O IGBT with three gate bias voltages, composite IGBT CT-1, and composite IGBT CT-2

shown by the black line. It is important to make sure that the capacitor C_{DC} is adequate to maintain the voltage at the desired value, 960 V in this case.

The red line was observed for the stand-alone Si A&O IGBT. The current reaches 72 A at the 1.5 μs time used to generate the high-voltage, high-current output characteristics in Fig. 18.7. However, the peak short-circuit current occurs at 4.8 μs with a much larger value of 130 A, leading to device failure with a short-circuit withstand time of 4.8 μs. These results confirm that an IGBT designed and optimized for low on-state voltage drop by manufacturers does not have good short-circuit withstand capability.

The green line in Fig. 18.8 was obtained for the composite IGBT CT-1. Its current reaches a peak value of 31 A at the 1.5 μs time used to generate the high-voltage, high-current output characteristics in Fig. 18.7. The current reduces gradually with time due to the temperature rise of the Si IGBT. The composite IGBT CT-1 failed when the pulse width was increased to a very large value of 36 μs—well beyond the desired value of 10 μs.

The blue line in Fig. 18.8 was obtained for the composite IGBT CT-2. Its current reaches a peak value of 56 A at the 1.5 μs time used to generate the high-voltage, high-current output characteristics in Fig. 18.7. The current reduces with time due to the temperature rise of the Si IGBT. The composite IGBT CT-2 failed when the pulse width was increased to a large value of 20.3 μs—well beyond the desired value of 10 μs. In all cases, it was found that the Si IGBT failed but the Si DMM

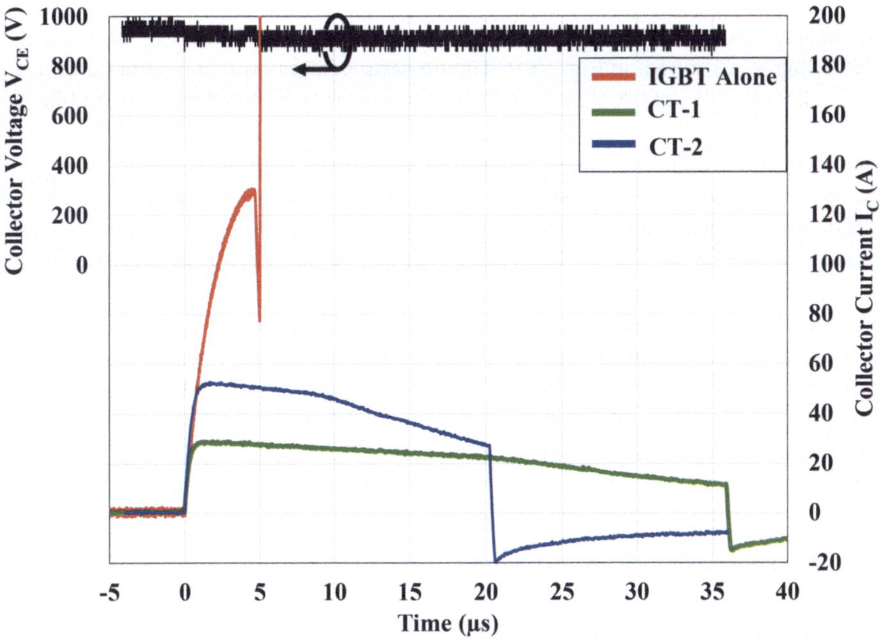

Fig. 18.8 Measured short-circuit waveforms: stand-alone Si A&O IGBT, composite IGBT CT-1, and composite IGBT CT-2

devices were not damaged. This is due to much lower power dissipation in the Si DMM devices because their drain voltage is only 6–8 V during the short-circuit event.

These short-circuit test results provide validation of the utility of the BaSIC topology for enhancing the short-circuit withstand capability of commercially available Type 2 IGBTs which are optimized for low on-state voltage drop. This topology was able to reduce the short-circuit current level and significantly increase the short-circuit withstand time for the 1.2 kV Si IGBT.

18.6 Normal Circuit Operation: A&O 1.2 kV IGBT

The BaSIC(DMM) topology discussed in this chapter is based on the addition of a Si depletion mode MOSFET series with the emitter electrode of the IGBT. The impact of the resistance of this element on the on-state voltage drop and switching behavior of the Si IGBT must be quantified. Trade-off curves between enhancing the short-circuit time and an increase on-state voltage drop and switching losses must be developed to fully comprehend the value of the BaSIC topology to Si IGBTs.

18.6.1 On-State Characteristics

The measured on-state characteristics of the stand-alone A&O IGBT at various gate bias voltages are shown in Fig. 18.9. The on-state voltage drop at a rated current of 20 A with gate bias of 15 V is 1.74 V. It increases to 1.80 V when the gate bias is reduced to 12 V that was used in Fig. 18.7 to generate a high-voltage, high-current characteristic. The device enters its current saturation mode well below 20 A at a gate bias of 8 V making on-state operation at this current level no longer possible.

The on-state characteristics of the composite IGBTs CT-1 and CT-2 are shown in Figs. 18.10 and 18.11. The on-state voltage drop at a rated current of 20 A with gate bias of 15 V are 3.24 and 2.46 V for devices CT-1 and CT-2. The single DMM has an on-resistance of 60 mΩ. Consequently, it adds a voltage drop of 1.2 V at a current of 20 A to the Si IGBT in the case of the composite IGBT CT-1. The two DMMs in parallel used for the composite IGBT CT-2 have a resistance of 30 mΩ. Consequently, they add a voltage drop of 0.6 V at a current of 20 A to the Si IGBT in the case of the composite IGBT CT-1.

18.6.2 Impact on Capacitances

The capacitances of an IGBT determine its switching performance [1]. The input, output, and reverse transfer capacitances of the stand-alone Si IGBT and the two composite IGBTs CT-1 and CT-2 were therefore obtained. These measured

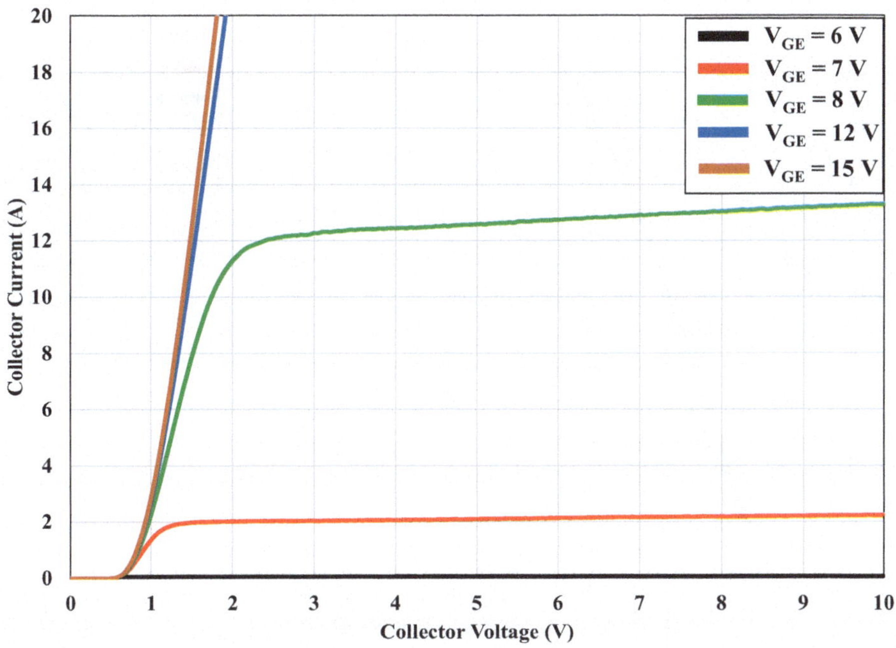

Fig. 18.9 Measured on-state characteristics of stand-alone Si A&O IGBT with various gate bias voltages

characteristics are displayed in Fig. 18.12. The black lines apply for the stand-alone A&O Si IGBT product. The measured values for the C_{ISS}, C_{OSS}, and C_{RSS} are very close to values provided in the datasheet for the device.

The red and green lines in Fig. 18.12 apply to the composite IGBT CT-1 and CT-2. They overlap the lines for the stand-alone A&O Si IGBT. This demonstrates that the BaSIC(DMM) topology does not alter the capacitances of the Si IGBT.

The reason for this can be explained by using the equivalent circuit for the BaSIC topology shown in Fig. 18.13. The Si IGBT has its usual gate-emitter $C_{IN,IGBT}$, collector-emitter $C_{O,IGBT}$, and gate-collector $C_{GD,IGBT}$ capacitances. In contrast, the silicon DMM device has only its usual drain-source $C_{O,DMM}$, and gate-drain $C_{GD,DMM}$ capacitances. The gate-source $C_{GS,DMM}$ capacitance is eliminated by the shorting of its gate and source.

Based on the equivalent circuit shown in Fig. 18.13, the input capacitance for the composite IGBT is given by:

$$C_{ISS,(CT)} = \frac{C_{ISS,IGBT} \cdot C_{OSS,DMM}}{C_{ISS,IGBT} + C_{OSS,DMM}} \tag{18.11}$$

The applied collector bias on the composite IGBT is supported by the Si IGBT because the Si DMM is a normally-on device. Its output capacitance, $C_{OSS,DMM}$ at low V_D, is 10,000 pF. This is highlighted in Fig. 18.14 where a figure from the

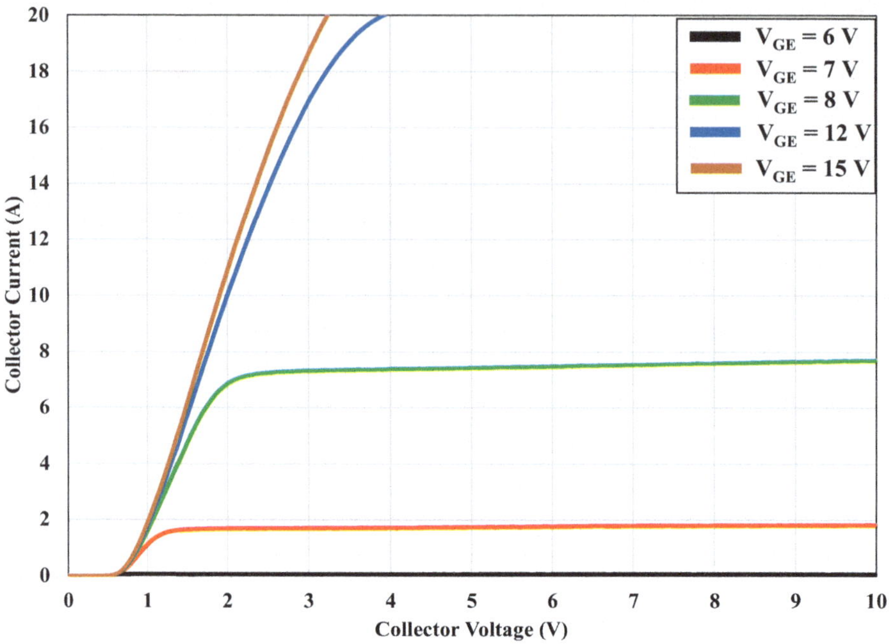

Fig. 18.10 Measured on-state characteristics of composite IGBT CT-1 with various gate bias voltages

datasheet for the Si DMM device is reproduced for reference. Since the $C_{OSS,DMM}$ value (10,000 pF) is much larger than the $C_{ISS,IGBT} = 1500$ pF for the IGBT, the input capacitance of the composite IGBT is very close to that of the stand-alone Si IGBT.

Based on Fig. 18.13, the output capacitance for the composite IGBT is given by:

$$C_{OSS,(CT)} = \frac{C_{OSS,IGBT} \cdot C_{OSS,DMM}}{C_{OSS,IGBT} + C_{OSS,DMM}} \tag{18.12}$$

As mentioned above, $C_{OSS,DMM}$ at low V_D (10,000 pF) is much larger than $C_{OSS,IGBT}$ (<20 pF), making the output capacitance of the composite IGBT very close to that of the stand-alone Si IGBT.

The reverse transfer capacitance, C_{RSS}, for the composite IGBT is measured with gate of the Si IGBT shorted to the source of the Si DMM device per standard practice in the industry. The shorting of the gate and emitter terminals of the composite IGBT eliminates any influence of the Si DMM on this capacitance. Consequently, the C_{RSS} for the composite device is equal to that of the stand-alone Si IGBT.

From the above discussion, it is apparent that the BaSIC(DMM) topology does not alter the capacitances of the Si IGBT. This is a favorable outcome from the point of view of the switching behavior.

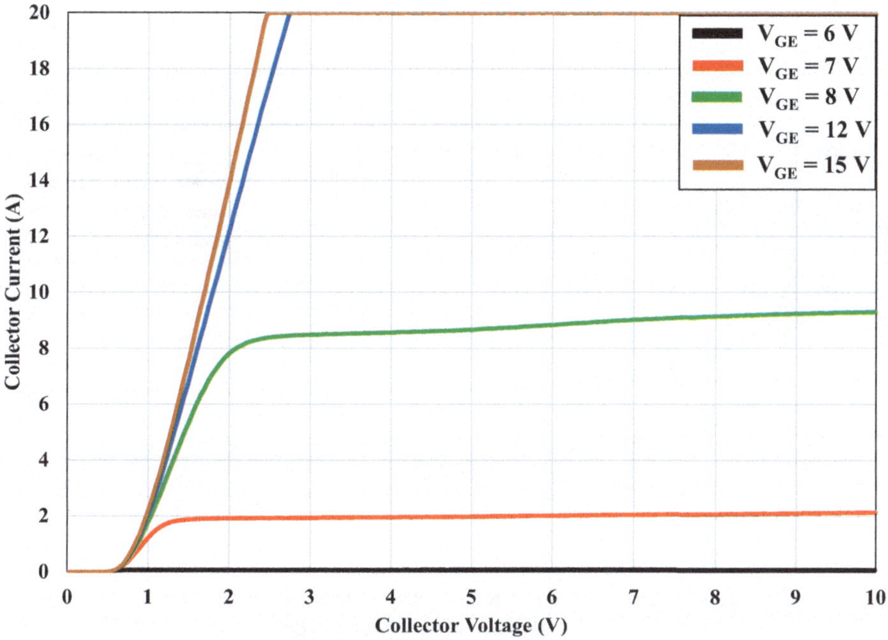

Fig. 18.11 Measured on-state characteristics of composite IGBT CT-2 with various gate bias voltages

18.6.3 Impact on Transconductance

The transconductance g_M of the IGBT plays a role in the switching transient for the drain voltage during operation with an inductive load [1]. Its value can be extracted from the transfer characteristics of the device. The transfer characteristics for the stand-alone and composite IGBTs were measured at a collector bias of 20 V. The transconductance was extracted by taking a derivative of the collector current with respect to the gate voltage. The results are displayed in Fig. 18.15. The values of the transconductance can be compared for the three cases at the same collector current level of 10 A by following the red dashed lines in the figure. It is given by the values at the three points indicated on the transconductance curve.

The stand-alone Si IGBT has a transconductance of 17.8 S, which is close to the value in its datasheet. The transconductances of the composite IGBTs CT-1 and CT-2 were 8.4 and 11.4 S, respectively. The observed reduction in the transconductance is associated with the resistance of the Si DMM device as described in Eq. (18.9). The transconductance value predicted by this equation is 8.6 S for the case of a series resistance R_S of 60 mΩ corresponding to a single DMM in the composite IGBT CT-1. The transconductance value predicted by this equation is 11.6 S for the case of a series resistance R_S of 30 mΩ corresponding to two DMMs in parallel in the composite IGBT CT-2. These calculated values are in excellent agreement with

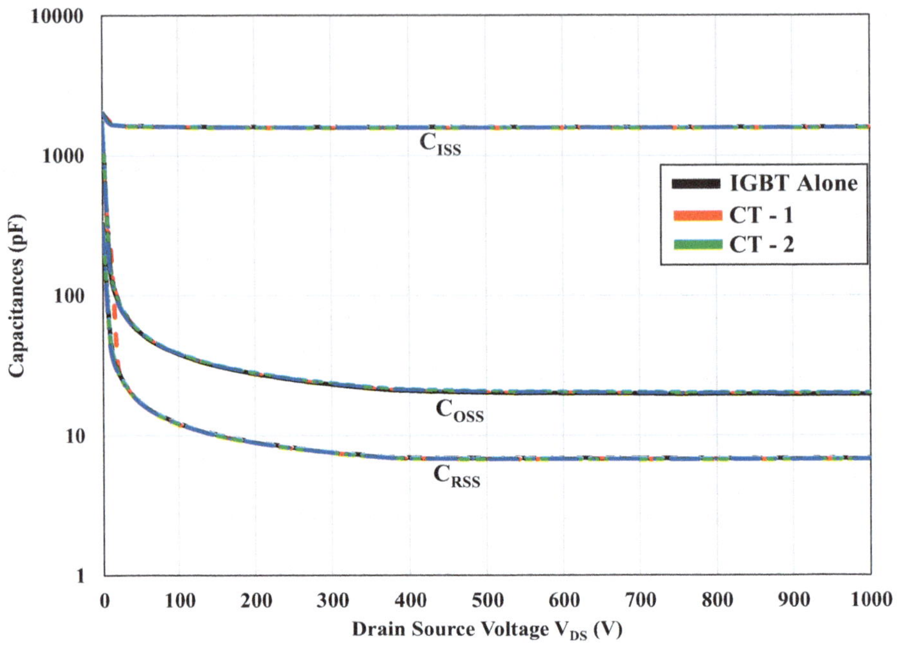

Fig. 18.12 Measured capacitances of the stand-alone A&O Si IGBT, the composite IGBT CT-1, and composite IGBT CT-2

Fig. 18.13 Equivalent circuit model for the capacitances of the composite IGBT

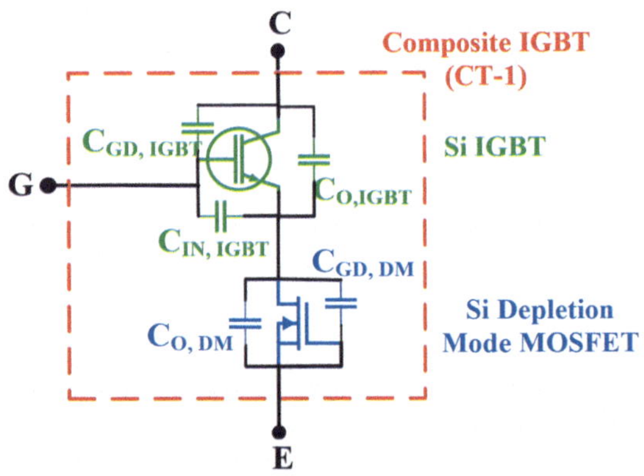

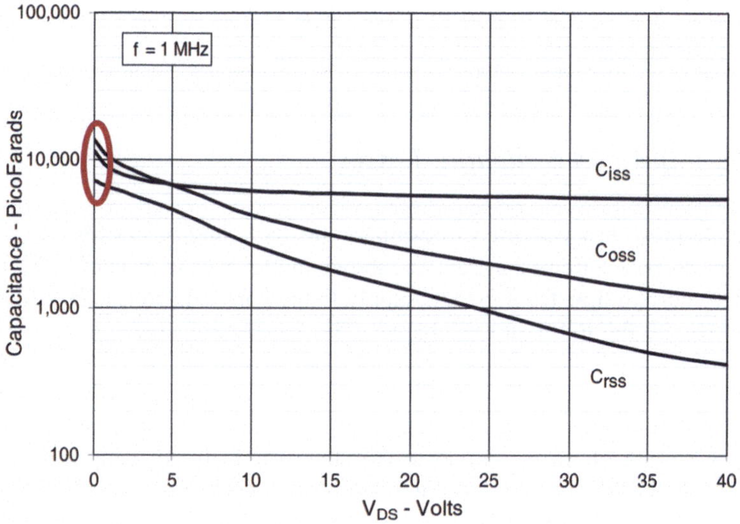

Fig. 18.14 Capacitances of the silicon DMM device

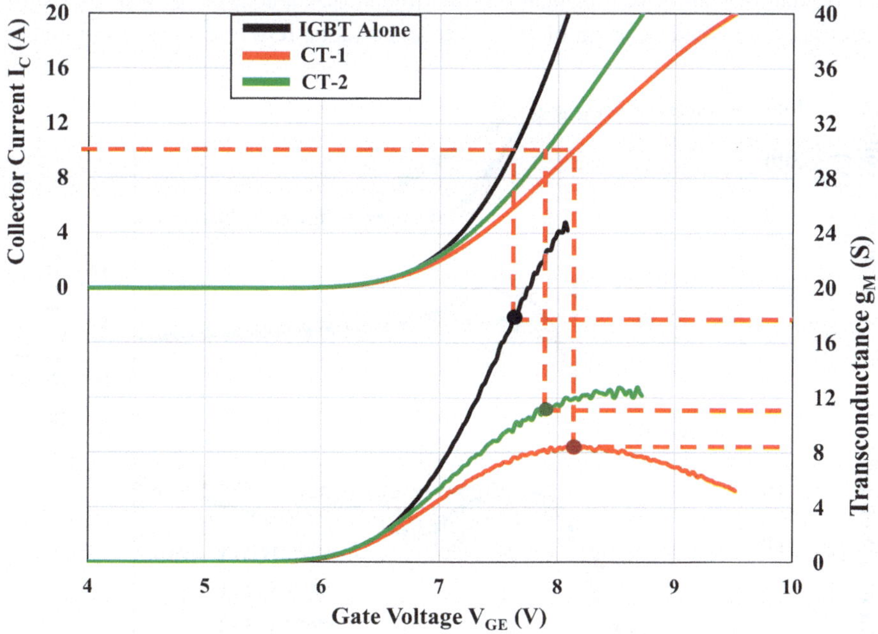

Fig. 18.15 Measured transconductances for the stand-alone A&O Si IGBT, the composite IGBT CT-1, and the composite IGBT CT-2

the measured values. The impact of the reduction of the transconductance on the switching losses is discussed in the next section.

18.6.4 Impact on Switching Losses

The double-pulse test was used to evaluate the switching power losses of power transistors. Its basic circuit diagram was shown in Fig. 5.15 with a power MOSFET, controlling current flow from the DC supply to the load inductance. This MOSFET is replaced with the IGBT in this chapter. The test setup includes a free-wheeling diode across the inductor. The anti-parallel diode within the A&O IGBT was used as the free-wheeling diode for the switching loss measurements in this chapter.

The double-pulse tests were conducted using a DC bus voltage of 600 V and a gate bias of 15 V to match the conditions used in the manufacturer's datasheet. In all the tests, an A&O IGBT was used as the high-side (HS) switch. The low-side (LS) switch for case 1 was a stand-alone A&O IGBT, and that for cases 2 and 3 was the composite transistors CT-1 and CT-2. Switching losses for all test cases were obtained by integrating the product of the voltage and current across the duration of their transitions.

The measured turn-on transients for the three cases are shown in Fig. 18.16 for the case of a collector current of 10 A. The stand-alone IGBT case (shown with the

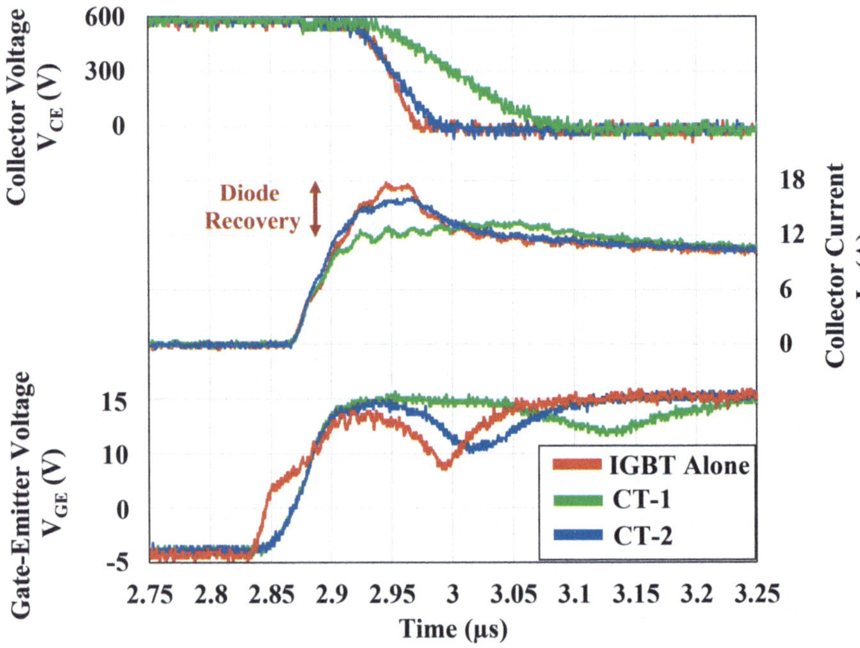

Fig. 18.16 Turn-on transients for the stand-alone A&O Si IGBT, the composite IGBT CT-1, and the composite IGBT CT-2

red lines) has the fastest increase in the current due to its large transconductance. This produces a large peak reverse recovery current of 11.7 A as determined by the design of the anti-parallel diode. In the case of the composite IGBT CT-2 (shown with the blue lines), the reduction of the transconductance slows down the current and voltage transients. This change is in accordance with Eq. (18.10). The lower *dI/dt* during the turn-on reduces the peak reverse recovery current [1]. The peak reverse recovery current in the case of the composite IGBT CT-2 is 9.5 A. An even further reduction of the peak reverse recovery current is observed in the case of the composite IGBT CT-1 (green lines) due to its lower transconductance. The peak reverse recovery current in the case of the composite IGBT CT-1 is 7.5 A.

The turn-on switching losses (E_{ON}) extracted from the waveforms shown in Fig. 18.16 were 0.58, 1.44, and 0.76 mJ for the stand-alone IGBT, the composite IGBT CT-1, and composite IGBT CT-2, respectively. The turn-off transients are not shown here because they are very similar for all three cases with turn-off energy losses (E_{OFF}) of 0.52 mJ. The total measured switching energy losses (E_{TOTAL}) were 1.10, 1.99, and 1.28 mJ, respectively. It can be seen that the BaSIC(DMM) topology produces significant increase in the switching energy losses when using the commercially available Si DMM device. However, this increase in switching power losses occurs with very large increase in the short-circuit withstand time, well beyond the desired 10 μs value, as shown in Sect. 10.5.2. The impact on switching energy loss is much smaller, as shown in a subsequent section by using the trade-off curve, to achieve the desired 10 μs value.

18.6.5 Sense Node Voltage

It was pointed out in Chap. 4 that the voltage across the non-linear resistance Q_{LV} can be used for monitoring the current level in the power transistor Q_{HV}. This is possible for the BaSIC(DMM) topology when applied to the Si IGBT because the Si DMM device operates in its linear regime at the on-state current level of the IGBT. The sense node voltage V_{SENSE} is therefore proportional to the current flowing through the IGBT. This is very beneficial feature of the BaSIC topology allowing power electronics engineers to make sure that the various IGBTs within the circuit do not exceed their reliable operating limits.

Experimental validation of using the sense node voltage to monitor the current flowing through the IGBT was obtained for the case of the A&O Si IGBT with a single silicon DMM device connected in series with its emitter terminal. The waveforms obtained with the double-pulse test circuit for this case of composite IGBT CT-1 are shown in Fig. 18.17. The current starts ramping up after application of the first gate drive pulse as shown by the red line. It reaches a value of 10 A at 2.3 μs. The sense node voltage V_{SENSE} is shown by the blue line. The sense node voltage V_{SENSE} increases in proportion to the current level and reaches a value of about 0.6 V when the current has a value of 10 A. The equivalent sense resistance is therefore 60

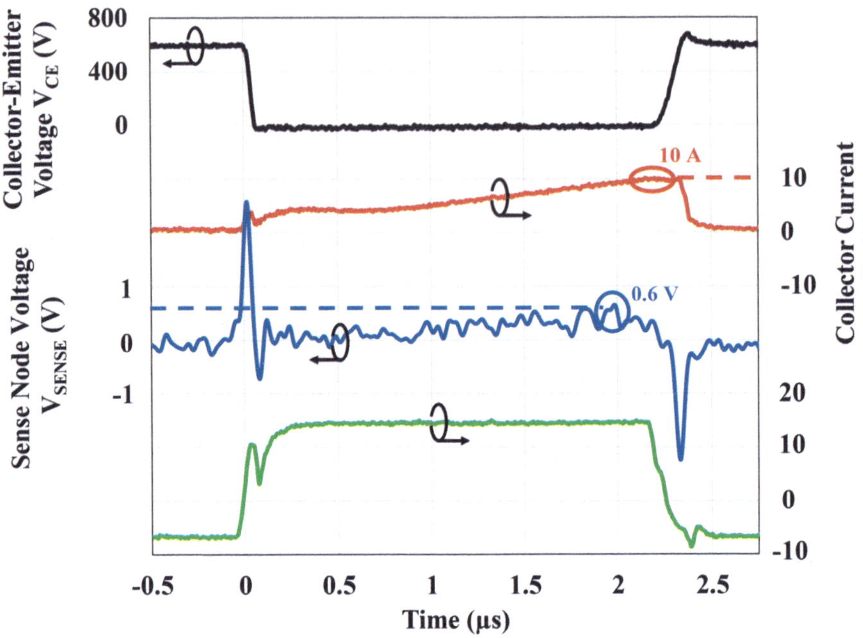

Fig. 18.17 Sense node voltage for the composite IGBT CT-1 during normal operation

mΩ as expected from the characteristics of the gate-source-shorted silicon DMM device described in Sect. 5.5.1.

The sense node voltage V_{SENSE} was monitored during the short-circuit tests as well for the case of the composite IGBT CT-1 with a single silicon DMM device connected in series with its emitter terminal. The resulting waveforms are shown in Fig. 18.18. The sense node voltage has a value of 6 V during the short-circuit event. The increase in the sense node voltage occurs in less than 0.5 μs allowing fast detection of the short-circuit problem. This time interval is much smaller than the 3 μs blanking time need for conventional IGBT DESAT gate control circuits. This difference is due to the sense node voltage V_{SENSE} operating at low voltages compared with the DESAT circuit where the collector terminal voltage of the IGBT is monitored.

The sense node voltage changes from 0.6 V in the normal on-state operating mode to 6 V under short-circuit conditions. This large change can be easily detected by the control circuits to identify the occurrence of the short circuit and then trigger turning-off the gate drive voltage to the Si IGBT. This approach is discussed in Chap. 17.

18.6.6 Trade-Off Curves

Trade-off curves between short-circuit withstand time and the on-state voltage drop or the switching losses are useful to understand the true impact of the BaSIC(DMM) topology on the Si IGBT. The trade-off curve between the short-circuit withstand time and the on-state voltage drop is shown in Fig. 18.19 for the 1.2 kV A&O IGBT.

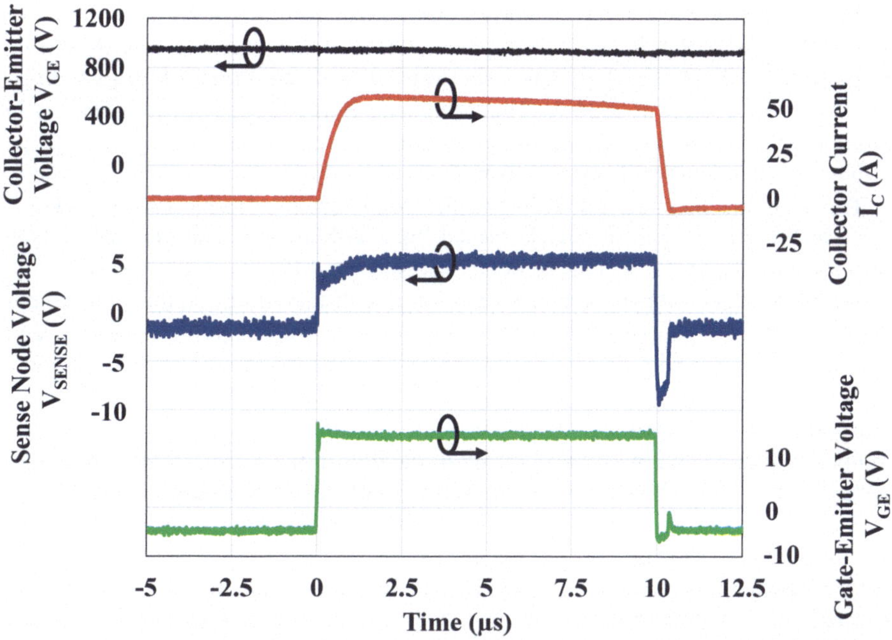

Fig. 18.18 Sense node voltage for the composite IGBT CT-1 during short-circuit operatio

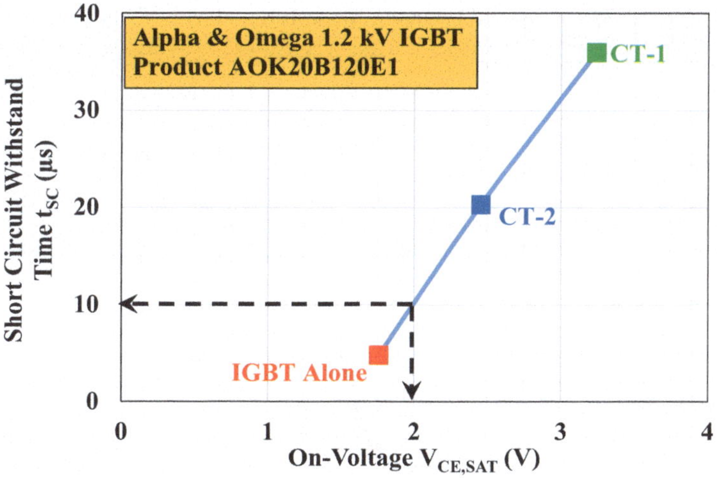

Fig. 18.19 Trade-off curve observed for the BaSIC(DMM) topology in the case of the A&O Si IGBT: on-state voltage drop

The stand-alone IGBT has a short-circuit withstand time of 4.8 μs with an on-state voltage drop of 1.76 V. The composite IGBT CT-1 with a single DMM device in the BaSIC(DMM) topology has a short-circuit withstand time of 36 μs with an on-state voltage drop of 3.24 V. The composite IGBT CT-2 with two DMMs in parallel in the BaSIC(DMM) topology has a short-circuit withstand time of 20.3 μs with an on-state voltage drop of 2.46 V.

The desirable short-circuit withstand time of 10 μs is indicated on the trade-off curve in Fig. 18.19 with the black dashed lines with arrows. The on-state voltage drop for this case is 2.0 V. This is an increase of only 14% to achieve the goal.

The trade-off curve between the short-circuit withstand time and the total switching energy loss is shown in Fig. 18.20. The stand-alone IGBT has a short-circuit withstand time of 4.8 μs with a total switching energy loss of 1.10 mJ. The composite IGBT CT-1 with a single DMM device in the BaSIC(DMM) topology has a short-circuit withstand time of 36 μs with total switching energy loss of 1.99 mJ. The composite IGBT CT-2 with two DMMs in parallel in the BaSIC(DMM) topology has a short-circuit withstand time of 20.3 μs with total switching energy loss of 1.28 mJ. The desirable short-circuit withstand time of 10 μs is indicated on the trade-off curve in Fig. 18.19 using the dashed lines with arrows. The total switching energy loss for this case is 1.15 mJ. This is an increase of only 4.5% to achieve the goal.

Based on the trade-off curves, it is clear that a short-circuit withstand time of 10 μs can be achieved by using the BaSIC(DMM) topology for the A&O 1.2 kV Si IGBT product with minimal impact on the on-state and switching losses. Even smaller impact can be achieved by using the optimized Si DMM devices described in Chap. 9.

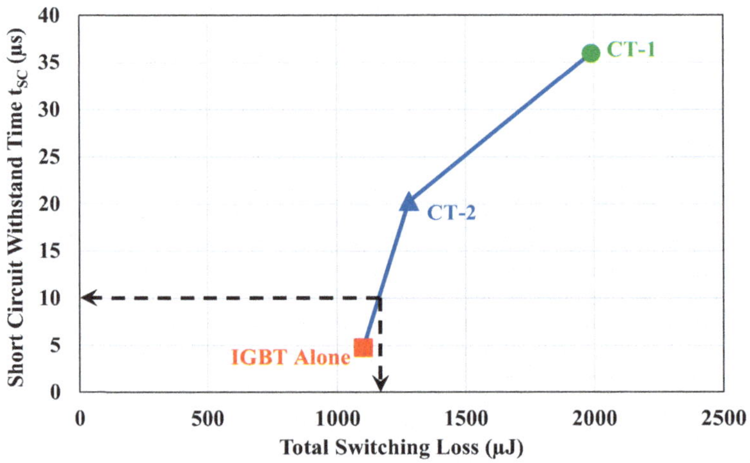

Fig. 18.20 Trade-off curve observed for the BaSIC(DMM) topology in the case of the A&O Si IGBT: switching energy loss

18.7 Experimental Results: Infineon and IXYS 1.7 kV IGBTs

The BaSIC(DMM) topology is scalable to higher blocking voltage IGBT products. The demonstrated of this is shown by using the Infineon IHW30N160R5 IGBT product with voltage rating of 1.6 kV and the IXYS IXGR16N170AH1 IGBT product with 1.7 kV blocking voltage capability. The Infineon IGBT is optimized for the induction cooking application according to the manufacturer's datasheet. This device is referred to as IGBT-1 in this chapter and represents the Type 2 devices mentioned in the introduction. It has a low on-state voltage drop $V_{CE,SAT}$ of 1.36 V at a collector current of 10 A device for a gate drive voltage of 15 V. The device has a monolithic anti-parallel diode, which reduces its turn-off losses [1]. No information on its short-circuit withstand capability is provided in the datasheet. This device was therefore chosen as a good candidate for the BaSIC(DMM) topology to examine if its short-circuit withstand capability can be enhanced with minimum impact on the on-state voltage drop and switching losses.

A second 1.7 kV rated IGBT, the IXYS IXGR16N170AH1 product, that is optimized for motor drive applications, was also selected for this study to contrast its performance with the BaSIC(DMM) topology. It has a short-circuit withstand time of 10 μs cited in the datasheet for the case of a collector bias of 1200 V and gate drive voltage of 15 V used for on-state operation. This device is referred to as IGBT-2 in this chapter and represents the Type 1 devices mentioned in the introduction.

The on-state characteristics of the Type 2 Infineon IHW30N160R5 1.6 kV IGBT at 25 °C are reproduced in Fig. 18.21 from the datasheet. The on-state operating point is indicated using the red dashed lines. The on-state voltage drop $V_{CE,SAT}$ at 10 A is 1.36 V at a gate bias of 15 V, a relatively low value due to sacrificing short-circuit withstand capability. It can be seen that the on-state voltage drop is not a very sensitive function of the gate drive voltage near 15 V. This is beneficial for operation of this device in the BaSIC(DMM) topology when the effective gate drive voltage is reduced by about 0.5 V due to the resistance of the DMM device. Three versions of the BaSIC(DMM) topology called CT-1, CT-2, and CT-3 were evaluated in this chapter. The versions CT-1, CT-2, and CT-3 were shown in Figs. 18.2, 18.4, and 18.5.

The on-state characteristics of the Type 1 IXYS IXGR16N170AH1 1.7 kV IGBT at 25 °C are reproduced in Fig. 18.22 from the datasheet. The on-state operating point is indicated using the red dashed lines. The on-state voltage drop $V_{CE,SAT}$ at 10 A is 3.8 V at a gate bias of 15 V, a relatively high value due to optimizing for good short-circuit withstand capability,

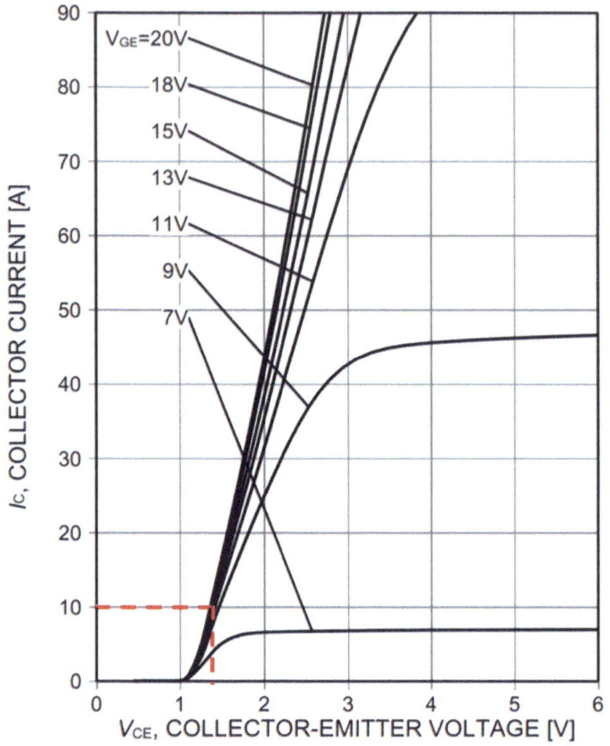

Fig. 18.21 The Infineon IHW30N160R5 IGBT product on-state characteristics from its datasheet

18.7.1 High Voltage Output Characteristics

As in the case of the 1.2 kV rated IGBTs, the operating point for the 1.7 kV rated devices during the short-circuit event is useful. This operating point is part of the IGBTs output characteristics at high collector currents and voltages. However, these characteristics are not provided in the Infineon and IXYS IGBT datasheets.

A novel method for extracting the high-voltage, high-current output characteristics of power transistors was developed by using the short-circuit test configuration [8]. The short-circuit test configuration shown in Fig. 18.1 was used with a small pulse width of 1.5 μs. During this time, the current increases to a peak value that is representative of the saturation current of the transistor at the DC supply voltage. The DC supply voltage can be varied in small steps to generate the high-voltage, high-current output characteristic for the chosen gate bias.

The experimentally extracted high-voltage, high-current output characteristic at a gate bias of 15 V for the stand-alone Infineon IHW30N160R5 Si IGBT-1 is shown in Fig. 18.23 with the red line. The measured saturation current value for IGBT-1 at a collector bias of 1100 V is 78 A as indicated by the red point on the graph. The experimentally extracted high-voltage, high-current output characteristic at a gate

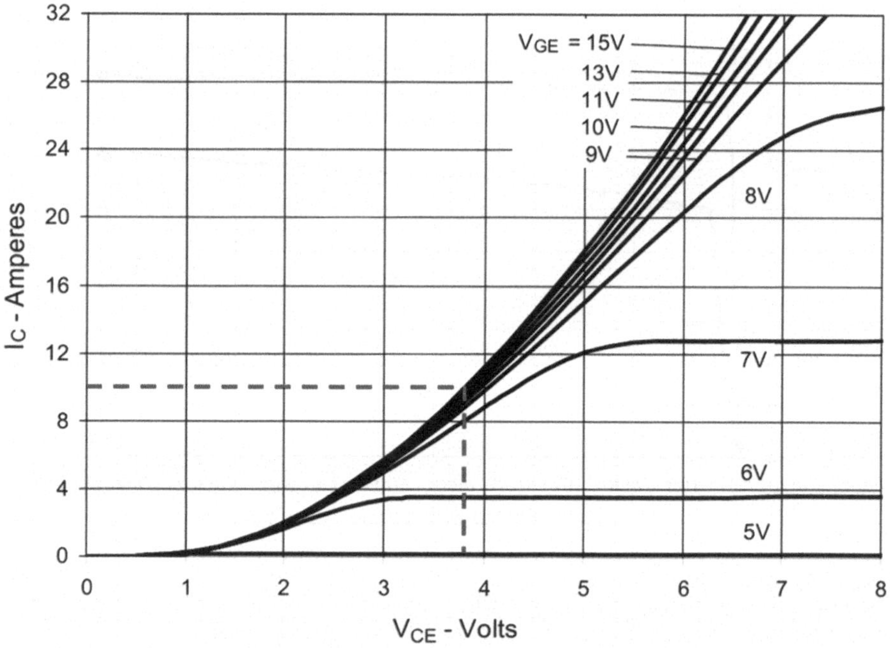

Fig. 18.22 The IXYS IXGR16N170AH1 IGBT product on-state characteristics from its datasheet

bias of 15 V for the stand-alone IXYS IXGR16N170AH1 Si IGBT-2 is shown in Fig. 18.23 with the purple line. The measured saturation current value for IGBT-2 at a collector bias of 1100 V is 61 A as indicated by the purple point on the graph.

The experimentally extracted high-voltage, high-current output characteristics at a gate bias of 15 V for the composite transistors CT-1, CT-2, and CT-3 are shown in Fig. 18.23 with the green, blue, and orange lines, respectively. The measured saturation current values at a collector bias of 1100 V, indicated by the green, blue, and orange points on the graph, are 18, 35, and 52 A for gate bias values of 15 for the composite IGBT CT-1, CT-2, and CT-3, respectively. The measured saturation currents for all three of the BaSIC(DMM) cases are lower than for the two stand-alone IGBT products.

18.7.2 Short-Circuit Test Results

The short-circuit withstand time for the IGBTs was obtained by using the test configuration shown in Figs. 18.1, 18.2, 18.4, and 18.5. The short-circuit pulse width is gradually increased while capturing the current waveform until the device failed. The DC supply voltage is monitored as shown by the black line in Fig. 18.24. It is

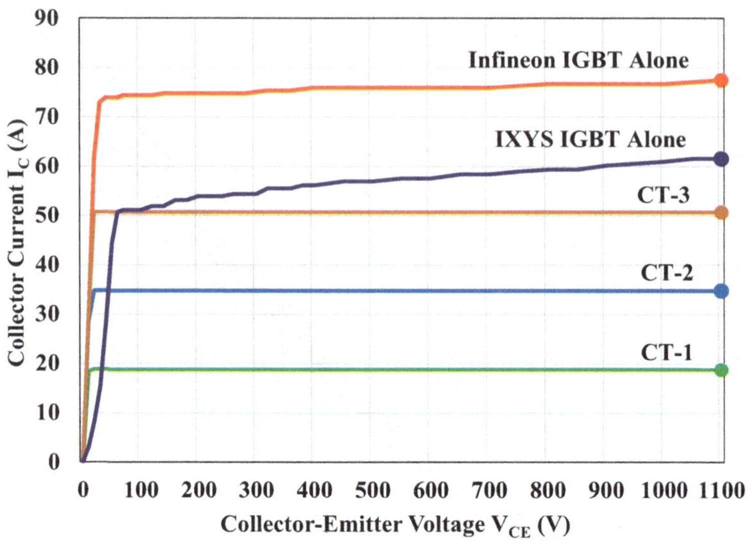

Fig. 18.23 High-voltage, high-current output characteristics: stand-alone Si Infineon 1.7 kV IGBT (red line), stand-alone IXYS 1.7 kV IGBT (purple line), composite IGBT CT-1 (green line), composite IGBT CT-2 (blue line), and composite IGBT CT-3 (orange line)

important to make sure that the capacitor C_{DC} is adequate to maintain the voltage at the desired value, 1100 V in this case.

The red line was observed for the stand-alone Infineon IHW30N160R5 Si IGBT-1. The current reaches 102 A at 3 μs. This is much larger than the 72 A reached at the 1.5 μs time used to generate the high-voltage, high-current output characteristics shown in Fig. 18.23. Device failure occurs at the time highlighted with the red arrow in the figure with a short-circuit withstand time of 5.7 μs. These results confirm that this Type 2 IGBT designed and optimized for low on-state voltage drop by the manufacturer does not have short-circuit withstand capability of 10 μs.

The purple line was observed for the Type 1 IXYS IXGR16N170AH1 Si IGBT-2. The current reaches a peak value of 61 A at 1.5 μs. This consequently matches the current on the output characteristics in Fig. 18.23. Device failure occurs at the time highlighted with the purple arrow in the figure with a short-circuit withstand time of 14 μs. These results confirm that this Type 1 IGBT has been designed and optimized by the manufacturer to have a good short-circuit withstand capability.

The green line in Fig. 18.24 was obtained for the composite IGBT CT-1 built using the Type 2 Infineon IHW30N160R5 Si IGBT. Its current reaches a peak value of 18 A at the 1.5 μs time used to generate the high-voltage, high-current output characteristics in Fig. 18.23. The current reduces gradually with time due to the temperature rise of the Si IGBT and DMM device. The composite IGBT CT-1 failed when the pulse width was increased to a very large value of 70 μs, well beyond the desired value of 10 μs.

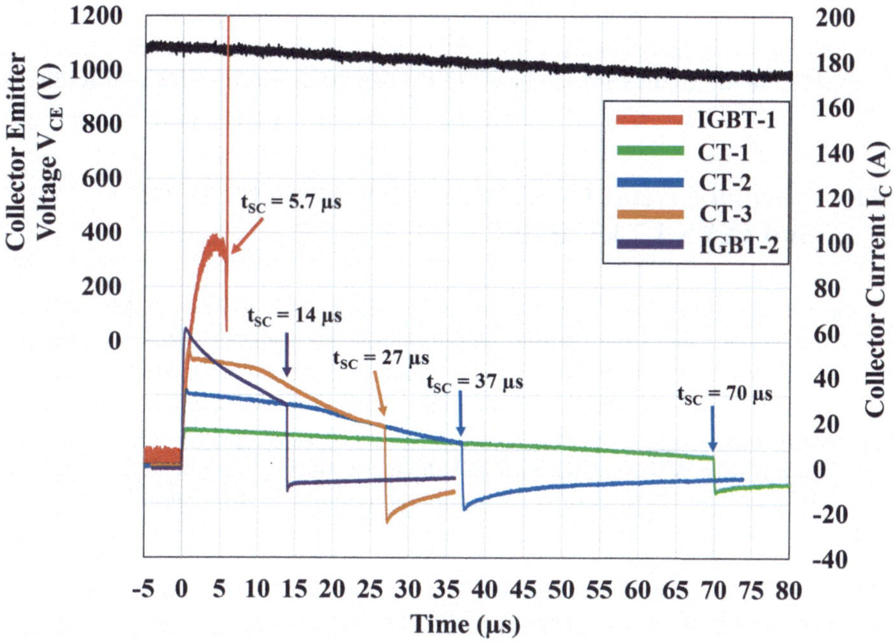

Fig. 18.24 Measured short-circuit waveforms: stand-alone Si Infineon 1.7 kV IGBT-1, stand-alone IXYS 1.7 kV IGBT-2, composite IGBT CT-1, composite IGBT CT-2, and composite IGBT CT-3

The blue line in Fig. 18.24 was obtained for the composite IGBT CT-2 built using the Type 2 Infineon IHW30N160R5 Si IGBT. Its current reaches a peak value of 35 A at the 1.5 μs time used to generate the high-voltage, high-current output characteristics in Fig. 18.23. The current reduces with time due to the temperature rise of the Si IGBT and DMM device. The composite IGBT CT-2 failed when the pulse width was increased to a large value of 37 μs, well beyond the desired value of 10 μs.

The orange line in Fig. 18.24 was obtained for the composite IGBT CT-3 built using the Type 2 Infineon IHW30N160R5 Si IGBT. Its current reaches a peak value of 52 A at the 1.5 μs time used to generate the high-voltage, high-current output characteristics in Fig. 18.23. The current reduces with time due to the temperature rise of the Si IGBT and DMM device. The composite IGBT CT-3 failed when the pulse width was increased to a large value of 27 μs, well beyond the desired value of 10 μs.

In all cases, it was found that the Infineon IHW30N160R5 Si IGBT failed but the Si DMM devices were not damaged. This is due to much lower power dissipation in the Si DMM devices because their drain voltage is only 6–8 V during the short-circuit event.

These short-circuit test results provide validation of the utility of the BaSIC(DMM) topology for enhancing the short-circuit withstand capability of commercially available Type 2 1.6 kV IGBTs which are optimized for low on-state voltage drop. This topology was able to reduce the short-circuit current level and significantly increase the short-circuit withstand time to beyond 10 μs for the 1.6 kV Si IGBT.

18.8 Normal Circuit Operation: Infineon 1.6 kV IGBT and IXYS 1.7 kV IGBT

For the BaSIC(DMM) topology, the impact of the resistance of the DMM device placed in series with the emitter electrode of the Type 2 1.6 kV Infineon IHW30N160R5 Si IGBT on the on-state voltage drop and switching behavior must be quantified. Trade-off curves between enhancing the short-circuit time and an increase on-state voltage drop and switching losses are developed for the case of the Type 2 Infineon IHW30N160R5 Si IGBT device to fully comprehend the value of the BaSIC topology. These results are compared with the performance of the IXYS IXGR16N170AH1 Si IGBT-2 as an example of a Type 1 device optimized with good short-circuit capability.

18.8.1 On-State Characteristics

The measured on-state characteristics of the stand-alone Type 2 1.6 kV Infineon IHW30N160R5 Si IGBT-1 and the stand-alone Type 1 1.7 kV IXYS IXGR16N170AH1 Si IGBT-2 at a gate bias voltage of 15 V are shown in Fig. 18.25 by the red and purple lines. The on-state voltage drop at a rated current of 10 A is 1.36 and 12.4 V for the two devices. The IXYS IXGR16N170AH1 Si IGBT-2 device evaluated in this work had a much larger on-state voltage drop of 12 V than shown in its datasheet (see Fig. 18.22).

The measured on-state characteristics of the composite IGBTs CT-1, CT-2, and CT-3 at a gate bias of 15 V are also provided in Fig. 18.25. The measured on-state voltage drops at a rated current of 10 A are 2.12, 1.60, and 1.56 V for devices CT-1, CT-2, and CT-3, respectively. The single DMM used with composite IGBT CT-1 has an on-resistance of 60 mΩ. Consequently, it adds a voltage drop of 0.6 V at a current of 10 A to the Si IGBT in the case of the composite IGBT CT-1. The two DMMs in parallel used for the composite IGBT CT-2 have a resistance of 30 mΩ. Consequently, they add a voltage drop of 0.3 V at a current of 10 A to the Si IGBT in the case of the composite IGBT CT-2. The three DMMs in parallel used for the composite IGBT CT-3 have a resistance of 20 mΩ. Consequently, they add a voltage drop of 0.2 V at a current of 10 A to the Si IGBT in the case of the composite IGBT CT-3.

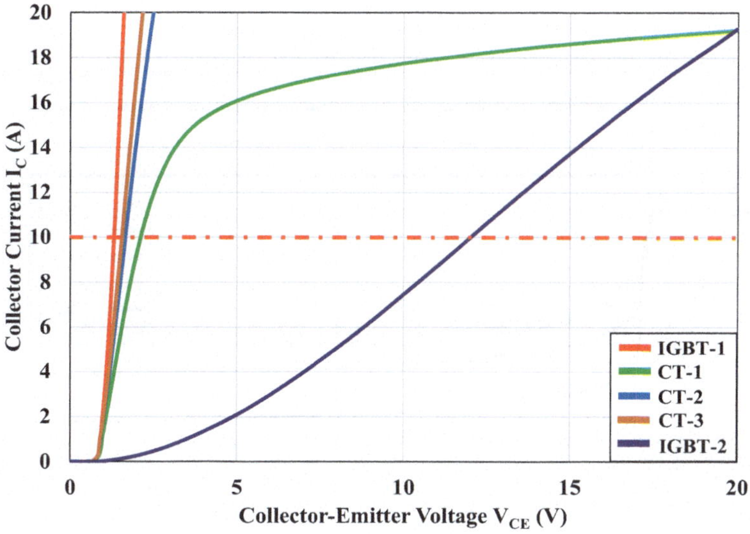

Fig. 18.25 Measured on-state characteristics: stand-alone Si Infineon 1.7 kV IGBT (IGBT-1), stand-alone IXYS 1.7 kV IGBT (IGBT-2), composite IGBT CT-1, composite IGBT CT-2, and composite IGBT CT-3

18.8.2 Impact on Capacitances

The capacitances of an IGBT determine its switching performance [1]. The input, output, and reverse transfer capacitances of the stand-alone Si IGBTs and the three composite IGBTs CT-1, CT-2, and CT-3 were therefore evaluated. These measured characteristics are displayed in Fig. 18.26. The red lines apply for the stand-alone Infineon IHW30N160R5 Si IGBT-1 product. The measured values for the C_{ISS}, C_{OSS}, and C_{RSS} are very close to values provided in the datasheet for the device. The purple lines apply for the stand-alone IXYS IXGR16N170AH1 Si IGBT-2 product.

The green, blue, and orange lines in Fig. 18.26 apply to the composite IGBTs CT-1, CT-2, and CT-3 which utilize the Infineon IHW30N160R5 Si IGBTs (IGBT-1) product in the BaSIC(DMM) implementations. They overlap the lines for the stand-alone Infineon IHW30N160R5 Si IGBT. This demonstrates that the BaSIC(DMM) topology does not alter the capacitances of the Si IGBT.

The reason for this can be explained by using the equivalent circuit for the BaSIC(DMM) topology shown in Fig. 18.13 and discussed in context of the 1.2 kV A&O IGBT. From the above measurements, it is apparent that the BaSIC(DMM) topology does not alter the capacitances of the 1.6 kV Si IGBT. This is a favorable outcome from the point of view of the switching behavior.

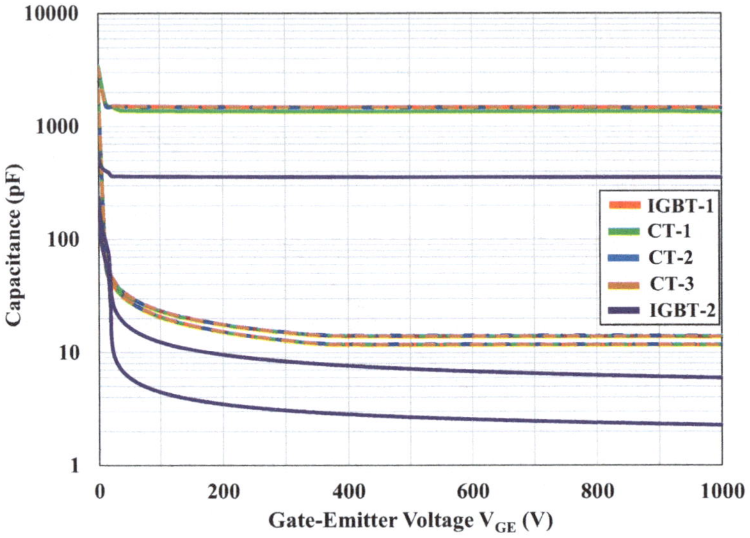

Fig. 18.26 Measured capacitances: stand-alone Si Infineon 1.7 kV IGBT (IGBT-1), stand-alone IXYS 1.7 kV IGBT (IGBT-2), composite IGBT CT-1, composite IGBT CT-2, and composite IGBT CT-3

18.8.3 Impact on Transconductance

The transconductance g_M of the IGBT plays a role in the switching transient for the drain voltage during operation with an inductive load [1]. Its value can be extracted from the transfer characteristics of the device. The transfer characteristics for the stand-alone and composite IGBTs were measured at a collector bias of 20 V. The transconductance was extracted by taking a derivative of the collector current with respect to the gate voltage. The results are displayed in Fig. 18.27. The values of the transconductance can be compared at the same collector current level of 10 A by following the red dashed lines in the figure. It is given by the values at the four points indicated on the transconductance curves.

The stand-alone Infineon IHW30N160R5 Si IGBT-1 has a measured transconductance of 15.8 S, which is consistent with its datasheet. The stand-alone IXYS IXGR16N170AH1 Si IGBT-2 product has a transconductance of 3.61 S.

The transconductances of the composite IGBTs CT-1, CT-2, and CT-3 were 4.8, 9.6, and 11.1 S, respectively. The observed reduction in the transconductance is associated with the resistance of the Si DMM device as described in Eq. (18.9). The transconductance value predicted by this equation is 8.1 S for the case of a series resistance R_S of 60 mΩ corresponding to a single DMM in the composite IGBT CT-1. The transconductance value predicted by this equation is 10.1 S for the case of a series resistance R_S of 30 mΩ corresponding to two DMMs in parallel in the composite IGBT CT-2. The transconductance value predicted by this equation is 12.0 S for the case of a series resistance R_S of 20 mΩ corresponding to three DMMs

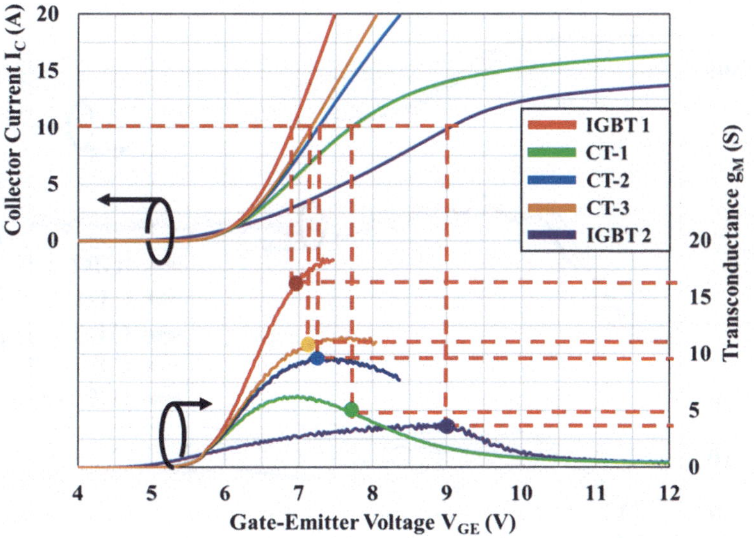

Fig. 18.27 Measured transfer characteristics and transconductances: stand-alone Si Infineon 1.6 kV IGBT (IGBT-1), stand-alone IXYS 1.7 kV IGBT (IGBT-2), composite IGBT CT-1, composite IGBT CT-2, and composite IGBT CT-3

in parallel in the composite IGBT CT-3. These calculated values are in agreement with the measured values. The impact of the reduction of the transconductance on the switching losses is discussed in the next section.

18.8.4 Impact on Switching Losses

The double-pulse test procedure was used to evaluate the switching power losses for all the devices. The tests were performed using a DC bus voltage of 850 V and collector current of 10 A. Switching losses for all test cases were obtained by integrating the product of the voltage and current across the duration of their transitions.

The measured turn-on transients for the five cases are shown in Fig. 18.28 for the case of a collector current of 10 A. The turn-on switching loss (E_{ON}) extracted from the waveforms for the stand-alone Infineon IHW30N160R5 Si IGBT (IGBT-1) was 0.50 mJ. The turn-on switching losses for the composite IGBT CT-1, CT-2, and CT-3 were 0.92, 0.55, and 0.51 mJ, respectively. The measured turn-off transients for the five cases are shown in Fig. 18.29. The turn-off energy loss (E_{OFF}) was 0.60 mJ for all four cases. The total measured switching energy loss (E_{TOTAL}) was 1.09 mJ for the Infineon IHW30N160R5 Si IGBT (IGBT-1) product. The total measured switching energy losses for the composite IGBT CT-1, CT-2, and CT-3 were 1.52, 1.14, and 1.10 mJ, respectively. It can be seen that the BaSIC(DMM) topology with a single commercially available Si DMM device produces significant increase

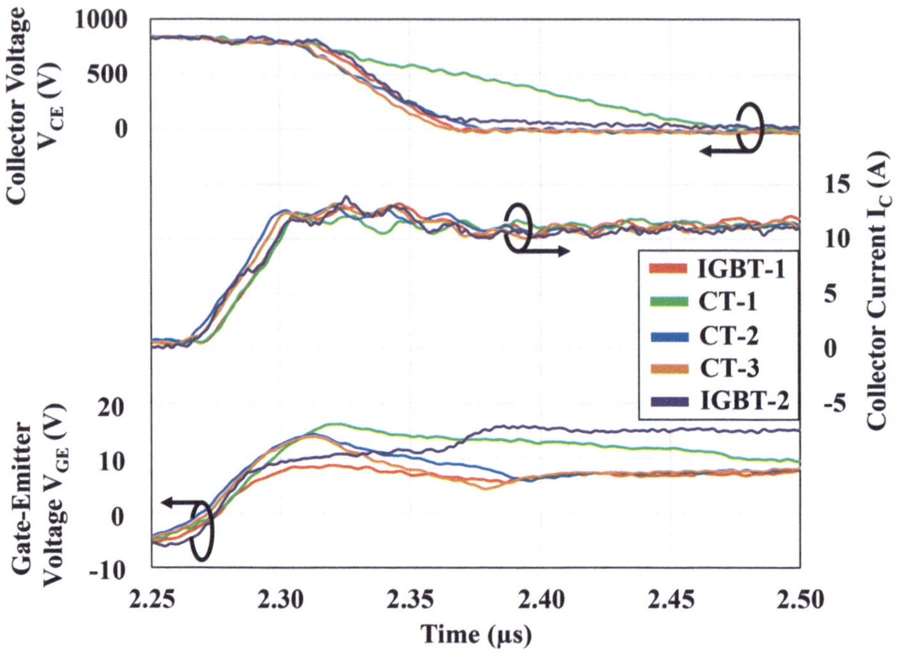

Fig. 18.28 Measured turn-on transients: stand-alone Si Infineon 1.6 kV IGBT (IGBT-1), stand-alone IXYS 1.7 kV IGBT (IGBT-2), composite IGBT CT-1, composite IGBT CT-2, and composite IGBT CT-3

in the switching energy losses. However, this increase in switching power losses occurs with a very large increase in the short-circuit withstand time, well beyond the desired 10 μs value, as shown in Sect. 18.7.2. The impact on switching energy loss is much smaller as shown in a subsequent section by using the trade-off curve to achieve the desired 10 μs value for the Infineon IHW30N160R5 Si IGBT (IGBT-1) product.

The turn-on switching losses extracted from the waveforms for the stand-alone IXYS IXGR16N170AH1 Si IGBT (IGBT-2) product was 0.59 mJ. Its turn-off switching loss was only 0.11 mJ. The total energy loss for the stand-alone IXYS IXGR16N170AH1 Si IGBT (IGBT-2) product was therefore 0.7 mJ. It is apparent that this Type 1 IGBT product has been tailored for low switching losses at the expense of a very high on-state voltage drop.

18.8.5 Trade-Off Curves

Trade-off curves between short-circuit withstand time and the on-state voltage drop or the switching losses are useful to understand the true impact of the BaSIC(DMM) topology on the 1.6 kV Infineon IHW30N160R5 Si IGBT (IGBT-1).

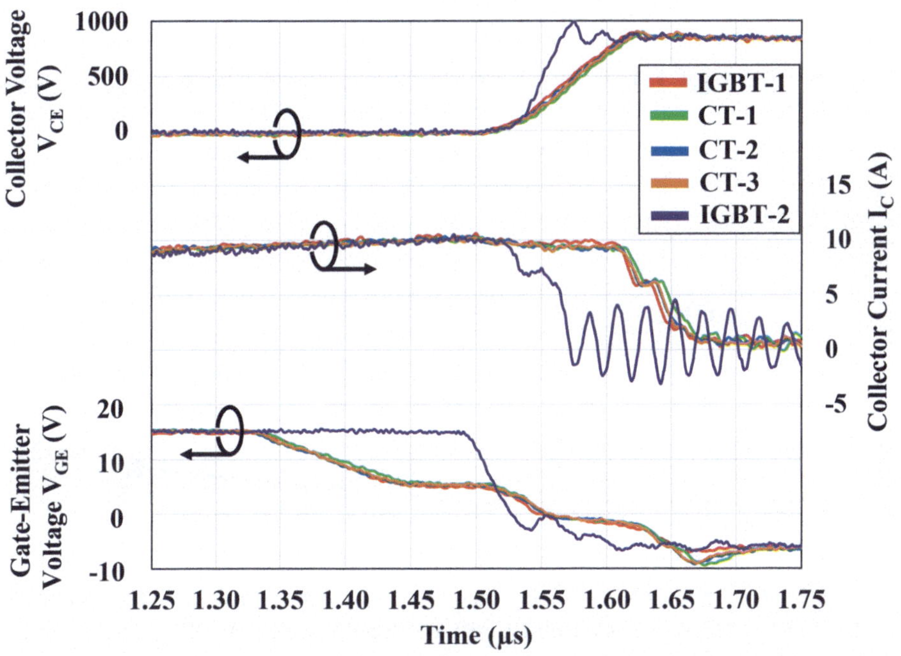

Fig. 18.29 Measured turn-off transients: stand-alone Si Infineon 1.6 kV IGBT (IGBT-1), stand-alone IXYS 1.7 kV IGBT (IGBT-2), composite IGBT CT-1, composite IGBT CT-2, and composite IGBT CT-3

The trade-off curve between the short-circuit withstand time and the on-state voltage drop is shown in Fig. 18.30 for the BaSIC(DMM) topology with the 1.6 kV Infineon IHW30N160R5 Si IGBT (IGBT-1). There is very steep increase in the short-circuit withstand time in relation to the on-state voltage drop. This is very desirable feature of the BaSIC(DMM) topology even with a non-optimized Si DMM device. The desirable short-circuit withstand time of 10 μs is indicated on the trade-off curve in Fig. 18.30 with the red dashed lines. The on-state voltage drop is increased by less than 5% to achieve this 10 μs short-circuit withstand time. In contrast, the on-state voltage drop is 6 V for the Type 1 IGBT with short-circuit withstand time of 10 μs, an increase of 440%. Consequently, the BaSIC(DMM) topology can be utilized to create devices for motor drive applications with much lower on-state voltage drop by using Type 2 IGBTs rather than the Type 1 IGBTs.

The trade-off curve between the short-circuit withstand time and the normalized total switching energy loss is shown in Fig. 18.31 for the BaSIC(DMM) topology for the 1.6 kV Infineon IHW30N160R5 Si IGBT-1. The desirable short-circuit withstand time of 10 μs is indicated on the trade-off curve in Fig. 18.31 with the red dashed line. The total switching energy loss for this case is essentially the same as for the stand-alone 1.7 kV Infineon IHW30N160R5 Si IGBT-1. The switching loss for the Type-1 IGBT-2 is much lower but it has a very high on-state voltage drop.

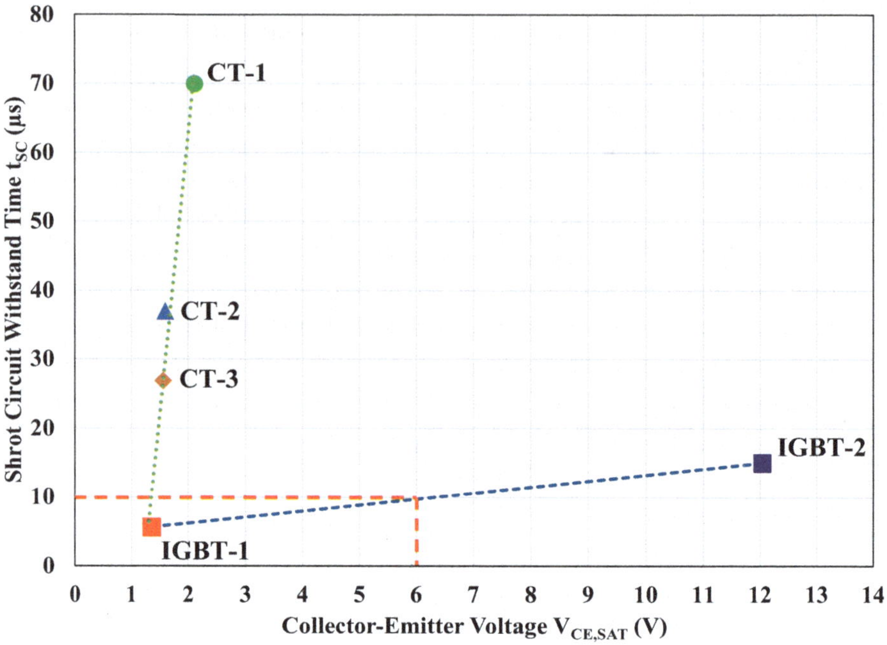

Fig. 18.30 Trade-off curve observed for the BaSIC(DMM) topology in the case of the Si Infineon 1.6 kV IGBT product (IGBT-1): on-state voltage drop

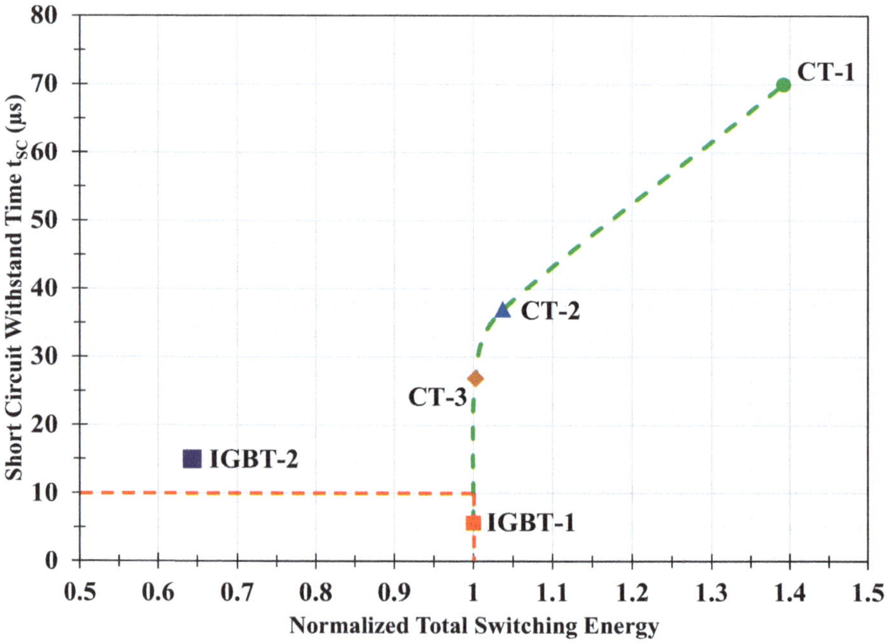

Fig. 18.31 Trade-off curve observed for the BaSIC(DMM) topology in the case of the Si Infineon 1.6 kV IGBT product (IGBT-1): switching energy loss

18.9 Summary

The benefits of using the BaSIC topology are demonstrated in this chapter for the case of silicon Type 2 IGBTs by using a gate-source-shorted silicon depletion mode MOSFET as the non-linear resistance. The BaSIC(DMM) topology allows achieving the desired short-circuit withstand time of 10 μs with essentially no change in the on-state voltage drop or switching losses of Type 2 IGBTs optimized for lower on-state voltage drop than the Type 1 IGBTs optimized for 10 μs short-circuit withstand time.

References

1. B.J. Baliga, *Fundamentals of Power Semiconductor Devices*, 2nd edn. (Springer Science, New York, 2019)
2. B.J. Baliga, Fast switching insulated gate transistors. IEEE Electron Device Lett. **4**, 452–454 (1983)
3. T. Yamada, G. Mazumdar, S. Mori, H. Hagino, H. Kondoh, T. Hirao, Next generation power module, in *IEEE International Symposium on Power Semiconductor Device and ICs* (1994), pp. 3–9
4. H. Iwamoto, H. Takahashi, M. Tabata, K. Satoh, New 1200 V power modules with sophisticated trench gate IGBT and superior soft recovery diode, in *IEEE International Conference on Power Electronic Drive Systems*, vol. 1 (1999), pp. 28–33
5. M. Mori, K. Oyama, Y. Kohno, J. Sakano, J. Uruno, K. Ishizaka, D. Kawase, A trench-gate high-conductivity IGBT with short-circuit capability. IEEE Trans. Electron Device **54**(8), 2011–2016 (2007)
6. R.S. Chokhawala, J. Catt, L. Kiraly, A discussion on IGBT short-circuit behavior and fault protection schemes. IEEE Trans. Ind. Electron. **31**, 256–263 (1995)
7. A. Kanale, B.J. Baliga, Enhancing short circuit capability of 1.2-kV Si IGBT using a gate-source shorted Si depletion mode MOSFET in series with the emitter. IEEE Trans. Power Electron. **35**, 6350–6361 (2020)
8. A. Kanale, K. Han, B.J. Baliga, S. Bhattacharya, Superior short circuit performance of 1.2kV SiC JBSFETs compared to 1.2kV SiC MOSFETs, in *12th European Conference on Silicon Carbide and Related Materials*, Paper WE.P.MO9 (2018)

Chapter 19
BaSIC(EMM) for Silicon IGBT

The efficacy of the BaSIC topology was first demonstrated for enhancing the short-circuit capability of silicon carbide power MOSFETs by using a Si depletion mode MOSFET (DMM). This effort was described in Chap. 5. A search for silicon depletion mode power MOSFETs produced only one device with the characteristics suitable for the BaSIC topology. It is an IXYS 100 V Si Depletion Mode MOSFET (GSS-DMM) with on-resistance of 64 mΩ at zero gate bias, i.e., with gate shorted to the source terminal. This device was found to have a measured breakdown voltage of 150 V, well above the 30 V required for the BaSIC topology. It has a relatively large on-resistance with zero gate bias.

The first application of the BaSIC topology to enhancing the performance of silicon IGBTs by using the above Si GSS-DMM device was discussed in Chap. 18. It was demonstrated that good short-circuit withstand time could be achieved for Si IGBTs optimized with low on-state voltage drop. Although successful, the limitations of the only commercially available Si GSS-DMM device constrain this approach.

However, the power semiconductor industry manufactures a large variety of Si enhancement mode MOSFETs (EMMs) due to their widespread applications. The benefits of utilizing one of these EMM devices for the BaSIC topology were described in Chap. 10 for the SiC power MOSFET [1]. This approach is therefore also useful for enhancing the short-circuit performance of Si IGBTs.

Experimental confirmation of the excellent performance of the BaSIC(EMM) topology for Si IGBTs is described in this chapter [2]. The trade-off between extending the short-circuit withstand time and the increase in on-resistance and switching losses is quantified here.

The BaSIC(EMM) implementation allows power electronics designers to achieve the desired short-circuit withstand time with Type 2 IGBTs that are optimized with low on-state voltage drop. They can use the same combination of the Type 2 IGBT and Si EMM device for a variety of applications that need different short-circuit withstand capabilities.

B. J. Baliga, A. Kanale, *The BaSIC Topology*,
https://doi.org/10.1007/978-3-031-86630-2_19

19.1 Short-Circuit Configuration: BaSIC(EMM) Topology

The short-circuit tests for the BaSIC(EMM) topology were performed using a single silicon EMM device connected in series with the emitter electrode of the silicon IGBT. The DC gate bias applied to the Si EMM device was varied to alter its saturated drain current and extend the short-circuit withstand time. Experimental demonstration of the proposed concept was performed with the commercially available IXYS 1.2 kV Si IGBT IXGH20N120A3. The device selected for this test is a Type 2 IGBT optimized for low on-state voltage drop. It has poor short-circuit withstand time of less than 10 μs as shown later.

The short-circuit test configuration is shown in Fig. 19.1 with the composite transistor, identified by the red dashed lines, labeled as CT. It is worth pointing out that the gate drive voltage is applied to the gate of the Si IGBT as usually done in its power circuit applications. The silicon EMM device only serves as the non-linear resistance Q_{LV} with a DC gate bias V_{GE}(EMM) applied to its gate terminal to control its saturation current level. The DC gate bias V_{GE}(EMM) applied to the Si EMM device can be programmed by the power electronics engineer based on the short-circuit withstand time required for a particular application. This provides a unique flexibility for the power electronics engineer while using a single commercially available Si IGBT optimized with low on-state voltage.

The short-circuit tests were performed for the BaSIC(EMM) topology by initially using a short duration gate pulse of 1.5 μs to avoid damaging the Si IGBT. The width of the gate pulse was gradually increased until the device failed while capturing the collector current waveform. This allowed determination of the short-circuit withstand capability of the stand-alone Si IGBT and the composite transistors.

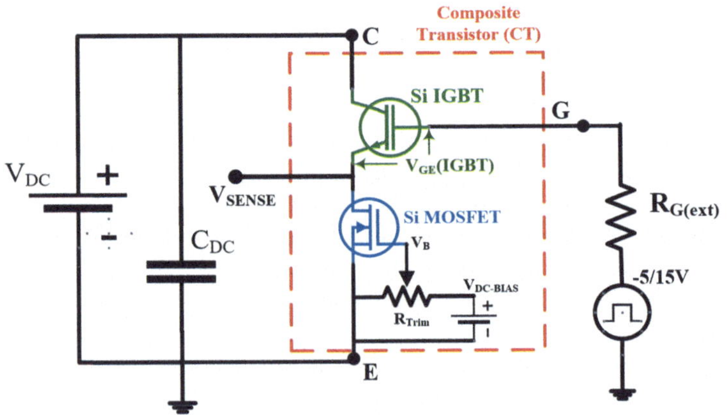

Fig. 19.1 Short-circuit test circuit with BaSIC(EMM) topology for Si IGBT

19.2 Silicon EMM Device

A Infineon Si OptiMOS enhancement mode MOSFET (EMM) IPI70N04S-06 with blocking voltage of 40 V was selected for implementation of the BaSIC(EMM) topology [3] due to its low on-state resistance of 6 mΩ at a gate bias of 10 V. Its output characteristics were reproduced from the datasheet at low drain bias voltages in Fig. 10.2. The device exhibits an on-resistance of 6 mΩ at a gate bias of 10 V with an extremely high drain saturation current level. The drain saturation current level can be brought down to the 25–50 A range needed to enhance the short-circuit withstand time of the Si IGBT when the gate bias voltage is reduced to the 5–6 V range.

The on-resistance of the silicon EMM device at 25 °C in its linear regime of operation is provided in its datasheet as a function of the gate bias and drain current [2]. This graph was reproduced in Fig. 10.3. It is appropriate to use the 25 °C data because the silicon EMM does not undergo switching losses under normal circuit operating conditions preventing a significant increase in its junction temperature. The on-state resistances of the silicon EMM device at the Si IGBT on-state operating current level of 10 A are highlighted using the red lines and dots. The on-state resistance has a value of 6.5, 7.2, 8.5, and 11.5 mΩ at gate bias values of 7.0, 6.5, 6.0, and 5.5 V, respectively. These resistances are much smaller than that of the Si DMM device used for the BaSIC(DMM) topology in previous chapters. Consequently, the Si EMM device used for the BaSIC(EMM) topology has a smaller impact on the on-state voltage drop and switching losses of the Si IGBT.

The voltage drop across the silicon EMM device increases to about 5 V during the short-circuit event. The effective gate drive voltage for the Si IGBT under short-circuit conditions is therefore reduced, which brings down its saturation current until it automatically matches that for the silicon EMM device. In other words, the voltage V_{GE}(IGBT) developed across the silicon EMM under short-circuit conditions in Fig. 19.1 automatically adjusts until the Si IGBT saturation current becomes equal to the saturation current of the silicon EMM. In general, the saturation characteristics of both the Si IGBT and silicon EMM device will vary from part to part for the same device due to manufacturing tolerances. The automatic adjustment of the operating point ensures seamless operation of the BaSIC(EMM) topology without the need for matching the characteristics of the Si IGBT and silicon EMM devices. This is an important feature of the BaSIC topology allowing its low cost implementation for power electronics applications.

19.3 High-Voltage Output Characteristics

As explained in Chap. 2, the power transistor operates in its current saturation mode while supporting a high voltage equal to the DC supply voltage for the load. This operating point is part of the Si IGBT output characteristics. Knowledge of the high-voltage, high-current output characteristics is therefore invaluable for understanding the short-circuit capability of these power transistors. These characteristics are not provided by the device manufacturer in the datasheets.

A novel method for extracting the high-voltage, high-current output characteristics of power transistors was developed by using the short-circuit test configuration [4]. The short-circuit test configuration shown in Fig. 19.1 was used with a small pulse width of 1.5 μs. During this time, the current increases to a peak value that is representative of the saturation current at the DC supply voltage. The DC supply voltage can be varied in small steps to generate the high-voltage, high-current output characteristic for the chosen gate bias.

The experimentally extracted high-voltage, high-current output characteristic at a gate bias of 15 V for the stand-alone IXYS 1.2 kV Si IGBT IXGH20N120A3, labeled as "IGBT Alone," is shown in Fig. 19.2 with the red line. The current saturates at a collector bias of 50 V and then increases gradually up to 1000 V. The saturated collector current is 88 A for the stand-alone Si IGBT at a collector bias of 800 V. The experimentally extracted high-voltage, high-current output characteristics for the composite MOSFET CM with gate biases of 6.0, 5.7, 5.4, and 5.0 V applied to the Si EMM device are also shown in Fig. 19.2 with the purple, orange, blue, and green lines, respectively. The current saturates at values of 73, 65, 57, and 40 A, respectively, at these DC gate bias values. These operating points for the five cases are highlighted with the brown circles.

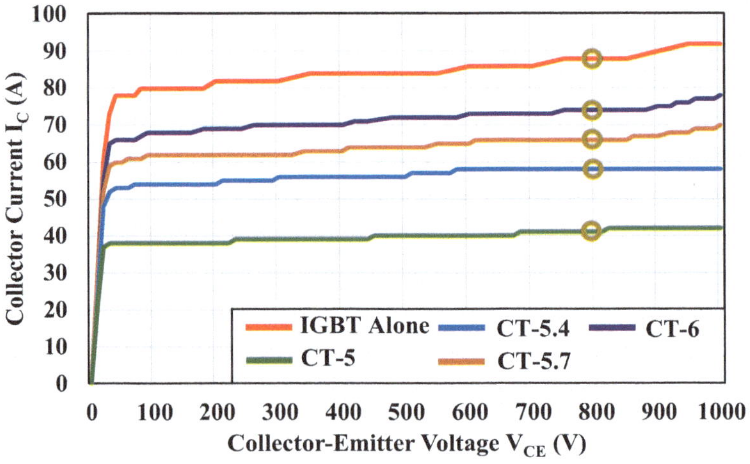

Fig. 19.2 High-voltage, high-current output characteristics: stand-alone IXYS 1.2 kV Si IGBT IXGH20N120A3, and composite transistor CT with DC gate bias voltages of 5.0, 5.4, 5.7, and 6.0 V applied to the Si EMM device

19.4 Short-Circuit Test Results

The short-circuit withstand time for the composite transistor CT was obtained by using the test configuration shown in Fig. 19.1 at a DC supply voltage of 800 V with various values programmed for the DC gate bias applied to the Si EMM device. The short-circuit pulse width is gradually increased while capturing the current waveform until the Si IGBT fails. The DC supply voltage is monitored as shown by the black line in Fig. 19.3. It is important to make sure that the capacitor C_{DC} is sufficiently large to maintain the voltage at the desired value, 800 V in this case.

The measured short-circuit waveforms for all the cases are shown in Fig. 19.3. The red line was obtained for the stand-alone Type 2 IXYS 1.2 kV Si IGBT IXGH20N120A3. The device failed when the pulse width was increased to 3.8 μs with a peak short-circuit current of 122 A. This value differs from the saturated current shown in Fig. 19.2 because the peak occurs after 1.5 μs. This comment is applicable to the composite transistors as well. The stand-alone IGBT selected for this test is a Type 2 device optimized for low on-state voltage drop and has a poor short-circuit withstand time of much less than 10 μs.

The green line labeled CT 5 in Fig. 19.3 was obtained for the composite transistor with a DC gate bias of 5.0 V applied to the Si EMM device. A peak short-circuit current of 42 A was observed for this case. The IXYS 1.2 kV Si IGBT

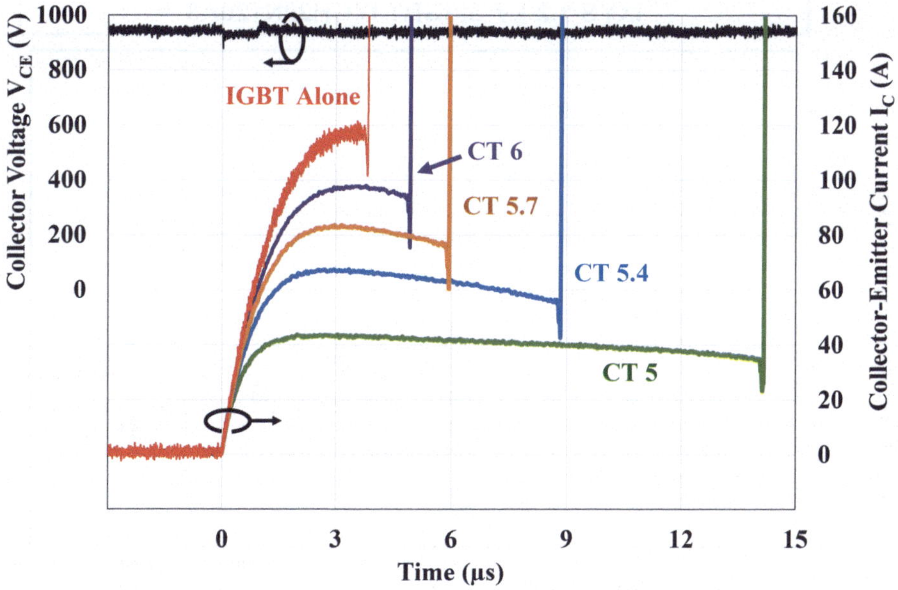

Fig. 19.3 Measured short-circuit waveforms: stand-alone IXYS 1.2 kV Si IGBT IXGH20N120A3, and composite transistor CT with DC gate bias voltages of 5.0, 5.4, 5.7, and 6.0 V applied to the Si EMM device

IXGH20N120A3 failed when the pulse width was increased to 14.1 μs. This demonstrates the ability of the BaSIC(EMM) topology to increase the short-circuit withstand time for a Type 2 IGBT to beyond 10 μs.

The blue line labeled CT 5.4 was obtained for the composite transistor with a DC gate bias of 5.4 V applied to the Si EMM device. A peak short-circuit current of 68 A was observed for this case. The IXYS 1.2 kV Si IGBT IXGH20N120A3 failed when the pulse width was increased to 8.8 μs in this case. The orange line labeled CT 5.7 was obtained for the composite transistor with a DC gate bias of 5.7 V applied to the Si EMM device. A peak short-circuit current of 83 A was observed for this case. The IXYS 1.2 kV Si IGBT IXGH20N120A3 failed when the pulse width was increased to 5.9 μs. The purple line labeled CT 6 was obtained for the composite transistor with a DC gate bias of 6.0 V applied to the Si EMM device. A peak short-circuit current of 98 A was observed for this case. The IXYS 1.2 kV Si IGBT IXGH20N120A3 failed when the pulse width was increased to 4.9 μs.

It can be concluded that the short-circuit withstand time increases with reduction in the peak short-circuit current level as expected from the theory provided in Chap. 2. This becomes obvious if the measured short-circuit withstand time is plotted against the measured peak short-circuit currents during the tests, as shown in Fig. 19.4. The red line shows an inverse dependence of the short-circuit time on the peak short-circuit current predicted by the theory. From the green dashed lines in the figure, it can be concluded that a peak short-circuit current of 62 A is needed to

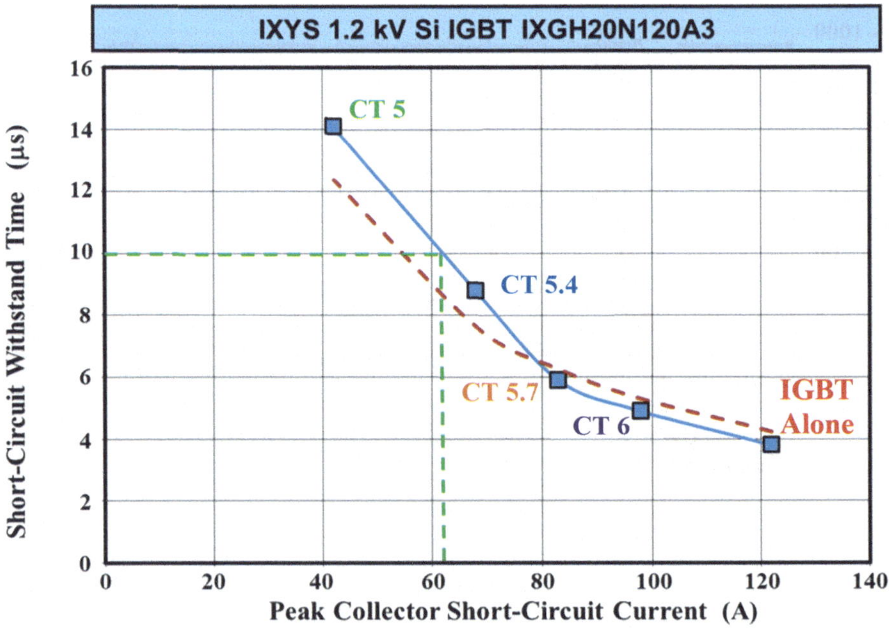

Fig. 19.4 Measured peak short-circuit current versus the observed short-circuit withstand time for the IXYS 1.2 kV Si IGBT IXGH20N120A3

achieve a short-circuit withstand time of 10 μs. The active area of the 1.2 kV, 20 A Si IGBT IXGH20N120A3 can be estimated to be 0.235 cm^2 based on an on-state current density of 85 A/cm^2 for the 1.2 kV Si IGBT [5]. The peak short-circuit current density to achieve a short-circuit withstand time of 10 μs is then found to be 263 A/cm^2 for the 1.2 kV Si IGBT IXGH20N120A3.

These short-circuit test results provide justification for using the BaSIC(EMM) topology to Si IGBTs. This topology was able to reduce the short-circuit current level and significantly increase the short-circuit withstand time for the Type 2 Si IGBT. These tests also demonstrate that the BaSIC(EMM) topology is able to increase the short-circuit withstand time for a commercially available Type 2 Si IGBT to beyond 10 μs. However, it is shown in the next section that the composite transistor has a very high on-state voltage drop in this case.

19.5 Impact on Normal Circuit Operation

The impact of the resistance of the Si EMM device on the on-state resistance and switching behavior of the composite transistor within the BaSIC(EMM) topology with a Si IGBT must be quantified. Trade-off curves between enhancing the short-circuit time and an increase on-state voltage drop and switching losses can then be developed to fully comprehend the benefits of the BaSIC(EMM) topology to enhancing the performance of Si IGBTs.

19.5.1 Impact on On-State Voltage Drop

The on-state characteristics of the IXYS 1.2 kV Si IGBT IXGH20N120A3 product were obtained at lower drain voltages by using a curve tracer with a gate drive voltage of 15 V. The on-state operating point was obtained at a current of 10 A as highlighted with the red horizontal line. The on-state voltage drop for the stand-alone Type 2 IXYS 1.2 kV Si IGBT IXGH20N120A3 at this operating point is 1.50 V consistent with its datasheet. This is a relatively low value when compared with Type 1 Si IGBTs [5].

The measured on-state characteristics for the composite transistors with four DC gate bias values applied to the Si EMM device are also shown in Fig. 19.5. The on-state operating point for each case was determined at a current of 10 A. The on-state voltage drop at this operating point is 1.65 V for case CT-6 with DC gate bias of 6.0 V applied to its Si EMM device. It increases to 1.70 V for case CT-5.7 with DC gate bias of 5.7 V applied to its Si EMM device. It becomes 1.87 V for case CT-5.4 with DC gate bias of 5.4 V applied to its Si EMM device. This is 25% increase compared with the stand-alone IGBT. The current begins to saturate in the case of the composite transistor CT-5 with DC gate bias of 5.0 V applied to its Si EMM

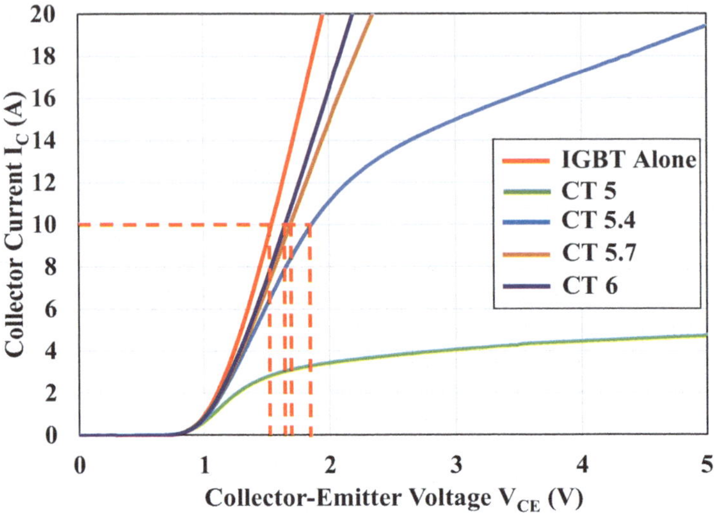

Fig. 19.5 Measured on-state characteristics at gate bias of 15 V: stand-alone IXYS 1.2 kV Si IGBT IXGH20N120A3, and composite transistor CT with DC gate bias voltages of 5.0, 5.4, 5.7, and 6.0 V applied to the Si EMM device

device. The on-state voltage drop in this case takes a large value of 13.6 V. This is an unacceptable operating point for the BaSIC(EMM) topology when using the selected Si EMM device.

In order to understand the on-state characteristics obtained for the composite transistors in the BaSIC(EMM) topology, it is necessary to examine the output characteristics of the Si EMM devices used for the BaSIC(EMM) topology with gate bias values in the range of interest to suppress the short-circuit current level. The results of this measurement are shown in Fig. 19.6 with gate bias values ranging from 5.0, 5.4, 5.7, and 6.0 V. The on-resistances were extracted at a drain current of 10 A, shown by the red dashed line in Fig. 19.6, to be relevant to the composite transistors. The Si EMM exhibits an on-resistance of 12 mΩ at a gate bias of 6.0 V. The on-resistance of the Si EMM increases to 15 mΩ at a gate bias of 5.7 V. These values produce a voltage drop of 0.12 and 0.15 V across the Si EMM devices in the composite transistors CT-6 and CT-5.7. This additional voltage drop is consistent with the measured on-state voltage drop for the composite transistors. For lower gate bias values, the Si EMM deviates from the linear regime and exhibits quasi-saturation at a current of 10 A. Consequently, the on-resistance of the Si EMM increases to 35 mΩ at a gate bias of 5.4 V. This produces the observed increase in on-state voltage drop by 0.37 V for the composite transistor CT-5.4 compared with the stand-alone IGBT. It can be observed that the Si EMM device enters its current saturation mode for a DC gate bias of 5.0 V with a high-voltage drop of 12 V at 10 A. This is responsible for the current saturation observed for the composite transistor CT-5 in Fig. 19.5, which produces a high on-state voltage drop of 13.6 V.

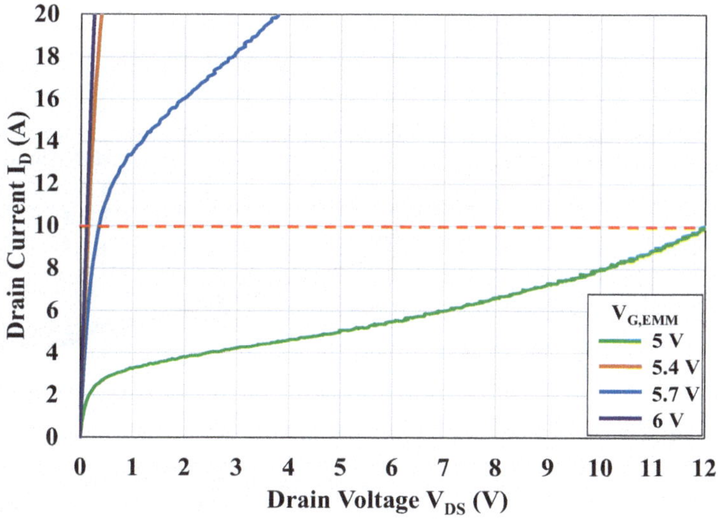

Fig. 19.6 Measured output characteristics of the Si EMM device (Infineon Si OptiMOS enhancement mode MOSFET IPI70N04S-06) at various gate bias voltages

19.5.2 Impact on Capacitances

The capacitances of the IGBT play a role in its switching performance [6]. The input, output, and reverse transfer capacitances of all the devices were therefore measured as a function of the collector voltage. These measured capacitances are displayed in Fig. 19.7. The curves for C_{ISS}, C_{OSS}, and C_{RSS} for all the five cases overlap. The explanation for this is the large capacitance of the Si EMM under normal circuit operating conditions as discussed in Chap. 10.

19.5.3 Impact on Transconductance

The transconductance g_M of the IGBT plays an important role in the switching transient for the collector voltage and current during operation with an inductive load [6]. Its value can be ascertained from the transfer characteristics of the device. The transfer characteristics for all the five transistor cases were measured at a collector bias of 10 V and are shown in Fig. 19.8. The transconductance, extracted by taking a derivative of the collector current with respect to the gate voltage, is also shown in the figure. The value of the transconductance can be compared for the four cases at the same drain current level of 10 A. It is given by the values at the points indicated on the transconductance curve.

The measured transconductance of the Si IGBT product was 11 S. The transconductances of the composite transistors were 10, 9.5, and 6.2 S for DC gate bias of

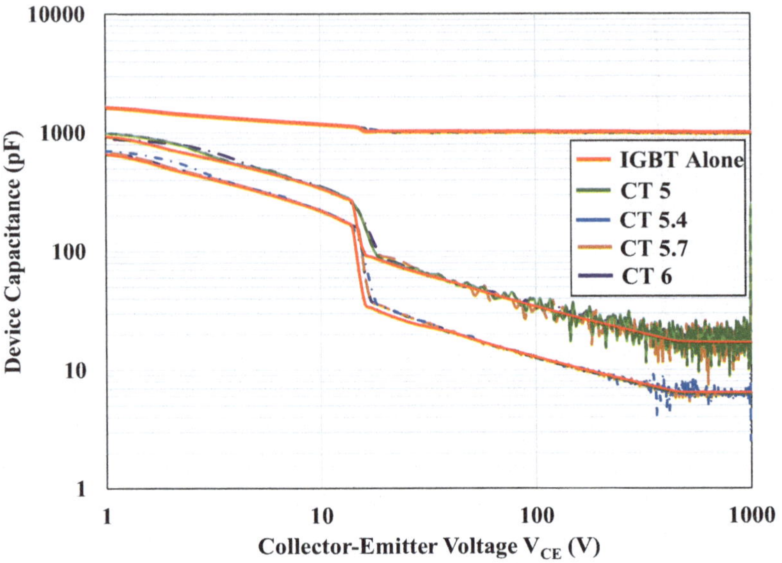

Fig. 19.7 Measured capacitances: stand-alone IXYS 1.2 kV Si IGBT IXGH20N120A3, and composite transistor CT with DC gate bias voltages of 5.0, 5.4, 5.7, and 6.0 V applied to the Si EMM device

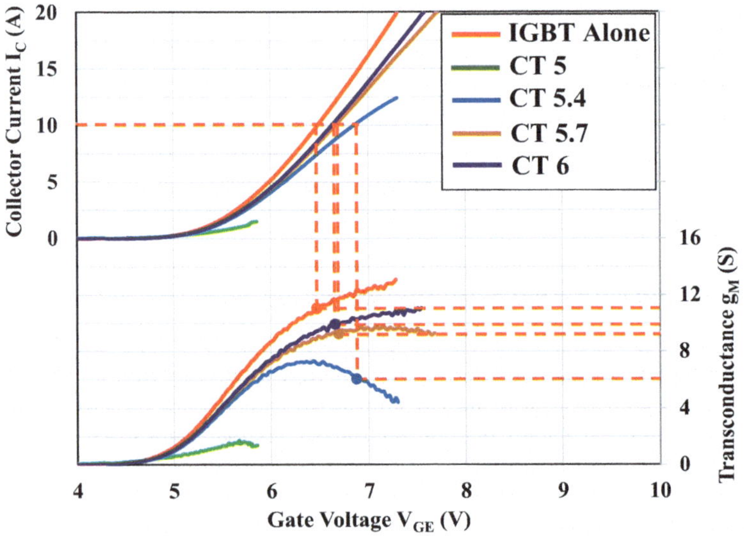

Fig. 19.8 Measured transconductances: stand-alone IXYS 1.2 kV Si IGBT IXGH20N120A3, and composite transistor CT with DC gate bias voltages of 5.0, 5.4, 5.7, and 6.0 V applied to the Si EMM device

6.0, 5.7, and 5.4 V, respectively. The reduction of the transconductance is consistent with calculated values based on the measured on-state resistances of the Si EMM device for the three DC gate bias values. The transconductance of the composite transistor CT-5 with DC gate bias of 5.0 V applied to the Si EMM could not be measured due to the Si EMM shifting to its drain current saturation mode.

19.5.4 Impact on Switching Losses

The double-pulse test, whose circuit diagram was shown in Fig. 5.15, was used to evaluate the switching power losses of all the power transistors. The double-pulse testing of all the power transistors was performed using SiC JBS rectifiers as free-wheeling diodes. The test was performed at a collector current of 20 A and collector supply voltage of 600 V to match the conditions in the IXYS 1.2 kV Si IGBT IXGH20N120A3 datasheet.

The measured turn-on transients for all the transistor configurations are shown in Fig. 19.9. The switching energy losses were determined by integrating the product of the voltage and current over the duration of the transient. All the five transistors show closely matching turn-on behavior. The turn-on loss for the stand-alone Si

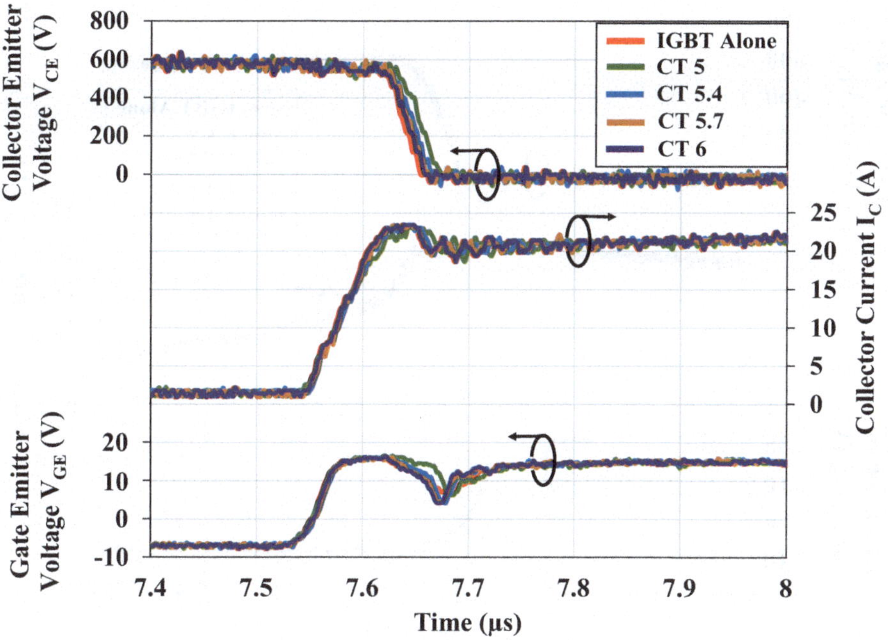

Fig. 19.9 Measured turn-on transients: stand-alone IXYS 1.2 kV Si IGBT IXGH20N120A3, and composite transistor CT with DC gate bias voltages of 5.0, 5.4, 5.7, and 6.0 V applied to the Si EMM device

IGBT was found to be 0.70 mJ. The composite transistors CT-6.0, CT-5.7, CT-5.4, and CT-5 had turn-on losses of 0.73, 0.75, 0.78, and 0.80 mJ for the case of DC gate bias of 6.0, 5.7, 5.4, and 5.0 V applied to the Si EMM device. The composite transistors exhibit slightly larger turn-on losses when the resistance of the Si EMM increases with smaller DC gate bias.

The measured turn-off transients for the five transistor configurations are shown in Fig. 19.10. A tail in the collector current occurs during the turn-off producing significant power losses as expected for Si IGBTs. However, very little difference in the turn-off waveforms is observed between all the five transistor cases. The turn-off loss for the stand-alone Si IGBT was found to be 5.30 mJ. The composite transistors CT-6.0, CT-5.7, CT-5.4, and CT-5 had turn-off losses of 5.30, 5.30, 5.50, and 5.50 mJ for the case of DC gate bias of 6.0, 5.7, 5.4, and 5.0 V applied to the Si EMM device. The total energy loss per cycle was obtaining by adding the turn-on and turn-off losses. It is found to be 6.00 mJ for the stand-alone Si IGBT. The composite transistors CT-6.0, CT-5.7, CT-5.4, and CT-5 had total turn-off losses of 6.03, 6.05, 6.28, and 6.30 mJ for the case of DC gate bias of 6.0, 5.7, 5.4, and 5.0 V applied to the Si EMM device. The energy loss per cycle for the composite transistors is larger than that for the stand-alone Si IGBT by 0.5, 0.8, 4.7, and 5.0% for the case of DC gate bias of 6.0, 5.7, 5.4, and 5.0 V applied to the Si EMM device. In conclusion, the double-pulse tests demonstrate that the composite transistors in the

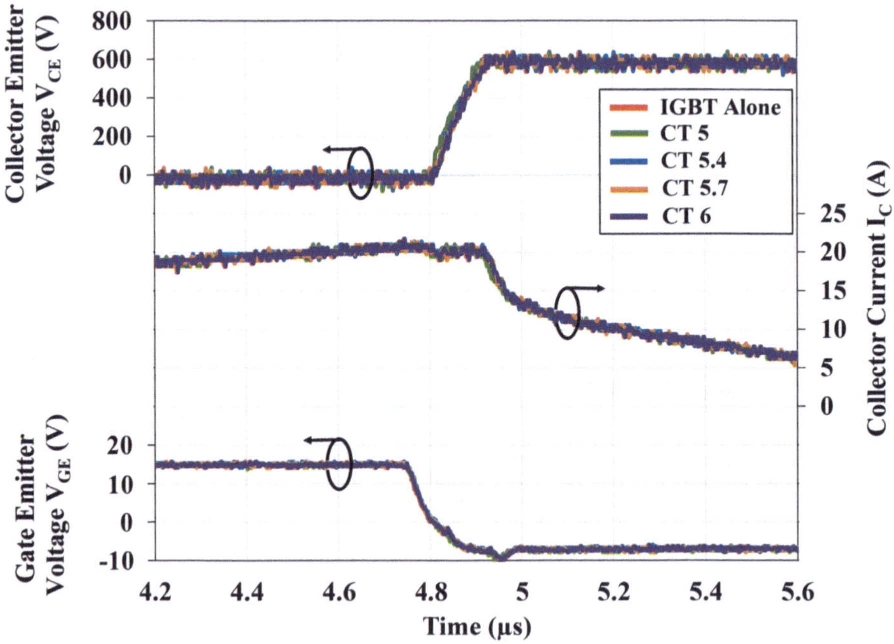

Fig. 19.10 Measured turn-off transients: stand-alone IXYS 1.2 kV Si IGBT IXGH20N120A3, and composite transistor CT with DC gate bias voltages of 5.0, 5.4, 5.7, and 6.0 V applied to the Si EMM device

BaSIC(EMM) topology have only slightly larger switching losses than the stand-alone Si IGBT.

19.6 Sense Node Voltage

The sense node voltage V_{SENSE} shown in Fig. 19.1 can be utilized to monitor the on-state current flowing through the Si IGBT and for detection of the short-circuit event. The detection of the short-circuit event was demonstrated for the case of the Si IGBT with a silicon EMM device at DC gate bias of 6.0 V connected in series with its emitter terminal. The short-circuit test waveforms are shown in Fig. 19.11 with a gate drive pulse width of 5 μs to avoid failure of the Si IGBT. The sense node voltage changes by 6 V under short-circuit conditions. This large change can be easily detected by control circuits used to identify the occurrence of the short-circuit and trigger turning-off the gate drive voltage to prevent failure of the Si IGBT.

A small spike in the sense node voltage is observed immediately after the short-circuit event is initiated. The sense node voltage then remains at the same value for 1 μs. A relatively short blanking time of 0.5 μs is therefore sufficient for the BaSIC(EMM) topology as indicated in Fig. 19.11. This time interval is much smaller than the typical 3 μs blanking time needed for the conventional IGBT DESAT gate control circuits. This difference is due to the sense node voltage V_{SENSE}

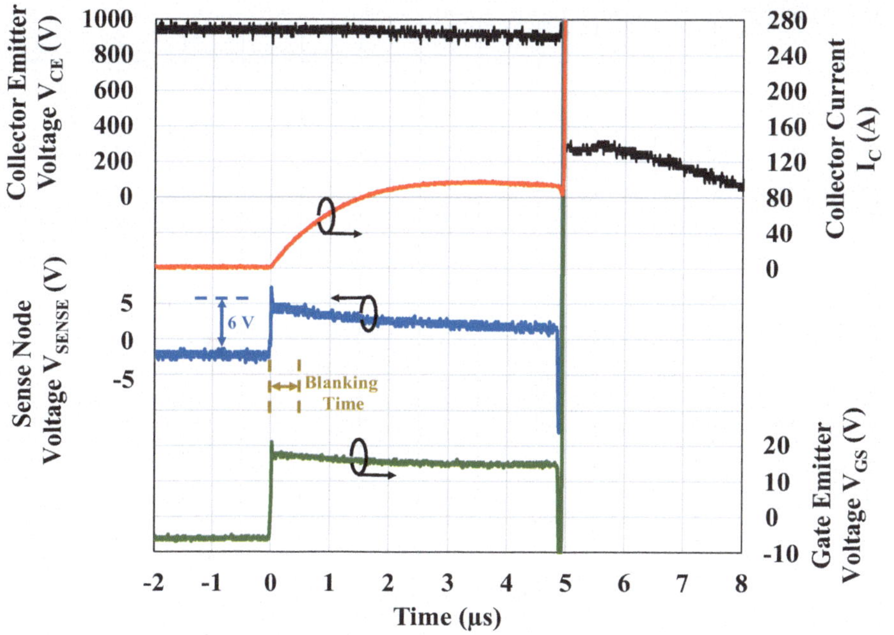

Fig. 19.11 Sense node voltage for the composite MOSFET CT-6 during short-circuit operation

operating at low voltages at all times during the short-circuit event compared with the conventional DESAT circuit where the collector terminal voltage of the IGBT is being monitored.

19.7 Trade-Off Curves

Trade-off curves between short-circuit withstand time and the on-state voltage drop or the switching losses are useful to understand the true impact of the BaSIC(EMM) topology on the performance of the Type 2 1.2 kV Si IGBT IXGH20N120A3.

The trade-off curve obtained between the short-circuit withstand time and the on-state voltage drop by using the measured data is shown in Fig. 19.12. There is very steep increase in the short-circuit withstand time in relation to the increase in the on-state voltage drop. This is very desirable feature of the BaSIC(EMM) topology. A black dotted trend line is shown in the figure extending to a short-circuit withstand time of 10 μs. The desirable short-circuit withstand time of 10 μs is obtained with an on-state voltage drop of 1.9 V. The on-state voltage drop is increased by 27% to achieve this 10 μs short-circuit withstand time. Consequently, the BaSIC(EMM) topology can be utilized to create devices for motor drive applications with much lower on-state voltage drop by using Type 2 IGBTs rather than using the Type 1 IGBTs.

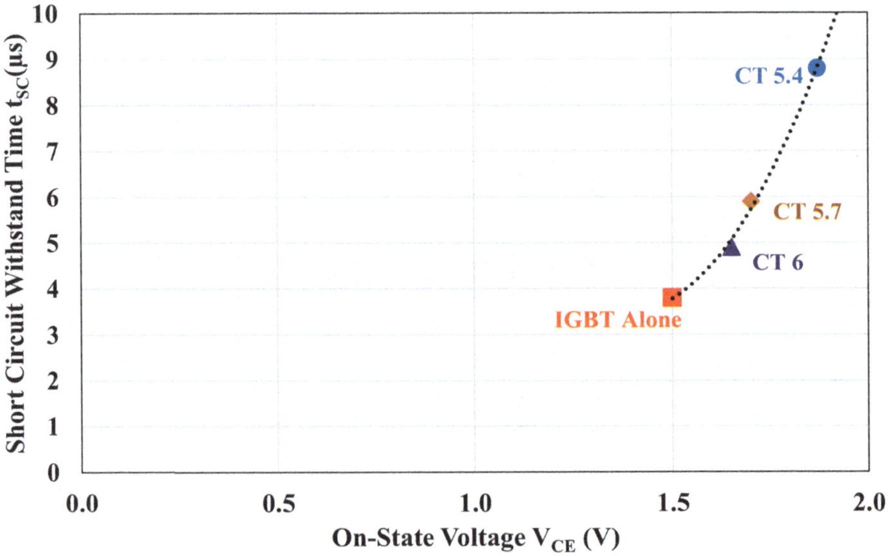

Fig. 19.12 Trade-off curve observed for the BaSIC(EMM) topology in the case of the IXYS 1.2 kV Si IGBT IXGH20N120A3 product: on-state voltage drop

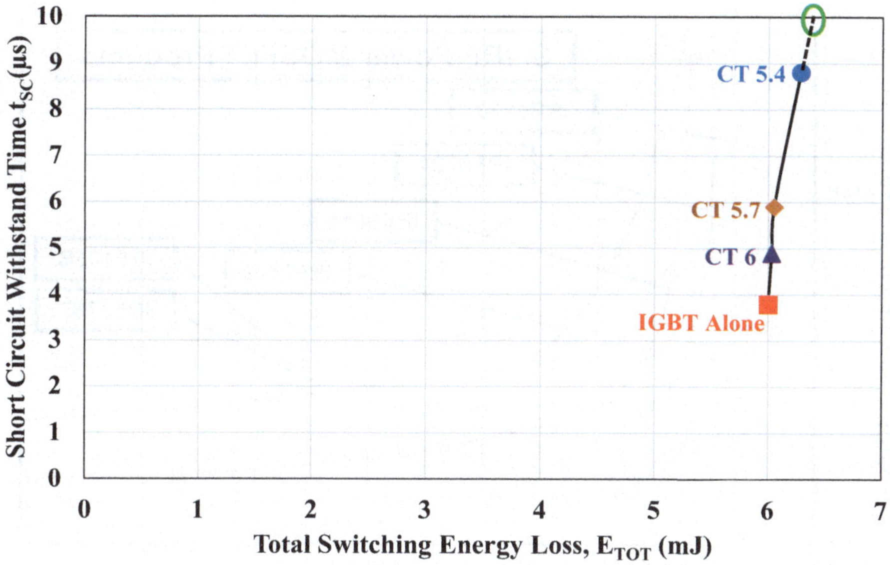

Fig. 19.13 Trade-off curve observed for the BaSIC(EMM) topology in the case of IXYS 1.2 kV Si IGBT IXGH20N120A3 product: switching energy loss

The trade-off curve between the short-circuit withstand time and the total switching energy loss is shown in Fig. 19.13 for the BaSIC(EMM) topology for the Type 2 1.2 kV Si IGBT IXGH20N120A3. A black dotted trend line is shown in the figure extending to a short-circuit withstand time of 10 µs circled in green. The total switching energy loss for this case is 6.35 mJ which is 6% larger than observed for the stand-alone Type 2 1.2 kV Si IGBT IXGH20N120A3 product. The impact on switching losses is therefore quite small when utilizing the BaSIC(EMM) topology for Si IGBTs.

19.8 Summary

The benefits of using the BaSIC topology with a silicon enhancement mode MOSFET (EMM) as the non-linear resistance are demonstrated in this chapter for the case of the Type 2 1.2 kV IXYS Si IGBT IXGH20N120A3 product. A wide variety of Si EMMs are manufactured by the industry making this approach practical. For the chosen commercially available Type 2 IGBT with low on-state voltage drop and poor short-circuit capability, an increase in the short-circuit withstand time by 230% was obtained with the BaSIC(EMM) topology with an increase in the on-state voltage drop by 25% and the switching losses by 5%. A short-circuit withstand time of 8.8 µs is observed for the above case with an Infineon Si OptiMOS enhancement mode MOSFET IPI70N04S-06 product.

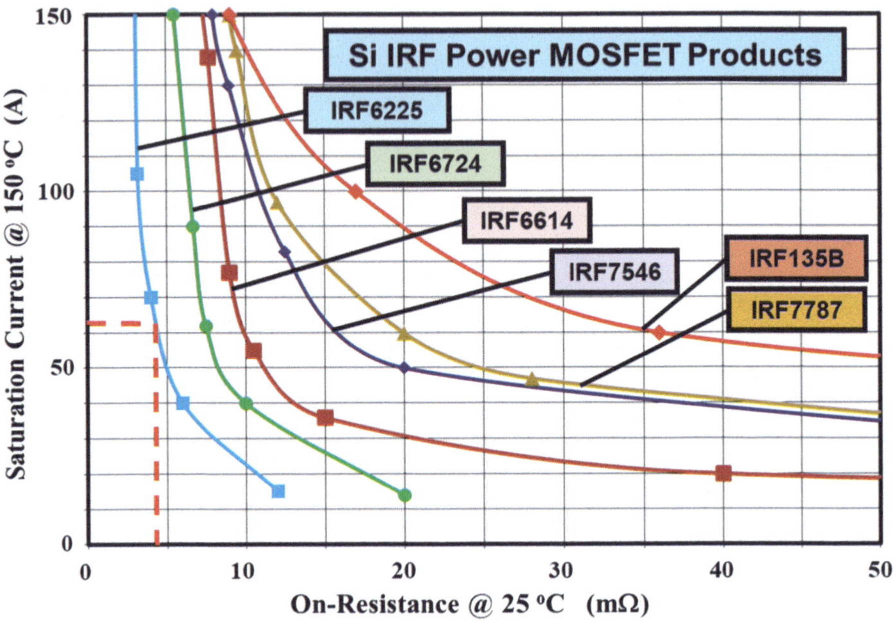

Fig. 19.14 Trade-off curve between on-resistance and drain saturation current for various Si EMM products

A selection methodology for the Si EMM device was discussed in Chap. 13. It allows identification of Si MOSFETs with superior trade-off curves between drain saturation current and on-resistance. This trade-off curve for various Si EMM products is reproduced in Fig. 19.14. The IRF6225 product was found to have the best trade-off curve among the evaluated products. The on-resistance for this Si EMM is 5 mΩ for a drain saturation current of 62 A required to achieve a short-circuit withstand time of 10 μs for the Type 2 Si IGBT as indicated by the red dashed line. The implementation of the BaSIC(EMM) topology with the IRF6225 product can therefore be predicted to increase the short-circuit withstand time of the Type 2 1.2 kV IXYS Si IGBT IXGH20N120A3 product to 10 μs while increasing its on-state voltage drop at 10 A by only 50 mV. This is an increase in on-state voltage drop by only 3.3%. This analysis demonstrates the importance of identifying the best Si EMM device for the BaSIC topology using the trade-off graph in Fig. 19.14.

References

1. A. Kanale, B.J. Baliga, A new user-configurable method to improve short-circuit ruggedness of 1.2-kV SiC power MOSFETs. IEEE Trans. Power Electron. **36**, 2059–2067 (2021)

2. A. Kanale, B.J. Baliga, User-programmable short-circuit capability enhancement for 1.2 kV Si IGBTs using a 40 V Si enhancement-mode MOSFET connected in series with the emitter, in *IEEE Applied Power Electronics Conference* (2020), pp. 680–685
3. https://www.infineon.com/cms/en/product/power/mosfet/12v-300v-n-channel-power-mosfet/irl60hs118/
4. A. Kanale, K. Han, B.J. Baliga, S. Bhattacharya, Superior short circuit performance of 1.2kV SiC JBSFETs compared to 1.2kV SiC MOSFETs, in *12th European Conference on Silicon Carbide and Related Materials*, Paper WE.P.MO9 (2018)
5. B.J. Baliga, *The IGBT Device*, 2nd edn. (Elsevier, Amsterdam, 2023)
6. B.J. Baliga, *Fundamentals of Power Semiconductor Devices*, 2nd edn. (Springer Science, New York, 2019)

Chapter 20
BaSIC(DMM) for GaN HEMT

Gallium Nitride (GaN) High Electron Mobility Field Effect Transistors (HEMTs) are a potential choice for power conversion applications due to their fast switching speeds and low on-state resistance [1]. A review of the performance of state-of-the-art GaN transistors has been recently published [2]. Many of these applications and especially motor drives require transistors that exhibit good short-circuit capability at the high DC bus supply voltage with on-state gate drive conditions. Type 1 silicon IGBT products have been optimized by manufacturers with short-circuit withstand time of over 10 μs at high DC bus voltages and on-state gate drive voltage for motor drive applications. However, GaN HEMT devices (Cascode, p-GaN, GaN MISHEMT) have been found to exhibit poor (<0.5 μs) short-circuit withstand time at high DC bus voltages with normal on-state gate drive [3–9]. This problem continues to impede the performance of recent devices [2]. Although adequate short-circuit withstand time has been observed at low DC bus voltages (<300 V) and reduced gate drive voltages, the short-circuit withstand time decreases drastically above a DC supply voltage of 300 V to <0.4 μs. These small short-circuit withstand times make protection of the GaN transistors very challenging.

The poor short-circuit withstand capability of GaN HEMTs can be improved by tuning the gate drive network to suppress the peak SC current [4] and terminating the short-circuit fault through soft-turn-off [10, 11]. This method of achieving short-circuit withstand capability is found to increase the on-state resistance and switching losses. Attempts to improve the short-circuit capability of Cascode GaN FETs by reducing the gate drive voltage has also been problematic due to observation of unstable operation [8].

The dramatic improvement in short-circuit withstand time that is achievable using the BaSIC topology is demonstrated in this chapter [12]. The BaSIC topology was implemented using a Gate-Source Shorted (GSS), low voltage Si depletion mode MOSFET (DMM) in series with the source of the GaN transistors.

B. J. Baliga, A. Kanale, *The BaSIC Topology*, https://doi.org/10.1007/978-3-031-86630-2_20

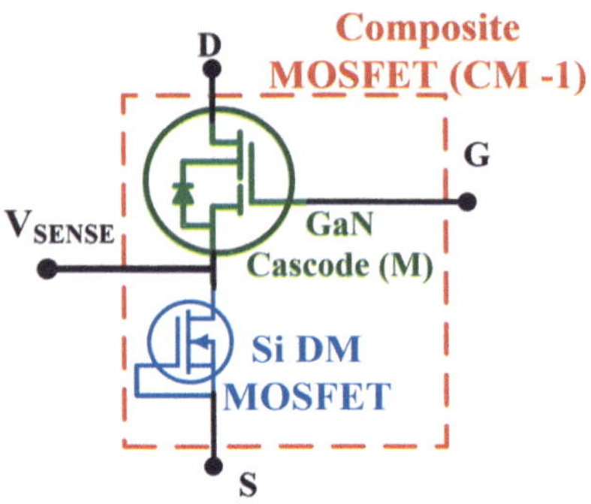

Fig. 20.1 The BaSIC(DMM) topology for the GaN HEMT device with composite MOSFET highlighted

20.1 GaN HEMT BaSIC(DMM) Topology

The BaSIC(DMM) topology for the GaN HEMT devices is illustrated in Fig. 20.1 using a single silicon DMM device connected in series with the source electrode of the GaN HEMT device (M). The GaN HEMT and Si DMM devices together form a composite MOSFET (CM-1) as indicated by the red box in Fig. 20.1, which can be treated as a single power switch in power circuits. The Si DMM device operates in its linear region with low on-resistance during normal on-state conditions. The GSS Si DMM enters its saturation region under short-circuit conditions limiting the current to a smaller level than for the stand-alone GaN HEMT device. The increased voltage drop across the Si DMM during the short-circuit event reduces the voltage appearing at the gate-source terminals of the GaN FET. This automatically reduces its saturation current until it matches the lower saturation current of the Si DMM device. The reduction in peak short-circuit current decreases the power dissipation in the GaN HEMT, which produces a larger short-circuit withstand time (t_{SC}).

The short-circuit tests were performed by initially using a short duration gate pulse of 0.1 μs to avoid damaging the GaN HEMT device. The width of the gate pulse was gradually increased until the device failed while capturing the drain current waveform. This allowed determination of the short-circuit withstand capability of the stand-alone GaN HEMT device.

20.2 Experimental Devices

The GaN HEMT devices chosen for the experimental demonstration were commercially available Transphorm 600 V GaN Cascode HEMT TPH3202PD [13] devices. It employs a depletion mode high voltage GaN HEMT transistor in series with a Si MOSFET to produce a normally-off device in accordance with the Baliga-Pair configuration [14, 15] originally proposed for SiC normally-on JFET devices. It has a rated blocking voltage of 600 V with on-state resistance of 350 mΩ at a gate bias of 8 V and drain current of 5.5 A.

The IXYS IXTH16N10D2 [16] device was used as the Si depletion mode MOSFET (DMM) as also employed in previous chapters. It has a rated blocking voltage of 100 V and on-state resistance of 65 mΩ. A Si DMM with lower blocking voltage is adequate for the BaSIC topology but there is a lack of commercially available DMMs with 20–30 V ratings.

20.3 Short-Circuit Test Configurations

The short-circuit test configuration is shown in Fig. 20.2 with the stand-alone GaN HEMT device. The DC supply voltage was set at 400 V due to the 600 V rating of the transistor. The gate drive voltage was set at 8 V since it is the recommended gate voltage for the on-state operation by the manufacturer. The capacitor C_{DC} must be sufficiently large in size to maintain the voltage across the transistor during the short-circuit event.

The BaSIC(DMM) topology for the GaN HEMT device has several configurations. The Si DMM devices can be paralleled to adjust the level of the short-circuit current observed for the BaSIC(DMM) topology. This allows examination of the trade-off between the improvement in short-circuit withstand time and the increase in on-state resistance produced by the addition of the Si DMM devices. In the first one shown in Fig. 20.3, a single Si DMM device was used as the non-linear resistance in series with the GaN HEMT device to create the composite MOSFET CM-1 highlighted by the red dashed box. In the second one shown in Fig. 20.4, two Si DMM devices were used in parallel as the non-linear resistance in series with the GaN HEMT device to create the composite MOSFET CM-2 highlighted by the red dashed box. In this case, the current during the short-circuit event is twice as large as for composite MOSFET CM-1 while the additional resistance of the Si DMM devices is reduced in half. In the third one shown in Fig. 20.5, three Si DMM devices were used in parallel as the non-linear resistance in series with the GaN HEMT device to create the composite MOSFET CM-3 highlighted by the red dashed box. In this case, the current during the short-circuit event is three times as large as for composite MOSFET CM-1 while the additional resistance of the Si DMM devices is reduced to one-third.

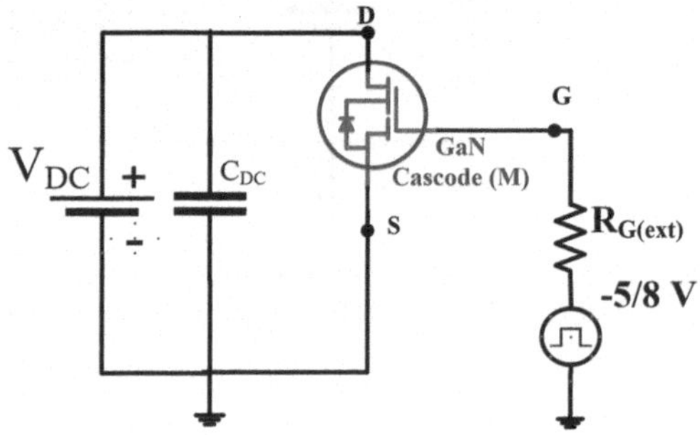

Fig. 20.2 The short-circuit test configuration with stand-alone GaN HEMT device

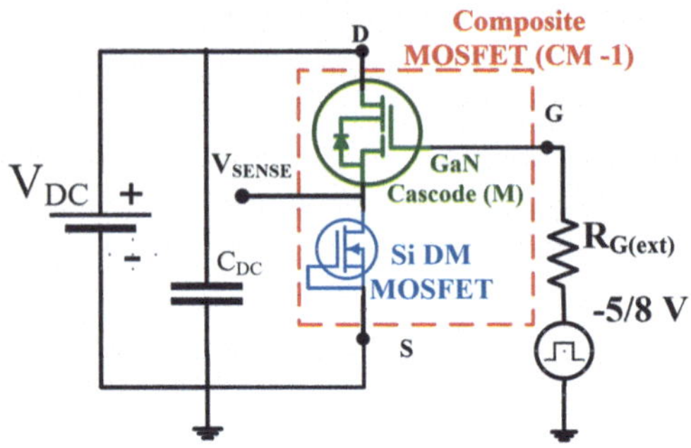

Fig. 20.3 The short-circuit test configuration with first BaSIC(DMM) topology for the GaN HEMT device: single Si DMM device is used to form the composite MOSFET CM-1

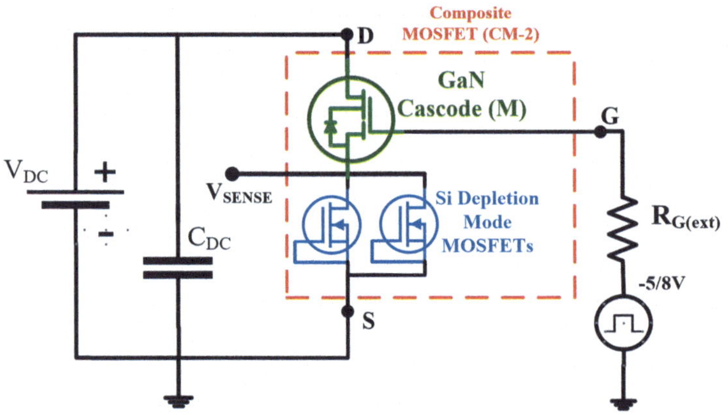

Fig. 20.4 The short-circuit test configuration with second BaSIC(DMM) topology for the GaN HEMT device: two Si DMM device are used in parallel to form the composite MOSFET CM-2

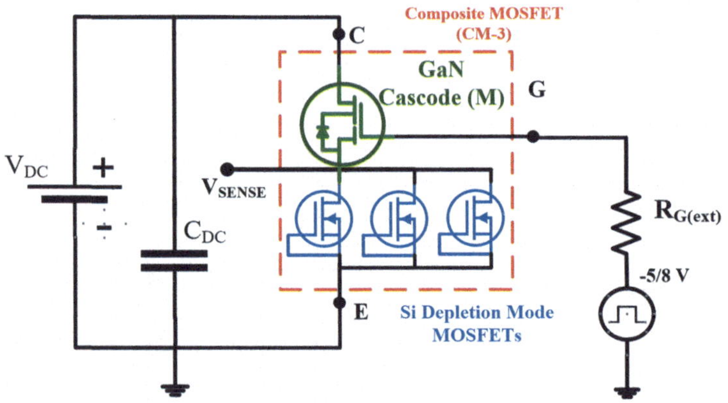

Fig. 20.5 The short-circuit test configuration with third BaSIC(DMM) topology for the GaN HEMT device: three Si DMM device are in parallel to form the composite MOSFET CM-3

20.4 Short-Circuit Test Results

The short-circuit withstand time for the stand-alone GaN HEMT device has been reported to be well below 1 μs when the DC supply voltage is 400 V and gate drive voltage is 8 V. This was observed for the Transphorm Cascode GaN HEMT TPH3202PD device used for the experiments.

The measured short-circuit waveforms for all the four test configurations are shown in Fig. 20.6. The red line (labeled M) was obtained for the stand-alone Transphorm 600 V GaN Cascode HEMTs TPH3202PD device. A peak short-circuit current of 84 A was observed with a decay as the device heated up. The device failed when the pulse width was increased to 0.33 μs. The orange line (labeled CM-3) was obtained for the composite MOSFET CM-3 in the BaSIC(DMM) topology. The peak short-circuit current was 53 A in this case. The device failed when the pulse width was increased to 3.30 μs, a ten-fold increase compared to the stand-alone GaN HEMT device. The blue line (labeled CM-2) was obtained for the composite MOSFET CM-2 in the BaSIC(DMM) topology. The peak short-circuit current was 35 A in this case. The device failed when the pulse width was increased to 4.35 μs, a 13-fold increase compared to the stand-alone GaN HEMT device. The green line (labeled CM-1) was obtained for the composite MOSFET CM-1 in the BaSIC(DMM) topology. The peak short-circuit current was 18 A in this case. The device failed when the pulse width was increased to 5.00 μs, a 15-fold increase compared to the stand-alone GaN HEMT device.

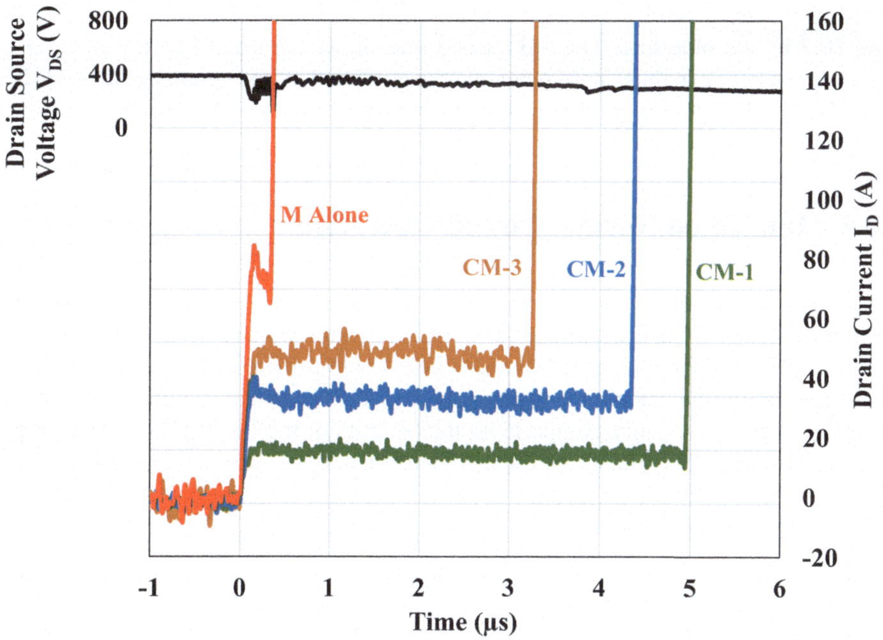

Fig. 20.6 The short-circuit test results for the stand-alone GaN HEMT device and the three composite MOSFETs of the BaSIC(DMM) topology

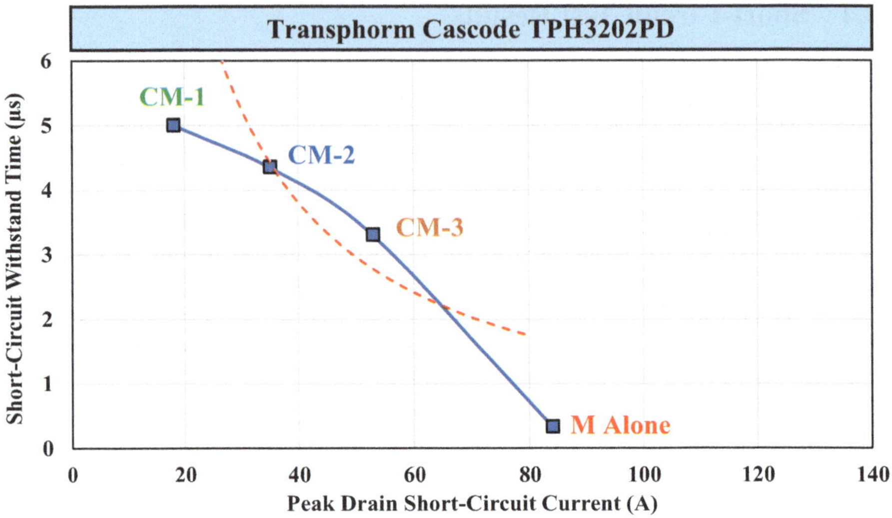

Fig. 20.7 The short-circuit withstand time versus short-circuit current for all devices

The short-circuit data demonstrates that the BaSIC(DMM) topology is able to substantially reduce the short-circuit current for the GaN HEMT device and consequently enhance the short-circuit withstand time. The relationship between the measured short-circuit withstand time and the peak shirt-circuit current is shown in Fig. 20.7 by the blue line. The red dashed line shows the trend line for an inverse relationship between these parameters that is predicted by the adiabatic theory [15]. The GaN HEMT devices do not follow this theoretically expected trend.

20.5 Impact on Normal Circuit Operation

The impact of the resistance of the Si EMM device on the on-state resistance and switching behavior of the composite transistor within the BaSIC(EMM) topology for the GaN HEMT device must be quantified. Trade-off curves between enhancing the short-circuit time and an increase on-state voltage drop and switching losses can then be developed to fully comprehend the benefits of the BaSIC(EMM) topology to enhancing the performance of the GaN HEMT devices.

20.5.1 Impact on On-State Voltage Drop

The on-state characteristics of the stand-alone Transphorm GaN HEMT TPH3202PD product was obtained at lower drain voltages by using a curve tracer with a gate drive voltage of 8 V as prescribed in its datasheet. The on-state operating point obtained at a current of 5.5 A as defined in the datasheet is highlighted with the red horizontal line. The on-state voltage drop for the Transphorm GaN HEMT TPH3202PD product at this operating point is 1.31 V. The measured on-state resistance is therefore 239 mΩ, which is close to the typical value of 290 mΩ in the datasheet.

The measured on-state characteristics for the composite MOSFETs are also shown in Fig. 20.8. The on-state operating point for each case was determined at a current of 5.5 A defined by the red dashed line. The on-state voltage drop at this operating point is 1.84 V for case CM-3 with a single Si(DMM) in series with the GaN HEMT device; 1.70 V for case CM-2 with two Si(DMM) devices in parallel used in series with the GaN HEMT device; and 1.62 V for case CM-3 with three Si(DMM) devices in parallel used in series with the GaN HEMT device. These values of on-state voltage drop correspond to on-state resistance value of 334 mΩ for case CM-3 with a single Si(DMM) in series with the GaN HEMT device; on-state resistance value of 309 mΩ for case CM-2 with two Si(DMM) devices in parallel used in series with the GaN HEMT device; and on-state resistance value of 294 mΩ for case CM-3 with three Si(DMM) devices in parallel used in series with the GaN HEMT device. The increase in on-state resistance is 95 mΩ for case CM-3 with a single Si(DMM) in series with the GaN HEMT device; 70 mΩ for case CM-2 with two Si(DMM) devices in parallel used in series with the GaN HEMT device; and 55 mΩ for case CM-3 with three Si(DMM) devices in parallel used in series

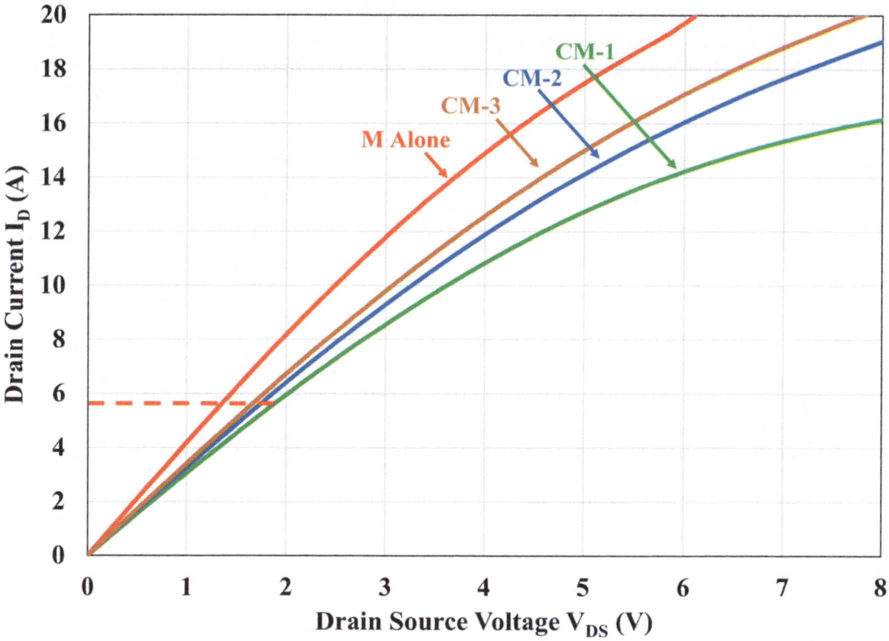

Fig. 20.8 Measured on-state characteristics: stand-alone Transphorm GaN HEMT TPH3202PD, and composite MOSFETs CM-1, CM-2, and CM-3

with the GaN HEMT device. These results demonstrate that the use of multiple parallel Si DMM devices in the BaSIC topology reduces the on-resistance of the composite MOSFET configuration as expected.

20.5.2 Impact on Capacitances

The capacitances of the GaN HEMT device play an important role in its switching performance. The input, output, and reverse transfer capacitances of all the devices were therefore measured as a function of the collector voltage. These measured capacitances are shown in Fig. 20.9. The curves for C_{ISS}, C_{OSS}, and C_{RSS} for all the four cases overlap. The explanation for this is the large capacitance of the Si EMM under normal circuit operating conditions as discussed in Chap. 10. The measured capacitance graphs are also in agreement with those shown in the datasheet for the GaN HEMT product. This indicates that the BaSIC(DMM) topology has no discernable impact on the device capacitances.

20.5.3 Impact on Transconductance

The transconductance g_M of the power switch plays an important role in the switching transient for the collector voltage and current during operation with an inductive load [15]. Its value can be derived from the transfer characteristics of the device.

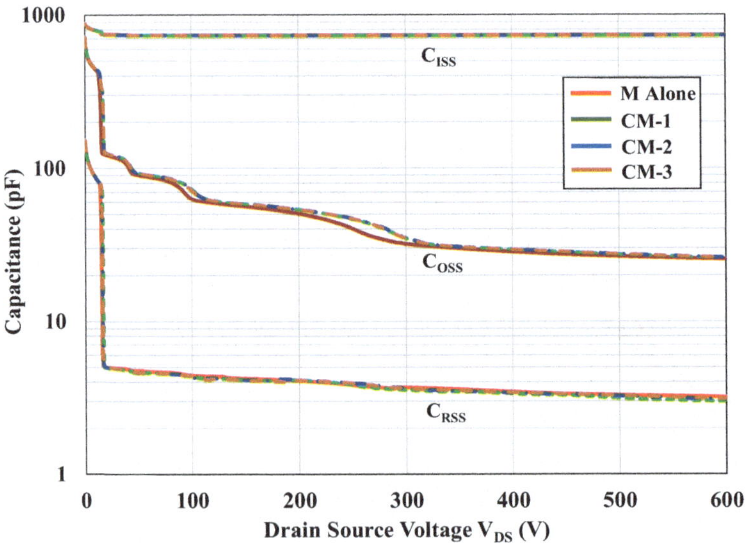

Fig. 20.9 Measured capacitances: stand-alone Transphorm GaN HEMT TPH3202PD, and composite MOSFETs CM-1, CM-2, and CM-3

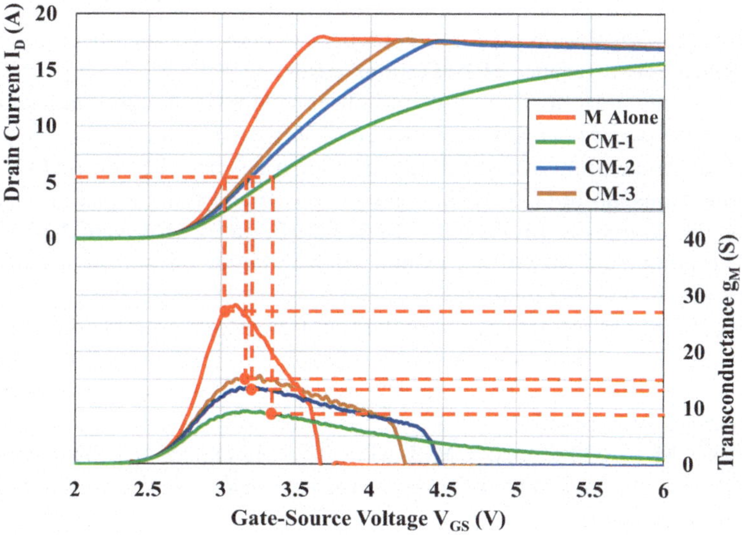

Fig. 20.10 Measured transconductances: stand-alone Transphorm GaN HEMT TPH3202PD, and composite MOSFETs CM-1, CM-2, and CM-3

The transfer characteristics for all four transistor cases were measured at a collector bias of 10 V and are shown in Fig. 20.10. The transconductance, extracted by taking a derivative of the collector current with respect to the gate voltage, is also shown in the figure. The value of the transconductance can be compared for the four cases at the same drain current level of 5.5 A. It is given by the values at the points indicated on the transconductance curve identified by the dashed red lines.

For a current of 5.5 A, the measured transconductance g_M for the stand-alone Transphorm GaN HEMT device was 26.9 S. This parameter is not provided in its datasheet for comparison. The measured transconductance g_M for the composite MOSFET CM-1 was 8.96 S. The measured transconductance g_M for the composite MOSFET CM-2 was 13.4 S. The measured transconductance g_M for the composite MOSFET CM-3 was 15.32 S. The reduction in transconductance is due to the influence of the resistance of the Si DMM devices in the source path. A reduced transconductance leads to slower switching transients for the drain current and voltage, which produces larger switching losses as discussed in the next section.

20.5.4 Impact on Switching Losses

The double-pulse test, whose circuit diagram is shown in Fig. 5.15, was used to evaluate the switching power losses of all the power transistors. The double-pulse testing of all the power transistors was performed using SiC JBS rectifiers as free-wheeling diodes. The tests were performed at a collector current of 5.5 A with a collector supply voltage of 400 V to match the conditions in the device datasheet. Gate pulses of −3/8 V were used in accordance the GaN FET datasheet recommendations with a gate resistance of 10 Ω for all cases.

The measured turn-on transients for all four transistor cases are shown in Fig. 20.11. The switching energy losses were determined by integrating the product of the voltage and current over the duration of the transient. The turn-on loss for the stand-alone Transphorm GaN HEMT device was found to be 88 μJ. The turn-on losses for the composite MOSFETs CM-1, CM-2, and CM-3 were 107, 88, and 86 μJ. The composite MOSFETs exhibit slightly larger turn-on losses when the resistance of the Si DMM increases.

The measured turn-off transients for the four transistor configurations are shown in Fig. 20.12. Very little difference in the turn-off waveforms is observed between all the four transistor cases. The turn-off loss for the stand-alone Transphorm GaN HEMT device was found to be 13 μJ. The turn-off losses for the composite MOSFETs CM-1, CM-2, and CM-3 were 16, 14, and 15 μJ. The composite MOSFETs have the same turn-off losses as the stand-alone device within the experimental margin of error.

The total energy loss per cycle for the stand-alone Transphorm GaN HEMT device is 101 μJ. The total energy losses per cycle for the composite MOSFETs CM-1, CM-2, and CM-3 are 123, 102, and 101 μJ. The total energy losses per cycle for the composite MOSFETs CM-1 are 20% larger than that observed for the stand-alone HEMT device. The total energy losses per cycle for the composite MOSFETs CM-2 and CM-3 are the same of that observed for the stand-alone HEMT device within the experimental margin of error.

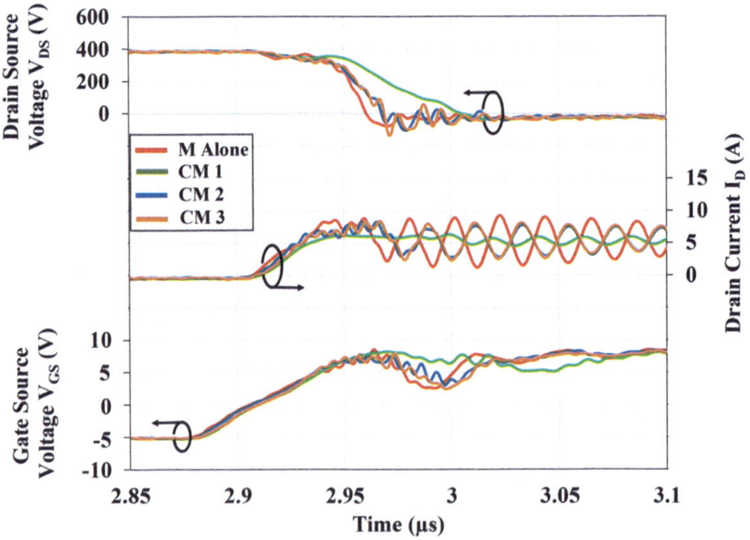

Fig. 20.11 Measured turn-on transients: stand-alone Transphorm GaN HEMT TPH3202PD and composite MOSFETs CM-1, CM-2, and CM-3

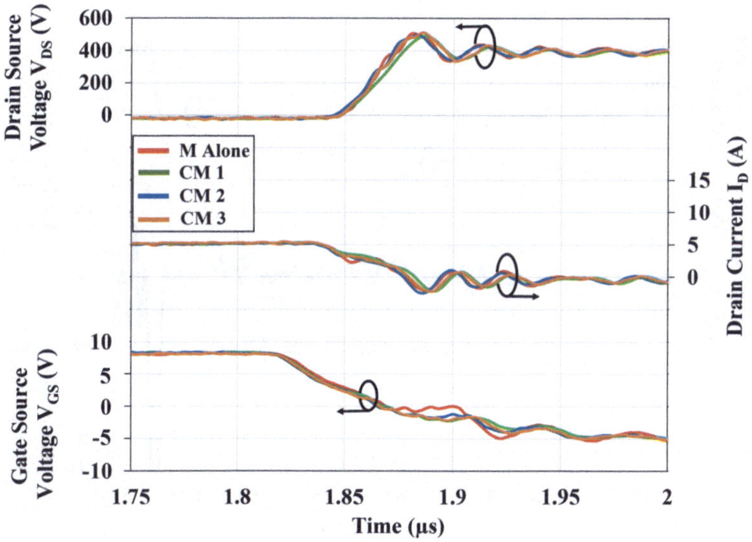

Fig. 20.12 Measured turn-off transients: stand-alone Transphorm GaN HEMT TPH3202PD and composite MOSFETs CM-1, CM-2, and CM-3

20.6 Sense Node Voltage

The current flowing through the GaN HEMT device can be monitored by using the drain of the Si DMM device as a fourth terminal. The sense voltage V_{SENSE}, shown in Figs. 20.3, 20.4, and 20.5, is proportional to the current flowing through the GaN HEMT device because the Si DMM device operates in its linear mode during on-state operation with a fixed resistance. As an example, the sense node voltage measured during on-state operation in shown in Fig. 20.13 for the composite MOSFET CM-2. V_{SENSE} has a value of ~0.4 V when the current reaches 5.5 A at the end of the gate pulse. This value is close to the product of the on-state current (5.5 A) and resistance (70 mΩ) of two Si DMM devices in parallel. This demonstrates that the sense node voltage is proportional to the GaN HEMT device current and can be used to monitor its value during normal circuit operation.

The Si DMM device enters its saturation mode under the larger short-circuit current conditions, and its voltage drop rises to value sufficient to reduce the gate-source voltage across the GaN HEMT device until its saturation current matches that of the Si DMM device. The sense node voltage behavior during the short-circuit event is shown in Fig. 20.14 for the composite MOSFET CM-2. V_{SENSE} reaches a value of 3.5 V within 0.2 μs after the onset of the short-circuit event and remains steady throughout the short-circuit event. It can therefore be used by the control circuit to detect the occurrence of a short-circuit event within 0.5 μs and initiate turning-off the gate drive voltage. This is a relatively shorter time than required for

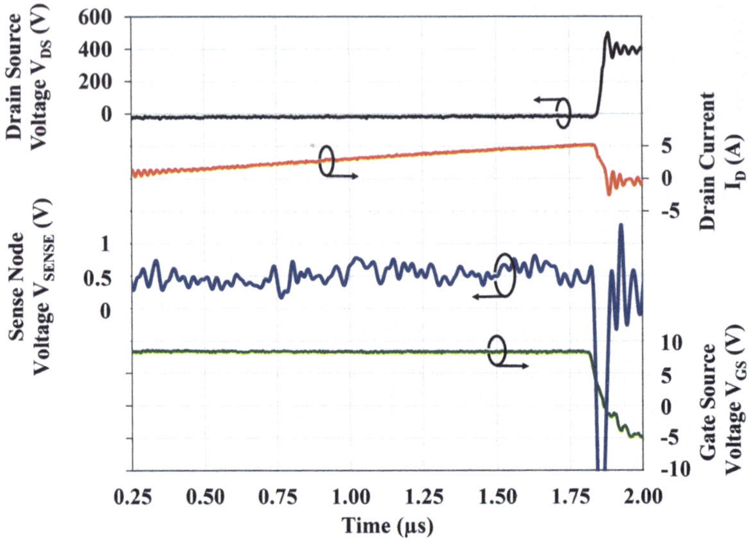

Fig. 20.13 Sense node voltage for the composite MOSFET CM-2 during normal operation

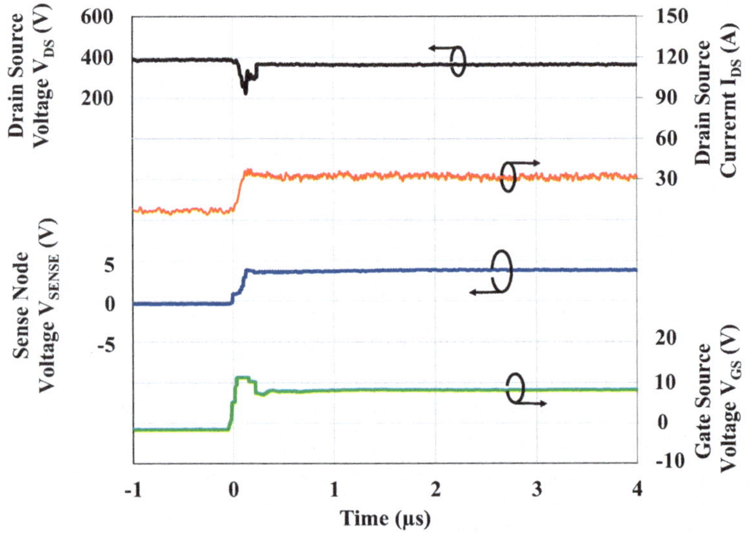

Fig. 20.14 Sense node voltage for the composite MOSFET CM-6 during short-circuit operation

the typical DESAT protection circuits, which require a blanking time of 3 μs because the sensing is performed at the drain terminal of the high voltage power device.

20.7 Trade-Off Curves

Trade-off curves between short-circuit withstand time and the on-state voltage drop or the switching losses are useful to understand the true impact of the BaSIC(EMM) topology on the performance of the Transphorm GaN HEMT TPH3202PD product.

The trade-off curve obtained between the short-circuit withstand time and the on-state state resistance by using the measured data is shown in Fig. 20.15. There is very steep increase in the short-circuit withstand time in relation to the on-state resistance. This is very desirable feature of the BaSIC(DMM) topology. An increase in short-circuit withstand time to 4.35 μs (13× or 1300% improvement) is achieved with 29% increase in the on-resistance.

The trade-off curve between the short-circuit withstand time and the normalized total switching energy loss is shown in Fig. 20.16 for the BaSIC(DMM) topology for the Transphorm GaN HEMT TPH3202PD product. A t_{SC} of 4.35 μs (13× or 1300% improvement) is achieved using the BaSIC(DMM) topology with no increase in total switching losses.

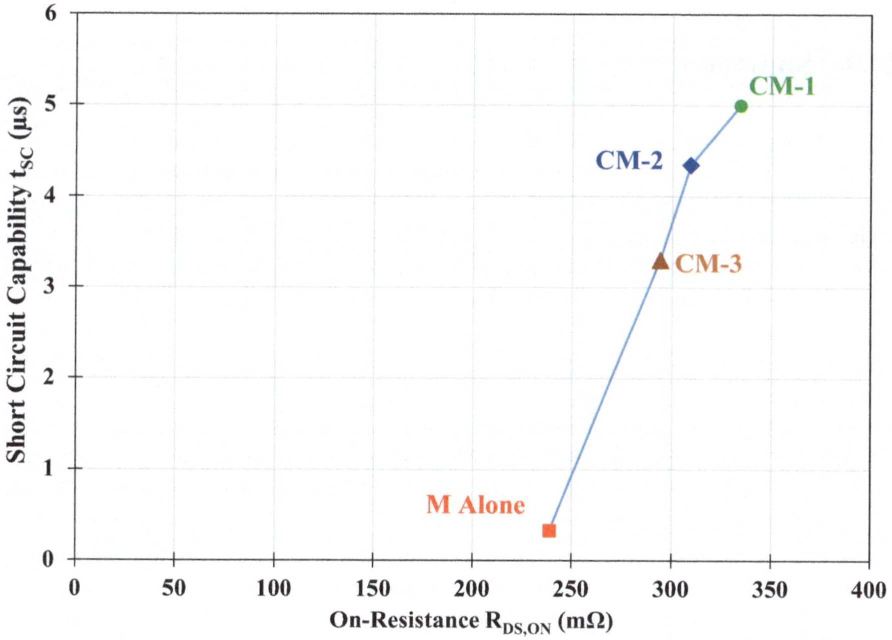

Fig. 20.15 Trade-off curve observed for the BaSIC(DMM) topology in the case of the Transphorm GaN HEMT TPH3202PD product: on-resistance

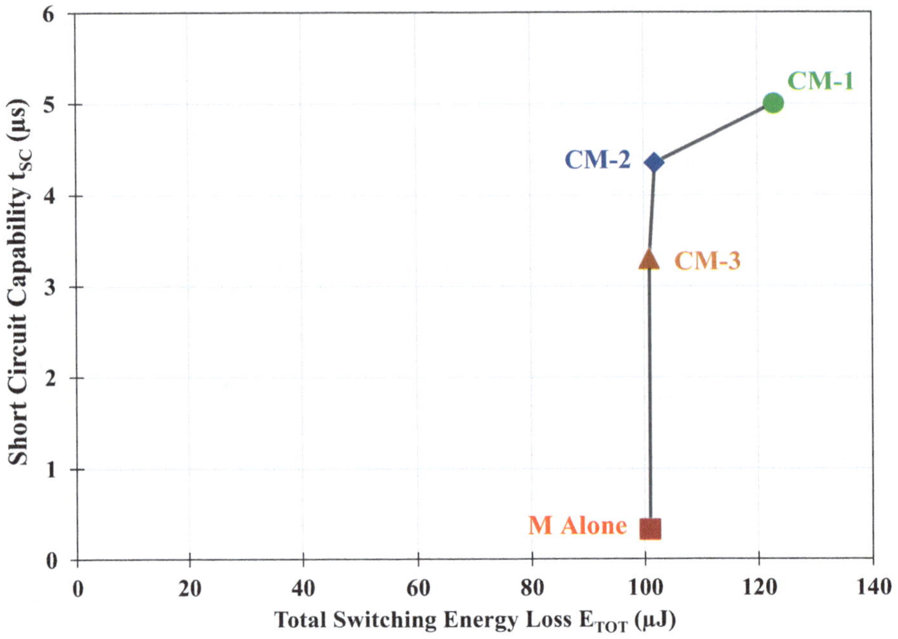

Fig. 20.16 Trade-off curve observed for the BaSIC(EMM) topology in the case of Transphorm GaN HEMT TPH3202PD product: switching energy loss

20.8 Summary

The benefits of using the BaSIC topology with a silicon depletion mode MOSFET (DMM) as the non-linear resistance have been demonstrated in this chapter for the case of the 600 V Transphorm GaN HEMT TPH3202PD product. The short-circuit withstand was improved from 0.33 μs for the stand-alone GaN HEMT device to 4.5 μs using the BaSIC(DMM) topology (case CM-2), an increase by 13× or 1300%. The on-state resistance increased for this case by 29% with no change in total switching energy loss per cycle. The performance of the BaSIC(DMM) topology was limited by the high on-state resistance of the only commercially available Si DMM device.

An even superior trade-off curve can be obtained by using the improved Si DMM devices that were discussed in Chap. 9. The trade-off curve between drain saturation current and the on-resistance for the UMOS DMM devices is reproduced in Fig. 20.17. The case of a drain saturation current of 35 A is highlighted in the figure by the red dashed line since this current produced a short-circuit withstand time of 4.5 μs for the 600 V Transphorm GaN HEMT TPH3202PD product. The on-state resistance for the optimum UMOS 30 DMM device can be observed to be only 5 mΩ for this saturated drain current of 35 A. Consequently, the implementation of the BaSIC(DMM) topology with this optimum UMOS 30 V DMM device will

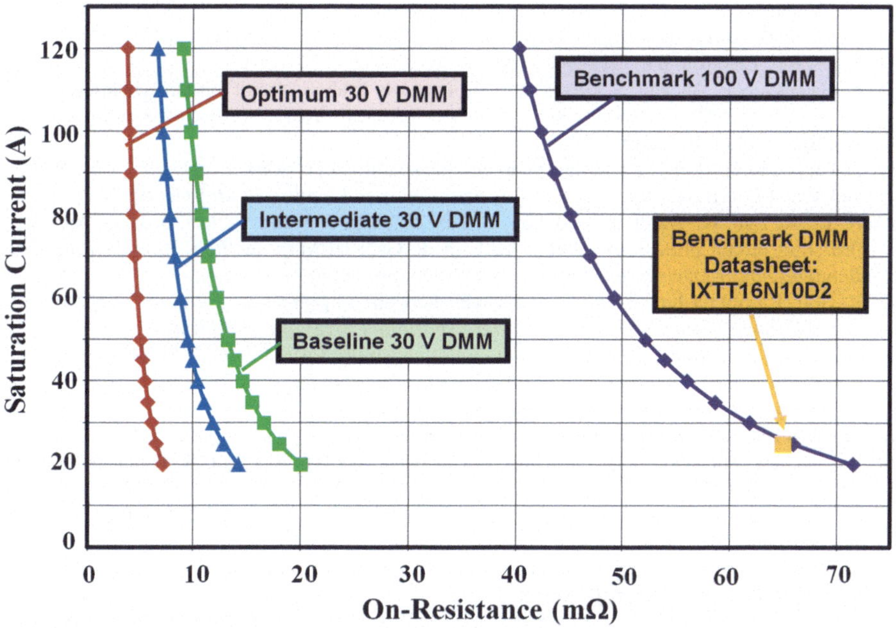

Fig. 20.17 Trade-off curve between on-resistance and drain saturation current for various Si DMM devices

produce an increase in on-state resistance by only 2% for the 600 V Transphorm GaN HEMT TPH3202PD product instead of the 29% observed using the commercially available benchmark Si DMM device.

References

1. E.A. Jones, F. Wang, D. Costinett, Review of commercial GaN power devices and GaN-based converter design challenges. IEEE J. Emerg. Sel. Top. Power Electron. **4**, 707–719 (2016)
2. J.P. Kozak, R. Zhang, M. Porter, Q. Song, J. Liu, B. Wang, R. Wang, W. Saito, Y. Zhang, Stability, reliability, and robustness of GaN power devices: A review. IEEE Trans. Power Electron. **38**, 8442–8471 (2023)
3. H. Li, X. Li, X. Wang, X. Lyu, H. Cai, Y.M. Alsmadi, L. Lui, S. Bala, J. Wang, Robustness of 650-V enhancement-mode GaN HEMTs under various short-circuit conditions. IEEE Trans. Ind. Electron. **55**, 1807–1816 (2019)
4. P. Williford, F. Wang, J. Xu, Short circuit study of 600 V GaN GITs, in *Proceedings of the 2019 IEEE 7th Workshop on Wide Bandgap Power Devices and Applications* (2019), pp. 36–42
5. F. D'Aniello, A. Fayyaz, A. Castellazzi, T. Oeder, M. Pfost, Damage accumulation in GaN GITs exposed to repetitive short-circuit, in *International Symposium on Power Semiconductor Devices and ICs* (2019), pp. 451–454
6. A. Castellazzi, A. Fayyaz, S. Zhu, T. Oeder, M. Pfost, Single pulse short-circuit robustness and repetitive stress aging of GaN GITs, in *IEEE International Reliability Physics Symposium* (2018), pp. 4E.1-1–4E.1-10

7. M. Fernandez, X. Perpina, J. Roig-Guitart, M. Vellvehi, F. Bauwens, M. Tack, X. Jorda, Short-circuit study in medium-voltage GaN cascodes, p-GaN HEMTs, and GaN MISHEMTs. IEEE Trans. Ind. Electron. **64**, 9012–9022 (2017)
8. N. Badawi, A.E. Awwad, S. Dieckerhoff, Robustness in short-circuit mode: Benchmarking of 600V GaN HEMTs with power Si and SiC MOSFETs, in *IEEE Energy Conversion Congress and Exposition* (2016), pp. 1–7
9. D. Pappis, K. Goebel, P. Zacharias, Short circuit capability of 650 V normally off GaN E-HEMT and MOSFET-HEMT cascode, in *Power Electronics and Intelligent Motion* (2017), pp. 1–8
10. H. Li, X. Lyu, K. Wang, Y. Abdullah, B. Hu, Z. Yang, J. Wang, L. Liu, S. Bala, An ultra-fast short circuit protection solution for E-mode GaN HEMTs, in *IEEE Workshop on Wide Bandgap Power Devices and Applications* (2018), pp. 187–192
11. R. Hou, J. Lu, D. Chen, An ultrafast discrete short-circuit protection circuit for GaN HEMTs, in *IEEE Energy Conversion Congress and Exposition* (2018), pp. 1920–1925
12. A. Kanale, B.J. Baliga, Achieving short circuit capability for 600 V GaN FETs using a gate-source-shorted Si depletion-mode MOSFET in series with the source, in *IEEE Workshop on Wide Bandgap Power Devices and Applications* (2018), pp. 187–192
13. https://www.transphormusa.com/en/document/600v-cascode-gan-fet-tph3202p/, Mar 2018
14. B.J. Baliga, Silicon carbide switching device with rectifying gate. U.S. Patent 5,396,085, Issued 7 Mar 1995
15. B.J. Baliga, *Fundamentals of Power Semiconductor Devices*, 2nd edn. (Springer International, 2019)
16. https://m.littelfuse.com/products/power-semiconductors/discrete-mosfets/n-channel-depletion-mode/d2/ixth16n10d2.aspx, 2017

Chapter 21
BaSIC Topology with Integrated JFET

It has been shown in the previous chapters that the Baliga Short-circuit Improvement Concept, or the BaSIC topology, is an elegant solution to enhance the short-circuit withstand capability of power semiconductor devices. The concept makes use of a low-voltage, non-linear element, Q_{LV}, that is connected in series with the high-voltage power device Q_{HV}. The non-linear element Q_{LV} must have a resistance that increases with increasing current until it limits the magnitude of the current. The non-linear element Q_{LV} must have a low resistance at the on-state current level to minimize its impact on on-state and switching losses.

Low blocking voltage rated silicon depletion and enhancement mode power MOSFETs were utilized in the previous chapters to serve as a non-linear element Q_{LV}. This creates a multi-chip implementation of the BaSIC topology with separately packaged power devices Q_{HV}, such as SiC power MOSFETs, GaN HEMT devices, or silicon IGBTs, and the low blocking voltage Si MOSFETs.

An alternate approach for creating the BaSIC topology is discussed in this chapter where the non-linear element Q_{LV} is integrated into the SiC power MOSFET structure. The non-linear element is formed with a lateral JFET structure located in series with the source region of the SiC power MOSFET.

21.1 Integrated Linear Element

The integration of a linear element, i.e., a resistor, into the SiC power MOSFET structure has been previously proposed and reported [1]. The embedded resistor is created by making a portion of the N^+ source region with reduced doping. A cross-section of this structure is illustrated in Fig. 21.1a. The series embedded resistance shown as a layer marked N^+ to indicate its lower doping that the source regions with higher doping N^{++}. In this approach, the doping concentration of the N^+ region is kept large (above 1×10^{18} cm^{-3}) to make its resistance small. The inclusion of the

B. J. Baliga, A. Kanale, *The BaSIC Topology*,
https://doi.org/10.1007/978-3-031-86630-2_21

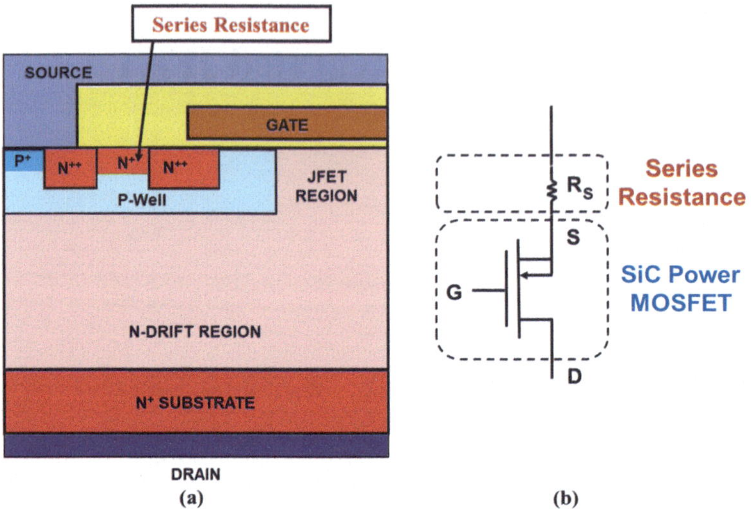

Fig. 21.1 SiC power MOSFET with embedded series resistance: (**a**) structure and (**b**) equivalent circuit

resistive N^+ region increases the cell pitch and contributes additional resistance, which leads to an increase in the specific on-resistance of the SiC power MOSFET.

The equivalent circuit for this case is the SiC power MOSFET with a series resistance R_S as illustrated in Fig. 21.1b. This situation resembles the SiC power MOSFET with an external ballast resistance as previously discussed in Chap. 7.

The authors [1] compared the SiC power MOSFET device with embedded source resistance to conventional SiC power MOSFETs whose short-circuit withstand time was enhanced by increasing the channel length. In both cases, the specific on-resistance is observed to increase with larger short-circuit withstand time. This trade-off can be compared for the two cases using Fig. 21.2. The stand-alone SiC power MOSFET with shortest channel length was reported to have a specific on-resistance of 3.2 mΩ cm^2. The percentage increase in the specific on-resistance for the other structures can be computed by using this value. It can be observed that the short-circuit withstand time for the conventional SiC power MOSFET structure can be enlarged to 12 μs by increasing the channel length. The specific on-resistance increases by 219% in this case. In comparison, the short-circuit withstand time for the SiC power MOSFET structure with embedded series resistance was enlarged to 8.7 μs by increasing the length of the embedded resistor. The specific on-resistance increases by 62.5% in this case. Consequently, the approach with embedded series resistance exhibits a superior trade-off curve between short-circuit withstand time and increase in specific on-resistance.

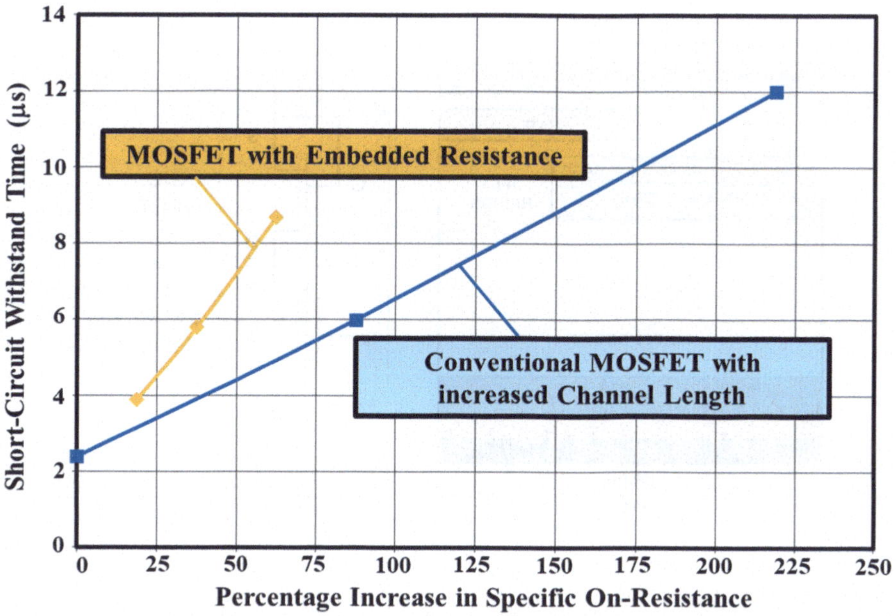

Fig. 21.2 Trade-off between short-circuit withstand time and increase in on-resistance for the SiC power MOSFET with embedded series resistance compared with impact of increasing the channel length

21.2 Integrated JFET Element

A cross-section of the SiC power MOSFET structure with an integrated lateral JFET [2] is illustrated in Fig. 21.3a with the series JFET shown by a region marked N(JFET). The equivalent circuit for this case is the SiC power MOSFET with a series SiC JFET as illustrated in Fig. 21.3b. In this approach, the doping concentration of the N(JFET) region is chosen to allow its depletion by voltage drop produced across it when the short-circuit current is flowing in the SiC power MOSFET. At the same time, the doping concentration of the N(JFET) region is chosen to be sufficiently high to keep its resistance low during current flow under normal operation of the SiC power MOSFET. The inclusion of the JFET region increases the cell pitch and adds additional resistance, which leads to an increase in the specific on-resistance of the SiC power MOSFET.

The objective of inclusion of the integrated JFET region is to reduce the current flowing through the SiC power MOSFET during short-circuit events. The integrated JFET region enters its current saturation-mode when the voltage across it increases due to current flow through the structure. The saturation current of the integrated JFET must be designed to be smaller than the drain saturation current of the SiC power MOSFET at its gate drive voltage. With this design, the current flowing

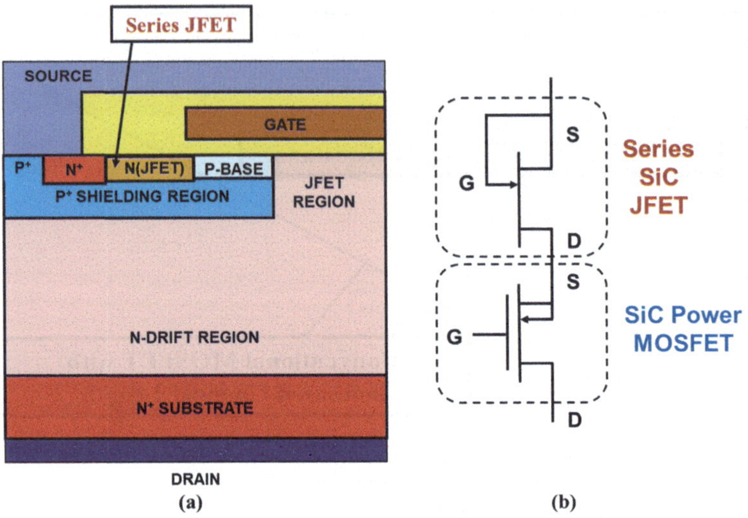

Fig. 21.3 SiC power MOSFET with integrated JFET: (**a**) structure and (**b**) equivalent circuit

during the short-circuit event is reduced allowing enlargement of the short-circuit withstand time.

21.2.1 JFET Region Physics

The electrical characteristics of the lateral JFET region can be modelled using analytical solutions that have been previously published [3]. The analytical solution for the JFET *i*–*v* characteristics is derived using the gradual channel approximation. The integrated lateral JFET structure is expanded in Fig. 21.4 to discuss its physics of operation. The JFET is formed between the P-base region and the N^+ source region of the SiC power MOSFET with its drain (D) located at the MOSFET P-base region and its source (S) located at the MOSFET N^+ source region. The gate of the JFET is formed by the P-N junction between the P^+ shielding region and the N(JFET) region. It extends along the entire length (L_{JFET}) of the lateral JFET.

The SiC power MOSFET current flows from the channel into the JFET region in the direction indicated by the red arrow. This current flow produces a voltage drop in the N(JFET) layer which increases the reverse voltage across the JFET gate P-N junction. The depletion layer formed at the P-N junction expands at point D with increasing current producing an increase in the resistance of the JFET region. As the drain current increases, the drain end D of the JFET is eventually pinched-off producing current saturation. The resistance of the JFET at low current levels and the saturated current are dictated by the doping concentration of the JFET region (N_{JFET}), its length (L_{JFET}), and its width ($2a$).

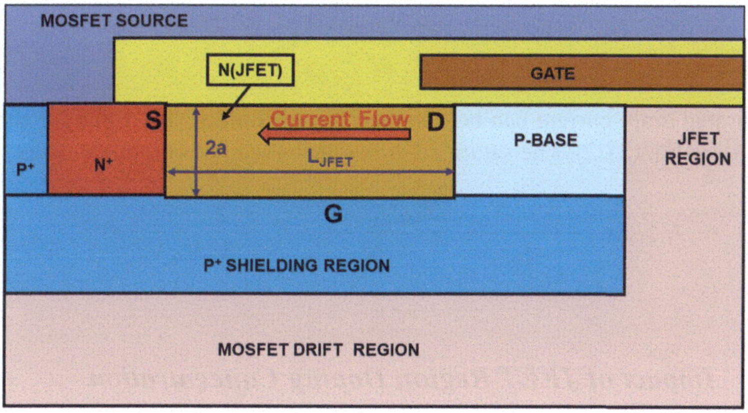

Fig. 21.4 Details of integrated JFET structure

21.2.2 Analytical Model for JFET Region

The integrated lateral JFET is designed with its gate shorted to the source via the P^+ plug located on the left-hand side in Fig. 21.4. The gate bias applied to the integrated JFET is consequently zero. This condition must be assumed when modeling the output characteristics of the integrated JFET.

The resistance of the JFET region can be obtained using [3]

$$R_{ON} = \frac{L_{JFET}}{q\mu_n N_{JFET} Z}\left[2a - \sqrt{\frac{2\varepsilon_S V_{bi}}{qN_{JFET}}}\right]^{-1} \tag{21.1}$$

where Z is the width of the cross-section. The built-in potential V_{bi} can be computed using the doping concentrations of the P^+ shielding region and the JFET region. The resistance decreases with increasing doping concentration N_{JFET} and thickness "a" of the JFET region. The resistance increases when the length of the JFET region (L_{JFET}) is increased.

The *i*–*v* characteristics of the JFET can be analytically modelled using the gradual channel approximation [3]:

$$I_D = \frac{2a\left(q\mu_n N_{JFET}\right)Z}{L_{JFET}}\left\{V_D - \frac{2}{3a}\sqrt{\frac{2\varepsilon_S}{qN_{JFET}}}\left[\left(V_D + V_{bi}\right)^{\frac{3}{2}} - V_{bi}^{\frac{3}{2}}\right]\right\} \tag{21.2}$$

This expression produces *i*–*v* characteristics that leads to saturation of the drain current I_D when the drain voltage V_D is increased.

The drain voltage at which the drain current reaches its saturated value is given by [3]:

$$V_{\mathrm{P}} = \frac{2qN_{\mathrm{JFET}}a^2}{\varepsilon_{\mathrm{S}}} - V_{\mathrm{bi}} \tag{21.3}$$

The saturated drain current can be obtained by substituting this value for the drain voltage V_{D} is Eq. (21.2). The saturated drain current can be computed using [3]:

$$I_{\mathrm{D,SAT}} = \frac{2a\left(q\mu_{\mathrm{n}}N_{\mathrm{JFET}}\right)Z}{L_{\mathrm{JFET}}} \frac{qN_{\mathrm{JFET}}a^2}{6\varepsilon_{\mathrm{S}}} - V_{\mathrm{bi}} + \frac{2}{3a}\sqrt{\frac{2\varepsilon_{\mathrm{S}}}{qN_{\mathrm{JFET}}}} V_{\mathrm{bi}}^{\frac{3}{2}} \tag{21.4}$$

21.2.3 Impact of JFET Region Doping Concentration

The doping of the lateral JFET region (N_{JFET}) is an important parameter that determines the characteristics of the device. The integrated lateral JFET saturation current increases with larger doping concentration of the integrated JFET while its resistance becomes smaller.

As an example, the computed *i*–*v* characteristics of the integrated lateral JFET are shown in Fig. 21.5 for various values of the doping concentration of the integrated JFET region. The doping concentration was varied from 0.8 to $1.5 \times 10^{17}\ \mathrm{cm}^{-3}$ for this example. A JFET region thickness of 0.23 μm and length of 0.5 μm were assumed to analyze this case. A width *Z* of 1 cm orthogonal to the cross-section was used for the MOSFET cell including the integrated lateral JFET. It can be observed

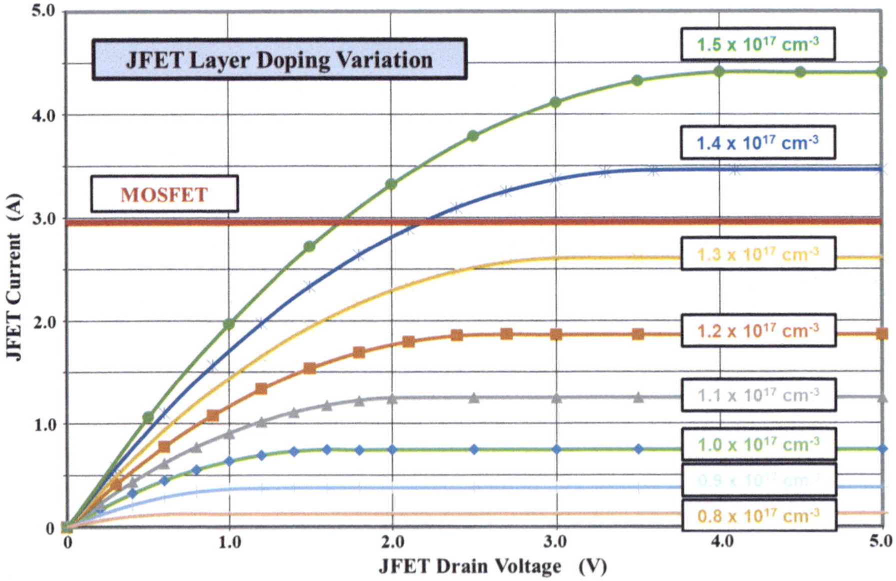

Fig. 21.5 Integrated JFET characteristics: impact of doping concentration

that the saturation current increases while the on-resistance decreases with increasing doping concentration. The saturation current for the SiC power MOSFET is also shown on the plot with a magnitude of 2.96 A by the horizontal red line. This was computed using a cross-sectional width of 3 μm and typical SiC power MOSFET parameters. The on-state gate bias was assumed to be 20 V with a threshold voltage of 3 V. The JFET saturation current becomes smaller than this value of 2.96 A when the JFET doping is reduced below 1.35×10^{17} cm^{-3}. The doping level of the integrated lateral JFET region can be precisely controlled by using ion-implantation of nitrogen into the SiC surface to the desired depth.

The on-state current level for the 1.2 kV rated SiC power MOSFET was computed assuming a specific on-resistance of 5 mΩ cm^2 and using a cell area of 3×10^{-4} cm^2 based on the cell size of 3 μm and width Z of 1 cm. It is found to be 0.12 A for an on-state voltage drop of 2 V. The ratio of the saturated drain current and the on-state current ($I_{D,SAT}/I_{ON}$) for the 1.2 kV rated SiC power MOSFET is then 24.65.

The computed changes in the ($I_{D,SAT}/I_{ON}$) ratio and the percentage increase in on-resistance contributed by the integrated JFET are plotted in Fig. 21.6. The ($I_{D,SAT}/I_{ON}$) ratio can be brought down below 50% of that for the SiC power MOSFET by using a JFET doping concentration of less than 1.1×10^{17} cm^{-3} as indicated by the red dashed lines. This will enlarge the short-circuit withstand time by a factor of 2 times. The on-resistance increases by only 5% for achieving this outcome.

The trade-off curve for the ($I_{D,SAT}/I_{ON}$) ratio versus the percentage increase in on-resistance contributed by the integrated lateral JFET, obtained by varying the doping concentration of the JFET region, is shown in Fig. 21.7. The red dashed lines indicate that the ($I_{D,SAT}/I_{ON}$) ratio can be reduced by 50% to increase the short-circuit

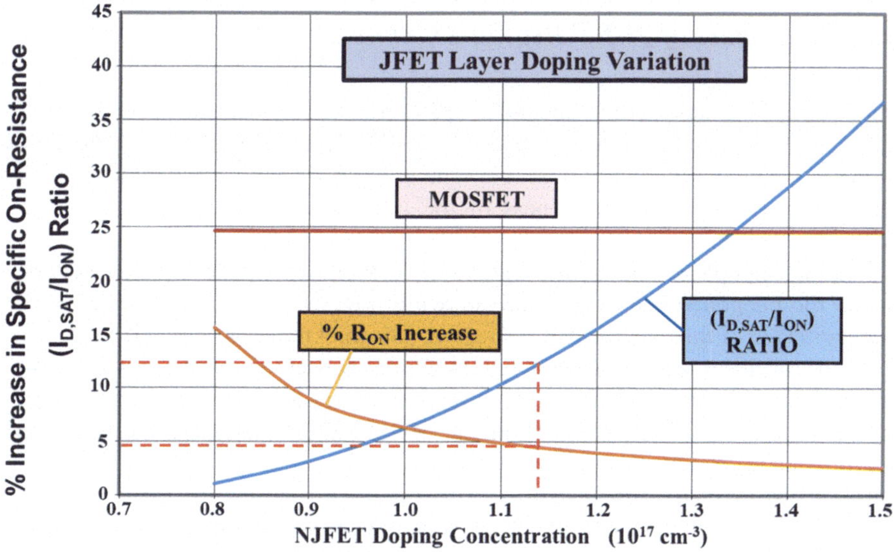

Fig. 21.6 Impact of JFET doping concentration on $I_{D,SAT}/I_{ON}$ ratio and % increase in R_{ON}

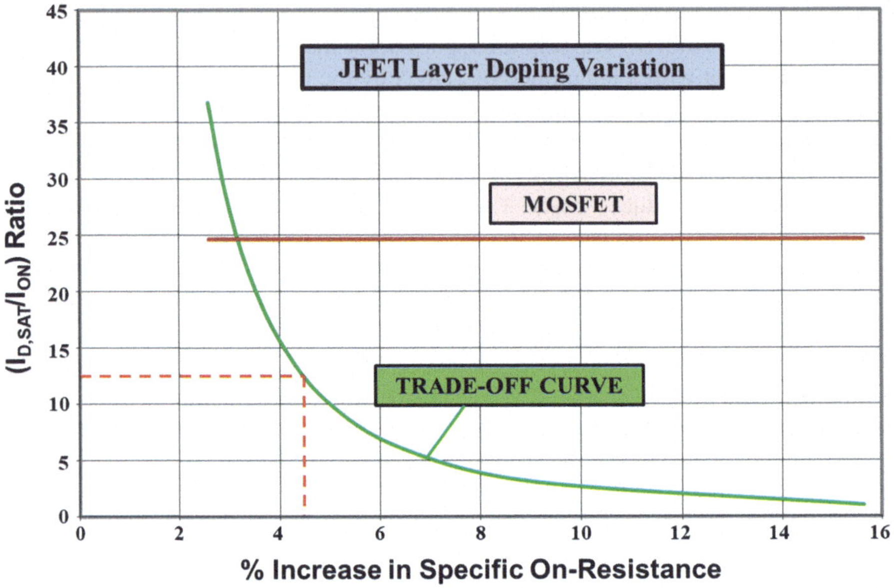

Fig. 21.7 Trade-off between $I_{D,SAT}/I_{ON}$ ratio and % increase in R_{ON}: impact of JFET doping concentration

withstand time by a factor of 2 times with less than 5% increase in the on-resistance.

21.2.4 Impact of JFET Region Thickness

The *i*–*v* characteristics of the integrated JFET are strongly dependent on the thickness (2*a*) of the JFET region. The depletion of JFET region by the built-in potential of the P-N gate plays an important role. The integrated lateral JFET saturation current increases with larger thickness of the integrated JFET while its resistance becomes smaller.

For the example discussed in the previous section, the computed *i*–*v* characteristics of the integrated lateral JFET are shown in Fig. 21.8 for various values of the thickness (2*a*) of the integrated JFET region. The thickness was varied from 0.21 to 0.28 μm for this example. A JFET region doping concentration of 1.0×10^{17} cm^{-3} and length of 0.5 μm was assumed to analyze this case. A width *Z* of 1 cm orthogonal to the cross-section was used for the MOSFET cell including the integrated JFET. It can be observed that the saturation current increases while the on-resistance decreases with increasing thickness. The saturation current for the SiC power MOSFET is also shown on the plot with a magnitude of 2.96 A by the horizontal red line. This was computed using a cross-section width of 3 μm and typical SiC power

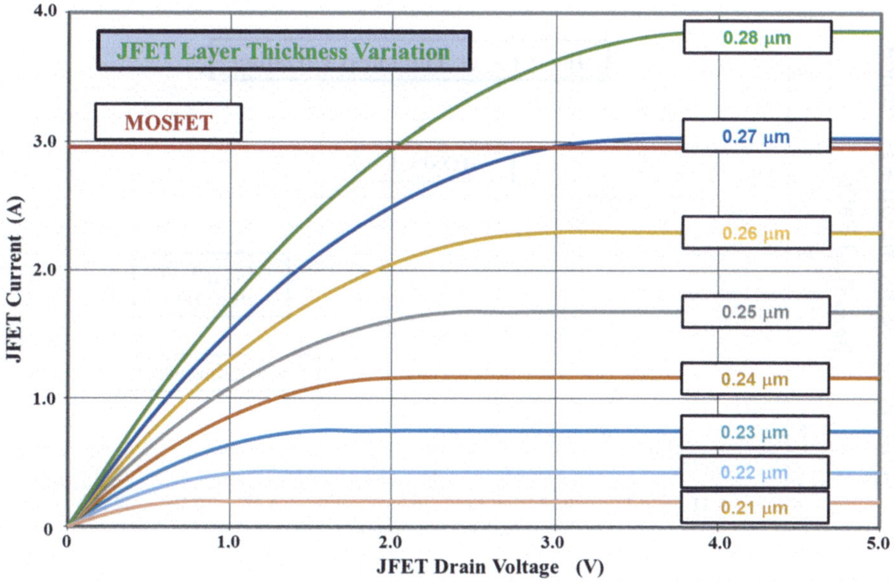

Fig. 21.8 Integrated JFET characteristics: impact of thickness

MOSFET parameters. The on-state gate bias was assumed to be 20 V with a threshold voltage of 3 V. The JFET saturation current becomes smaller than this value of 2.96 A when the JFET thickness is reduced below 0.27 µm. The thickness of the integrated JFET region can be precisely controlled by adjusting the energy of nitrogen ion-implantation into the SiC surface.

The on-state current level for the 1.2 kV rated SiC power MOSFET was computed assuming a specific on-resistance of 5 mΩ cm^2 and using a cell area of 3×10^{-4} cm^2 based on the cell size of 3 µm and width Z of 1 cm. It is found to be 0.12 A for an on-state voltage drop of 2 V. The ratio of the saturated drain current and the on-state current ($I_{D,SAT}/I_{ON}$) for the 1.2 kV rated SiC power MOSFET is then 24.65.

The computed changes in the ($I_{D,SAT}/I_{ON}$) ratio and the percentage increase in on-resistance contributed by the integrated JFET are plotted in Fig. 21.9. The ($I_{D,SAT}/I_{ON}$) ratio can be brought down below 50% of that for the SiC power MOSFET by using a JFET thickness of less than 0.25 µm as indicated by the red dashed lines. This will enlarge the short-circuit withstand time by a factor of 2 times. The on-resistance increases by only 5% for achieving this outcome.

The trade-off curve for the ($I_{D,SAT}/I_{ON}$) ratio versus the percentage increase in on-resistance contributed by the integrated JFET obtained by varying the thickness of the JFET region is shown in Fig. 21.10. The red dashed lines indicate that the ($I_{D,SAT}/I_{ON}$) ratio can be reduced by 50% to increase the short-circuit withstand time by a factor of 2 times with less than 5% increase in the on-resistance.

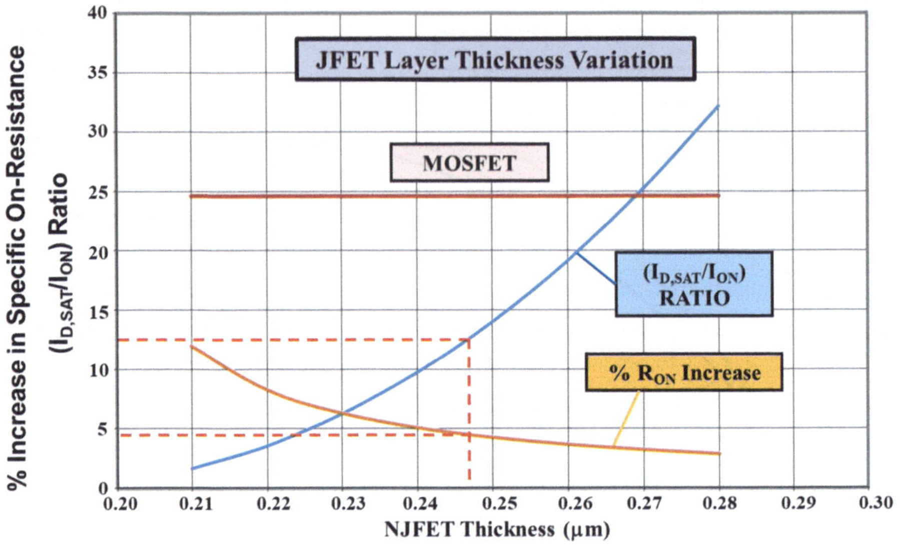

Fig. 21.9 Impact of JFET thickness on $I_{D,SAT}/I_{ON}$ ratio and % increase in R_{ON}

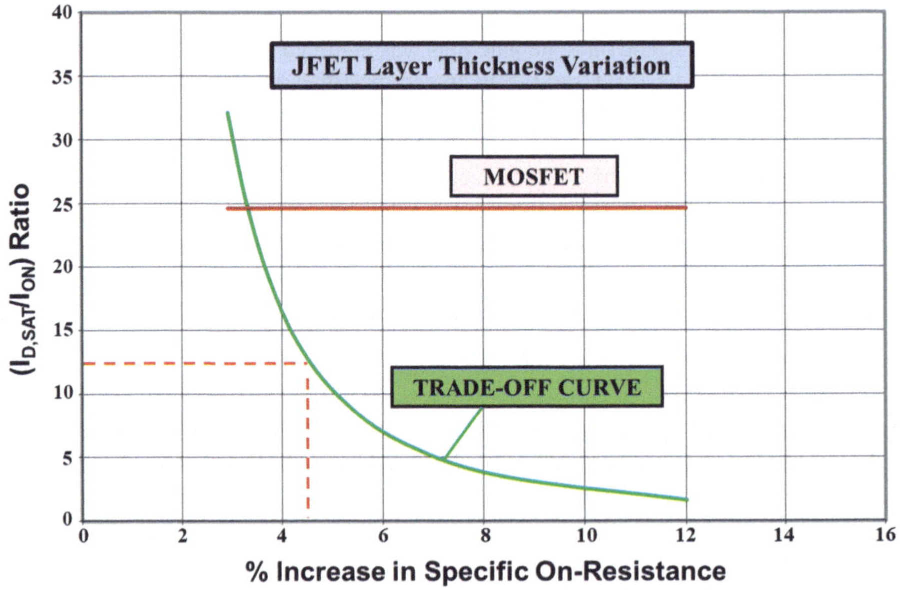

Fig. 21.10 Trade-off between $I_{D,SAT}/I_{ON}$ ratio and % increase in R_{ON}: impact of JFET thickness

21.2.5 Impact of JFET Region Length

The *i–v* characteristics of the integrated JFET are also dependent on the length (L_{JFET}) of the JFET region. The drain voltage at which the current saturates is independent of the JFET region length according to Eq. (21.3). The integrated JFET saturation current decreases with larger length of the integrated JFET because of the increase in its resistance.

For the example discussed in the previous sections, the computed *i–v* characteristics of the integrated JFET are shown in Fig. 21.11 for various values of the length (L_{JFET}) of the integrated JFET region. The length was varied from 0.5 to 1.0 μm for this example in steps of 0.1 μm. A JFET region doping concentration of 1.0×10^{17} cm^{-3} and thickness of 0.23 μm was assumed to analyze this case. A width *Z* of 1 cm orthogonal to the cross-section was used for the MOSFET cell including the integrated JFET. It can be observed that the saturation current decreases while the on-resistance increases with increasing length. The saturation current for the JFETs in this example is well below the 2.96 A value for the SiC power MOSFET. The length of the integrated JFET region can be precisely controlled during the layout of the cell.

The on-state current level for the 1.2 kV rated SiC power MOSFET was computed assuming a specific on-resistance of 5 mΩ cm^2 and using a cell area of 3×10^{-4} cm^2 based on the cell size of 3 μm and width *Z* of 1 cm. It is found to be 0.12 A for an on-state voltage drop of 2 V. The ratio of the saturated drain current and the on-state current ($I_{D,SAT}/I_{ON}$) for the 1.2 kV rated SiC power MOSFET is then 24.65.

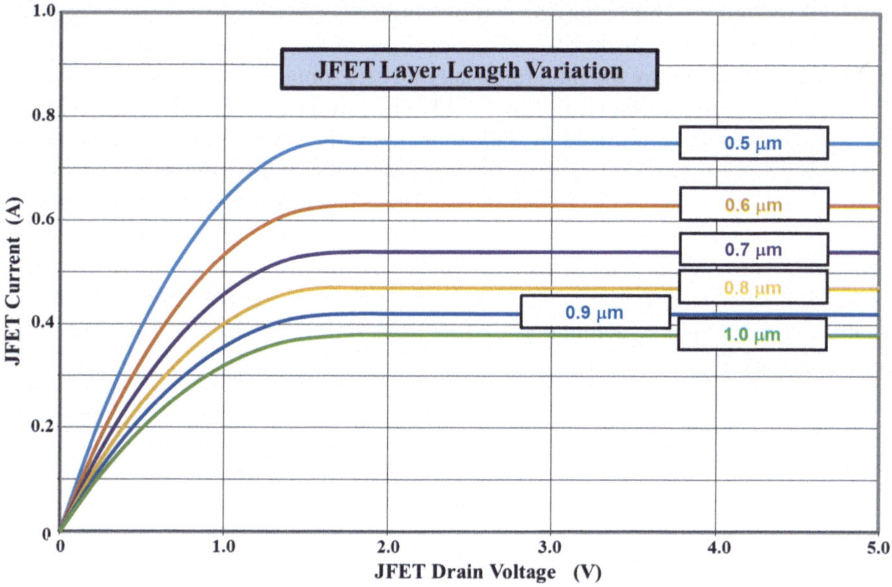

Fig. 21.11 Integrated JFET characteristics: impact of length

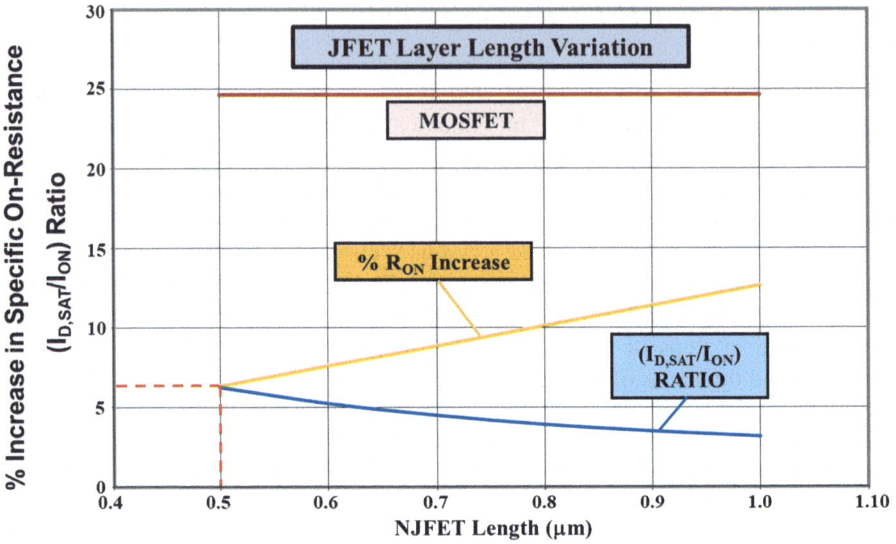

Fig. 21.12 Impact of JFET length on $I_{D,SAT}/I_{ON}$ ratio and % increase in R_{ON}

The computed changes in the ($I_{D,SAT}/I_{ON}$) ratio and the percentage increase in on-resistance contributed by the integrated lateral JFET are plotted in Fig. 21.12. The ($I_{D,SAT}/I_{ON}$) ratio can be brought down below 25% of that for the SiC power MOSFET by using a JFET region length of 0.5 μm as indicated by the red dashed lines. This will enlarge the short-circuit withstand time by a factor of 4 times. The on-resistance increases by only 6% for achieving this outcome.

The trade-off curve for the ($I_{D,SAT}/I_{ON}$) ratio versus the percentage increase in on-resistance contributed by the integrated lateral JFET, obtained by varying the length of the JFET region, is shown in Fig. 21.13. The red dashed lines indicate that the ($I_{D,SAT}/I_{ON}$) ratio can be reduced by 25% to increase the short-circuit withstand time by a factor of 4 times with less than 7% increase in the on-resistance.

Based on the analysis discussed above, it can be concluded that the integrated lateral JFET region length should be reduced to the smallest value compatible with device fabrication guidelines. This is typically 0.5 μm in size for the JFET length. A shorter JFET length also makes the cell pitch shown in the cross-section of Fig. 21.4 smaller which favors achieving a low specific on-resistance for the SiC power MOSFET.

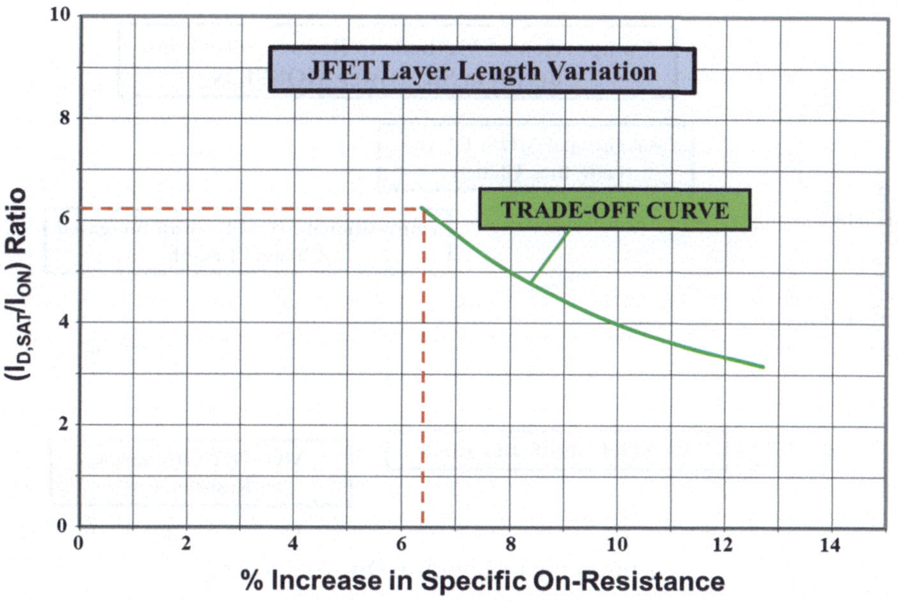

Fig. 21.13 Trade-off between $I_{D,SAT}/I_{ON}$ ratio and % increase in R_{ON}: impact of JFET length

21.3 Comparison of Approaches

The advantages of using an integrated lateral JFET within the SiC power MOSFET structure can be brought out by using trade-off curves between the ($I_{D,SAT}/I_{ON}$) ratio versus the percentage increase in on-resistance for various approaches. These trade-off graphs are shown in Fig. 21.14 for various approaches to reduce the ($I_{D,SAT}/I_{ON}$) ratio in order to extend the short-circuit withstand time.

It can be observed that increasing the channel length of the SiC power MOSFET produces the worst trade-off curve shown by the blue line. The ($I_{D,SAT}/I_{ON}$) ratio must be reduced by a factor of 4 times to improve the short-circuit time for the SiC power MOSFET from about 3.5 μs to over 10 μs. This corresponds to the horizontal red dashed line at a ($I_{D,SAT}/I_{ON}$) ratio of 5.5 shown in the figure. An increase in on-resistance by more than 120% occurs when the channel length is increased to achieve this improvement. The large increase in on-resistance is due to increased cell pitch and channel resistance, which makes the largest contribution to the on-resistance relative to other components in SiC power MOSFETs.

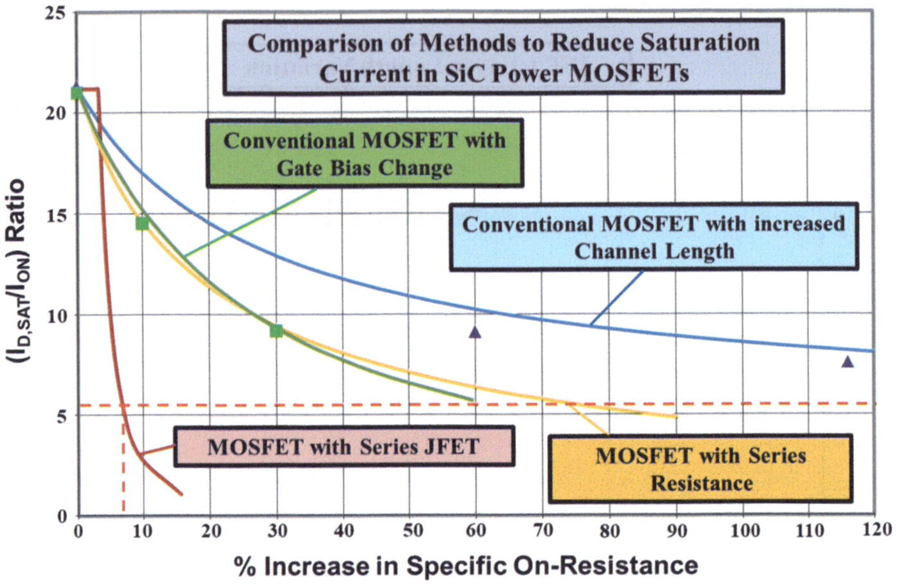

Fig. 21.14 Trade-off curves between $I_{D,SAT}/I_{ON}$ ratio and % increase in R_{ON}

An alternate approach to reduce the ($I_{D,SAT}/I_{ON}$) ratio is by decreasing the gate drive voltage. This produces the green trade-off curve which is superior to the approach of reducing the channel length. Based on the horizontal red dashed line at a ($I_{D,SAT}/I_{ON}$) ratio of 5.5 shown in the figure, an increase in on-resistance by 60% occurs when the gate bias is reduced to achieve this value.

The results obtained by using an integrated series resistance in the SiC power MOSFET structure, as shown in Fig. 21.1, are shown in Fig. 21.14 by the yellow line. The trade-off curve for this case closely matches the case of reducing the gate drive voltage. Based on the horizontal red dashed line at a ($I_{D,SAT}/I_{ON}$) ratio of 5.5 shown in the figure, an increase in on-resistance by 73% occurs when the gate bias is reduced to achieve this value.

The trade-off curve obtained for the SiC power MOSFET with the integrated lateral JFET region, as shown in Fig. 21.3, is depicted in Fig. 21.14 using the red line. This trade-off curve was generated by varying the JFET doping concentration. The same trade-off curve was previously shown in Fig. 21.7 using a different scale. It can be observed that this trade-off curve is much superior to the other approaches. Based on the horizontal red dashed line at a ($I_{D,SAT}/I_{ON}$) ratio of 5.5 shown in the figure, an increase in on-resistance by only 7% is observed in this case.

21.4 Summary

The improvement of the short-circuit withstand capability of SiC power MOSFETs by the integration of lateral JFET region within the cell structure has been discussed in this chapter. This approach is found to allow reduction in the drain saturation current by a factor of 4 times while increasing the on-resistance by only 7%. This approach has been shown to be significantly superior to adding a resistance in series with the N^+ source region, as well as other options such as reducing the gate bias voltage or increasing the channel length.

The JFET region can be formed by creating a lower doped N-type region adjacent to the N^+ source region in the SiC power MOSFET structure. This can be achieved using one additional masking step with ion-implantation of nitrogen as the n-type dopant during device fabrication. The ion implant energy and dose must be carefully selected to obtain the optimal depth and doping concentration for the JFET region.

The integrated lateral JFET region was placed within the SiC power MOSFET cross-section in Fig. 21.3. This enlarges the cell pitch leading to an increase in the specific on-resistance of the device. An alternative approach is to place the integrated lateral JFET orthogonal to the cross-section shown in Fig. 21.3. This provides the benefits of the BaSIC topology without increasing the intrinsic SiC MOSFET specific on-resistance.

The integrated JFET approach eliminates the need for using the external Si depletion mode or enhancement mode power MOSFETs in the BaSIC topology that were discussed in previous chapters. This makes the approach simpler in terms of packaging and implementation. However, this approach does not allow the current sensing that was discussed in previous chapters to monitor the on-state current level and to detect the on-set of short-circuit events.

References

1. H. Hatta, T. Tominaga, S. Hino, N. Miura, S. Tomohisa, S. Yamakawa, Suppression of short-circuit current with embedded source resistance in SiC MOSFET, in *International Conference on Silicon Carbide and Related Materials* (2017), p. 2755335
2. B.J. Baliga, Power devices having tunable saturation current clamps therein that support improved short-circuit capability and methods of operating same. U.S. Patent Application, Feb 2019
3. B.J. Baliga, *Modern Power Devices*, Chapter 4 (John Wiley & Sons, New York, 1987)

Chapter 22
Synopsys

The Baliga Short-circuit Improvement Concept, or the BaSIC topology, was proposed as an elegant solution to enhance the short-circuit withstand capability of power semiconductor devices across a wide range of voltage and power ratings. The concept employs a non-linear element, Q_{LV} that is connected in series with the high voltage power device Q_{HV}. The resistance of the non-linear element Q_{LV} must be designed to increase with increasing current until it limits the magnitude of the current to a value less than the drain saturation current of the high-voltage power device Q_{HV}. The non-linear element operates within the gate loop of the high- voltage power device Q_{HV}. Consequently, the highest voltage across it remains below 20 V under all operating modes. This allows creation of non-linear elements with low resistance at the on-state current level because they do not have to be designed to withstand high voltages.

During the short-circuit event, the voltage drop across the non-linear element Q_{LV} increases and opposes the gate drive voltage applied to the high voltage power device Q_{HV}. This reduces the saturation current for the high voltage power device while it supports the DC supply voltage during the short-circuit event. The smaller saturation current greatly reduces the power dissipation in the high voltage device Q_{HV} leading to an increase in the short-circuit withstand time.

The small magnitude of the resistance of the non-linear element Q_{LV} during normal circuit operating conditions minimizes its impact on the on-state voltage drop of the high voltage device Q_{HV}. The low resistance of the non-linear element Q_{LV} during normal circuit operating conditions also allows switching the composite transistor created in the BaSIC topology with power losses close to that of the standalone device.

The voltage across the non-linear element Q_{LV}, called sense voltage V_{SENSE}, can be used to monitor the magnitude of the current flowing through the high voltage power device Q_{HV}. This current sensing is an asset of the BaSIC topology when operating the high voltage device Q_{HV} in power circuits.

B. J. Baliga, A. Kanale, *The BaSIC Topology*,
https://doi.org/10.1007/978-3-031-86630-2_22

The voltage at the non-linear element Q_{LV}, called sense voltage V_{SENSE}, can also be used to detect the short-circuit event because it increases from less than 0.5 V to about 5.0 V. This voltage can be utilized with existing DESAT protection drivers for power MOS-gated devices to prevent failure of the high voltage power device Q_{HV} during the short-circuit event with a response time of less than 0.5 μs. This is possible because the high voltage isolation diode required for the conventional DESAT protection method is not needed for the BaSIC topology.

22.1 Non-linear Element Characteristics

In general, any electrical element that exhibits a non-linear characteristic with resistance increasing with current level is suitable for the BaSIC topology. The ideal desired characteristic for the non-linear element Q_{LV} is shown in Fig. 22.1 by the blue line. This characteristic has a current saturation behavior, which limits the current to $I_{SAT}(Q_{LV})$. However, this behavior is not essential for the non-linear element Q_{LV}. Any characteristic that exhibits an increase in resistance with current level is also acceptable as shown by the purple line.

In both cases, the non-linear elements have a low resistance during on-state operation at point A with a small voltage drop $V_{ON}(Q_{LV})$ at the on-state current level $I_{ON}(Q_{LV})$. This minimizes the impact of the non-linear element Q_{LV} on normal operation of the high voltage power device Q_{HV}.

In both cases, the non-linear resistances operate at point C during the short-circuit event with a voltage drop $V_{SC}(Q_{LV})$. This voltage drop subtracts from the gate drive voltage for the high voltage power device Q_{HV}, which reduces its drain saturation current to the same level $I_{SAT}(Q_{LV})$. Consequently, the short-circuit withstand time for the high voltage power device Q_{HV} is the same in both cases. The beauty of the BaSIC topology is that the voltage $V_{SC}(Q_{LV})$ automatically adjusts until the saturation current of the high voltage power device Q_{HV} becomes equal to the saturation current $I_{SAT}(Q_{LV})$ of the non-linear resistance at this voltage $V_{SC}(Q_{LV})$. The BaSIC topology therefore works despite variations in the characteristics of the high voltage

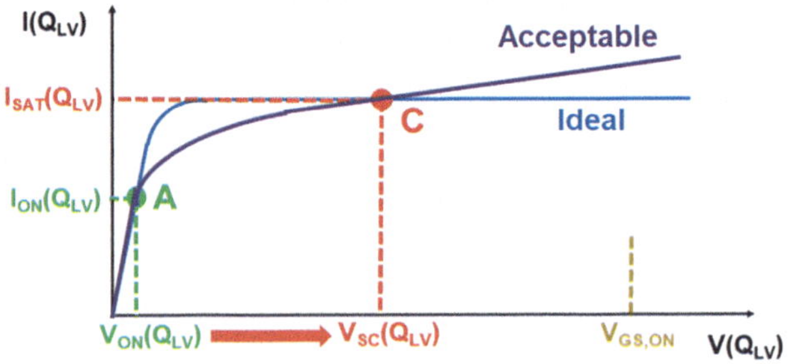

Fig. 22.1 The desired characteristics of the non-linear element Q_{LV}

power device Q_{HV}, and the non-linear element due to manufacturing tolerances. There is no need to perform any matching of the devices.

22.2 Non-linear Element Options

The experiments documented in this book for the implementation of the BaSIC topology with discrete components were conducted using silicon depletion mode and enhancement mode power MOSFETs as the non-linear resistances because of their commercial availability. They could be utilized with high voltage silicon carbide power MOSFETs, gallium nitride HEMT devices, and even silicon IGBTs to enhance the short-circuit withstand time.

However, many other options are also suitable:

(a) Silicon Junction-Field-Effect-Transistors (JFETs) that are either vertical or lateral in configuration [1]
(b) Gallium Arsenide Junction-Field-Effect-Transistors (JFETs) or Metal-Semiconductor-Field-Effect-Transistors (MESFETs) that are either vertical or lateral in configuration [2]
(c) Gallium Nitride lateral high-electron-mobility-transistors [3]

In addition, it is possible to integrate a lateral JFET into any of the high voltage power MOSFET structures to serve as the non-linear element.

22.3 Current Sensing

One of the unique features of the BaSIC topology is the ability to sense the current level flowing through the high voltage power device Q_{HV}. This can be achieved by sensing the voltage across the non-linear element Q_{LV}. The voltage across the non-linear element is proportional to the current level during normal on-state and switching conditions because it operates in the linear regime at low current levels. The ability to sense the current flowing in the high voltage power device Q_{HV} is valuable for balancing the current when paralleling devices. It can also be employed to maximize the output power of converters while maintaining safe operating conditions for the high voltage power device Q_{HV}.

22.4 Short-Circuit Detection

The voltage across the non-linear increases from the 0.5 V level to above 5 V during the short-circuit event. This increase occurs in less than 0.5 μs. In addition, the voltage across the non-linear element remains below 10 V under all operating conditions. This voltage can therefore be fed to the commonly available DESAT protection circuits for silicon IGBTs without using high voltage isolation diodes that are normally required. The rapid detection of the short circuit reduces the magnitude of the short-circuit current which decrease the thermal stress on the high voltage power device Q_{HV}.

22.5 Repetitive Short-Circuit Failure

High voltage power devices, such as SiC power MOSFETs, may survive single short-circuit events but are vulnerable to failure when repetitive short-circuit events are imposed on them. This failure mode occurs due to the melting of the aluminum top metallization during the short-circuit pulse. The aluminum penetrates through the source region of the SiC power MOSFET during repetitive short-circuit pulses leading to the observed failure of devices. It has been shown that the BaSIC topology can sufficiently reduce the temperature rise in the SiC power MOSFET so that the aluminum metallization does not melt during the short-circuit pulse. This eliminates failure of devices during repetitive short-circuit events. The BaSIC topology is the only approach that has been demonstrated to prevent failure of the SiC power MOSFETs during repetitive short-circuit events.

22.6 Summary

This book has provided a detailed discussion of the BaSIC topology. The principles of operation have been explained. Its implementation to improve the short-circuit withstand time has been experimentally confirmed for silicon carbide power MOSFETs, gallium nitride HEMT devices, and silicon IGBTs. The benefits of current monitoring and more rapid short-circuit detection have been experimentally demonstrated.

The power device industry can take advantage of the BaSIC topology to avoid the impediment imposed by the poor short-circuit withstand capability of manufactured devices. The non-linear element can be mounted on top of the source pad of the high voltage power devices to create an elegant compact solution for power electronics applications.

References

1. B.J. Baliga, *Modern Power Devices* (John Wiley & Sons, 1987)
2. P.M. Campbell, W. Garwacki, A.R. Sears, P. Menditto, B.J. Baliga, Trapezoidal-groove Schottky-gate vertical-channel GaAs FET. IEEE Electron Device Lett. **EDL-6**, 304–306 (1985)
3. U. Mishra, J. Singh, *Semiconductor Device Physics and Design* (Springer, 2007)

Index

B. J. Baliga, A. Kanale, *The BaSIC Topology*,
https://doi.org/10.1007/978-3-031-86630-2

The manufacturer's authorised representative in the EU is Springer Nature Customer Service Centre GmbH, Europaplatz 3, 69115 Heidelberg, Germany. If you have any concerns regarding our products, please contact ProductSafety@springernature.com

Printed and bound by CPI Group (UK) Ltd, Croydon, CR0 4YY
01/10/2026
02225949-0011